MODERN ATOMIC PHYSICS

MODERN ATOMIC PHYSICS

Vasant Natarajan
Indian Institute of Science
Bangalore, India

CRC Press is an imprint of the
Taylor & Francis Group, an **informa** business
A CHAPMAN & HALL BOOK

CRC Press
Taylor & Francis Group
6000 Broken Sound Parkway NW, Suite 300
Boca Raton, FL 33487-2742

First issued in paperback 2018

ISBN-13: 978-1-4822-4203-4 (hbk)
ISBN-13: 978-0-367-07713-6 (pbk)

Visit the Taylor & Francis Web site at
http://www.taylorandfrancis.com

and the CRC Press Web site at
http://www.crcpress.com

To

Akanksha

The philosopher should be a man willing to listen to every suggestion, but determined to judge for himself. He should not be biased by appearances; have no favorite hypothesis; be of no school; and in doctrine have no master. He should not be a respecter of persons, but of things. Truth should be his primary object. If to these qualities be added industry, he may indeed hope to walk within the veil of the temple of nature.

— Michael Faraday

Contents

Appendixes

Abbreviations

AC	—	Alternating Current
AOM	—	Acousto-Optic Modulator
BEC	—	Bose–Einstein Condensation
CCD	—	Charge-Coupled Device
C-G	—	Clebsch–Gordan
CPT	—	Coherent Population Trapping
DC	—	Direct Current
ECDL	—	External Cavity Diode Laser
EDM	—	Electric Dipole Moment
EIA	—	Electromagnetically Induced Absorption
EIT	—	Electromagnetically Induced Transparency
EOM	—	Electro-Optic Modulator
FM	—	Frequency Modulation
FWHM	—	Full-Width-at-Half-Maximum
FORT	—	Far Off Resonance Trap
JWKB	—	Jeffreys–Wentzel–Kramers–Brillouin
MOR	—	Magneto-Optic Rotation
NMOR	—	Nonlinear Magneto-Optic Rotation
NMR	—	Nuclear Magnetic Resonance
PBS	—	Polarizing Beam Splitter cube
PZT	—	Piezo-electric transducer (stands for lead zirconate titanate)
QED	—	Quantum Electro-Dynamics
QPD	—	Quadrant Photo Diode
RF	—	Radio Frequency
SAS	—	Saturated Absorption Spectroscopy
SM	—	Standard Model
SNR	—	Signal-to-Noise Ratio

Preface

ATOMIC physics is the mother of all physics. It is no exaggeration to say that much of our understanding of physics in the 20th century comes from understanding atoms, photons, and their interactions. Starting with the discrete energy levels of atoms, which led to the birth of *quantum mechanics*, to the measurement of the Lamb shift in hydrogen, which led to the birth of *quantum electrodynamics*, atomic physics has played a key role in the development of physics. *Lasers*, which have impacted modern life in countless ways, are a manifestation of coherent radiation emitted by atoms. The recent excitement in laser cooling and Bose-Einstein condensation arises from exploiting atom-photon interactions. Frontier research in quantum optics and quantum computation requires a thorough understanding of atomic physics. Five Nobel Prizes have been awarded in this area since 1995 alone, highlighting the relevance of atomic physics to modern society.

Even though there are a few textbooks in this important subject, I decided to write one for several reasons. The first and foremost is that, when students ask me to clear some doubt in atomic physics, there is no single book to which I can refer them. I know all the information is there, but it is scattered in various textbooks and trade books, many of which are not available in college libraries. In other words, I felt there was a real need for a textbook on *modern*—what has been done since 1980 or so—atomic physics. The second reason is that, after looking at textbooks in various subjects, I have come to realize that each author brings a *unique* perspective to a book, different from what you will find in other textbooks on the same subject. This is because the author is conditioned by his (or her) experiences and education, which is by definition different from that of others. Therefore each textbook is valuable, even if there are several of them already available. The third reason is that when I took a similar course from my PhD thesis advisor (Dave Pritchard) at Massachusetts Institute of Technology (MIT) in 1991, he had prepared *elaborate notes* for the course. Combined with the class notes that I took, they formed a valuable resource for students, and I felt it would be a criminal waste not to make them available to a wider audience. In fact, I have used them extensively in the Atomic Physics course that I have taught here at Indian Institute of Science

(IISc) since 1997, especially due to the lack of a comparable book covering these topics. Much of the material in the initial chapters of this book is therefore reproduced from those notes. Last, but certainly not least, I have come to view a book as an *immortal legacy* for students. What better joy for a teacher than to think that some creation of his has the potential to impact students, well beyond his limited domain in space and time.

I take this opportunity to express gratitude for the wonderful time I had learning physics at MIT. The people and the place are incomparable. Coming from an engineering background, I learned to think "physics" for the first time. The intuitive way of thinking and speaking physics—the MIT style—cannot be learned from courses. You *imbibe* it from the experience of being there. Two people at MIT have greatly influenced my way of thinking. One, of course, is my thesis advisor **Dave Pritchard**. He took a young engineer from India, and molded him into a physicist trained in precision measurements. He has taught me about things beyond just physics, and influenced me in ways that he does not know. The second person to influence me is **Eric Cornell**, my co-graduate student when I started my PhD. Ask him to explain something, and he would do so starting from the very basics. It was from him that I learned that if you understand the basics well, the advanced stuff is a breeze. I also learned from him that everything we do in the lab has a reason, including something as simple as twisting a pair of signal-carrying wires to reduce noise. My two-year stint at AT&T Bell Labs rounded off my education in a way that a post-doc position at a university would not have.

Since joining the Physics Department at IISc in 1996, I have taught many different courses in physics. I am grateful to the students in these courses for the varied questions that they have asked, questions that have helped me formulate this book in the most pedagogical manner. I thank my past and present graduate students, again for asking the right questions. Of my current batch, I thank in particular the following: Dipankar Kaundilya for urging me to write this book; Ketan Rathod for making such beautiful figures (using Libre Office); Apurba Paul for help with the typesetting; and Sumanta Khan, Pushpander Kumar, Aaron Markowitz, Lal Muanzuala, K. P. Nagarjun, Harish Ravishankar, Alok Singh, and Vineet Bharti for a critical reading of the manuscript. I am grateful to my assistant S. Raghuveer for help with the typing. I also thank Donald Knuth (whom I have never met) for inventing TEX, which has made typesetting this book such a pleasure. Finally, I thank my family and friends for being supportive during this long (ad)venture.

Some points about the book itself. The contents have been chosen to give enough preparation to make the latest research papers accessible to the student. To facilitate its use as a textbook by students, each chapter has several worked out problems, chosen to emphasize the concepts presented in the chapter. However, the USP of the book, in my opinion, are the **appendixes**. The four appendixes cover some of the topics in the main book, but are written in a way that avoids jargon. Many of them contain historical notes and personal anecdotes. They should therefore be accessible to a wider audience of non-atomic-physics students, and are intended to convey the excitement of atomic physics to all readers.

Bangalore, 2014 Vasant Natarajan

About the Author

Vasant Natarajan earned his B.Tech. from Indian Institute Technology, Madras; his M.S. from Rensselaer Polytechnic Institute, Troy, New York; and his Ph.D. from MIT, Cambridge, Massachusetts. He then worked for two years at AT&T Bell Labs, Murray Hill, New Jersey. He joined the Physics Department at IISc in 1996, where he has been ever since. His research interests are in laser cooling and trapping of atoms; quantum optics; optical tweezers; quantum computation in ion traps; and tests of time-reversal symmetry violation in the fundamental laws of physics using laser-cooled atoms.

Chapter 1

Metrology

THE science of measurement, called **metrology**, is indispensable to the science of physics because the accuracy of measurement limits the accuracy of understanding. In fact, the construction, inter-comparison, and maintenance of a system of units is really an art, often dependent on the latest advances in the art of physics—e.g. quantum Hall effect, laser cooling and trapping, trapped-ion frequency standards, etc. As a result, metrological precision typically marches forward a good fraction of an order of magnitude per decade. Importantly, measurements of the same quantity (e.g. the fine structure constant α) in different fields of physics (e.g. atomic structure, QED , and solid state) provide one of the few cross-disciplinary checks available in a world of increasing specialization. Precise null experiments frequently rule out alternative theories, or set limits on present ones. Examples include tests of local Lorentz invariance and the equivalence principle, searches for atomic lines forbidden by the exclusion principle, and searches for electric dipole moments in fundamental particles which indicate violation of time-reversal symmetry.

A big payoff, often involving new physics, sometimes comes just from attempts to achieve routine progress. In the past, activities like further splitting of the line and increased precision have led to the discoveries of fine and hyperfine structure in hydrogen, anomalous Zeeman effect, and the Lamb shift; trying to understand residual noise in a microwave antenna has resulted in the discovery of the cosmic background radiation; etc. One hopes that the future will bring similar surprises. Thus, we see that precision experiments especially involving fundamental constants or metrology not only solidify the foundation of physical measurement and theories, but occasionally open new frontiers.

This chapter deals briefly with SI units (and its ancestor, the mks system); systems of units which are more natural to a physicist; and then introduces atomic units.

A. Measurement systems

1. MKS units

Most of us were introduced in high school to the **mks** system—meter, kilogram, and second. This simple designation emphasizes an important fact: three dimensionally independent units are sufficient to span the space of all physical quantities. The dimensions are respectively: L — length, M — mass, and T — time. These three units suffice because when a new physical quantity is discovered it always obeys an equation which permits its definition in terms of m, k, and s. For example, consider the Coulomb force between two charges q_1 and q_2 separated by a distance r

$$F = K\frac{q_1 q_2}{r^2}$$

One possibility is to take K as being a dimensionless constant equal to 1, and define the dimensions of charge in terms of the three basic dimensions as

$$[Q]^2 = [M]^1[L]^3[T]^{-2}$$

In fact, **dimensional analysis**, i.e. finding the right combination of the basic units, is quite useful because it can give a rough estimate of the value of an unknown quantity, or tell us if some derived dependencies are correct.

One can argue that only two dimensions are necessary because Einstein's relativity teaches us the space and time are the same physical quantity, and transform into each other depending on your state of motion—one man's space is another man's time. The parameter that allows us to measure space and time in the same units to enable such transformation is c—the speed of light. In fact, it is more appropriate to call c as the relativity parameter rather than the speed of light; it is called speed of light only because E&M was the first relativistically covariant theory that was discovered. If Einstein's relativistic theory of gravity (general relativity) had been discovered before Maxwell's equations, then c would have been called the speed of gravity! And space and time may be interchangeable, but their character is different. Therefore, it is natural to keep both meter and second as independent units, which is what is followed in all practical measurement systems. But it is important to note that our faith in the correctness of relativity leads to the fact that the current definition of the meter is dependent on the definition of the second in such a way that the speed of light is exactly equal to 299792458 m/s.

2. SI units (and why Gaussian units are better)

The modern *avatar* of the mks system is the SI units system (abbreviated from French: *Le Système International d'Unités*). All the details about the system are available on their website: http://www.bipm.org/en/si/.

Briefly, the three basic units are defined as follows.* The second is defined as 9192631770 periods of the ^{133}Cs hyperfine oscillation in zero magnetic field. This definition is so reliable and "democratic" that secondary standards (clocks) with 10^{-12} accuracy are commercially available. The meter is defined as the distance light travels in 1/299792458 of a second, making the speed of light a defined quantity, as mentioned earlier. As metrological precision improves with time, it is the realization of the meter that will improve without changing the value of c. The kilogram, which is the only unit still defined in terms of an artifact, is the weight of a platinum-iridium cylinder kept in a clean ambient at the Bureau de Poids et Measures in Severes, France. The dangers of mass change due to cleaning, contamination, handling, or accident are so perilous that this cylinder has been compared with the dozen secondary standards that reside in the various national measurement laboratories only two times in the last century. Clearly one of the major challenges for metrology is replacement of the artifact kilogram with an atomic definition. This could be done analogously with the definition of length by making Avogadro's number a defined quantity; however, there is currently no sufficiently accurate method of realizing this definition.

There are four more base units in the SI system—the ampere, kelvin, mole, and the candela—for a total of seven. While three are sufficient (or more than sufficient) to do physics, the other four reflect the current situation that electrical quantities, atomic mass, temperature, and luminous intensity, are regularly measured with respect to auxiliary standards at levels of accuracy greater than what can be expressed in terms of the above three base units. Thus measurements of Avogadro's number, the Boltzmann constant, or the mechanical equivalents of electrical units play a role in inter-relating the base units of mole (defined as the number of atoms of ^{12}C in 0.012 kg of ^{12}C), kelvin, or the new volt and ohm (defined in terms of Josephson and quantized Hall effects, respectively). In fact, independent and accurate measurement systems exist for other quantities such as X-ray wavelengths (using diffraction from calcite or other standard crystals), but these other measurement scales are not formally sanctified by the SI system.

While SI units are the world-wide accepted system for making real measurements, they are not the best system for understanding the underlying physics when compared with Gaussian or centimeter-gram-second (cgs) units. Some examples will highlight this fact.

*For a description of the evolution of standards, the reader is referred to the essay in Appendix A, "Standards."

(i) The constant K appearing in the Coulomb force is not unity in SI units, and the equation becomes

$$F = \frac{1}{4\pi\epsilon_\circ} \frac{q_1 q_2}{r^2}$$

where $\epsilon_\circ$ is "the permittivity of free space." Isn't free space supposed to be the absence of any matter? Then how can we talk about it being polarizable, about having a permittivity. Gaussian units do not complicate matters by using $\epsilon_\circ$.

(ii) The Coulomb force directly leads to the potential

$$V = \frac{1}{4\pi\epsilon_\circ} \frac{q_1 q_2}{r}$$

The physics is all contained in the fact that it is a $1/r$ potential; the presence of $\epsilon_\circ$ merely complicates the understanding. In Gaussian units, the potential is just $q_1 q_2/r$, or e^2/r in the hydrogen atom.

(iii) Electric and magnetic fields should have the same dimensions, which is not the case in SI units. Maxwell's equations tell us that the two fields transform into one another. Consider a stationary charge in one frame. There is thus a pure E field surrounding it. But viewed from a moving frame, the charge appears as a current, and therefore has a B field associated with it. The easiest way to see this is to say that the $3 + 3$ components of the E and B vector fields are actually the 6 components of the totally anti-symmetric $F_{\mu\nu}$ tensor. But such a viewpoint necessitates that E and B fields are measured in the same units, just like space and time in any relativistically covariant theory.

To see that this is not true in SI units, consider that the Lorentz force on a charged particle moving with a velocity $\vec{v}$ is

$$\vec{F} = q\left(\vec{E} + \vec{v} \times \vec{B}\right) \qquad \text{SI}$$

This shows that E has dimensions of vB. Whereas the same force expressed in Gaussian units is

$$\vec{F} = q\left(\vec{E} + \frac{\vec{v}}{c} \times \vec{B}\right) \qquad \text{Gaussian}$$

which shows that E and B have the same dimensions.

(iv) In many cases, the SI units may not be the best system for a physicist to get a "feel" for some number. For example, B fields in the lab are measured in gauss (G), using what is called a gaussmeter. But the SI unit for this is tesla (T). The Bohr magneton is familiar to experimentalists in megahertz per gauss (MHz/G), whereas the correct SI unit is hertz per tesla (Hz/T). X-ray wavelengths are more familiar in angstrom units rather than nanometers. Electric fields are known in V/cm and not V/m.

The above examples show that while the functional dependencies are the same in both SI and Gaussian systems, the expressions are less transparent in the SI system. Since the idea of this book is to show the underlying physics through dependencies on various parameters, expressions will be given in Gaussian, but the values in SI. For instance, it is important to note that the Bohr radius of the hydrogen atom depends on atomic parameters as $\hbar^2/me^2$ and its value is about 0.5 Å; it is immaterial that the SI expression has $4\pi\epsilon_o$ in it.

B. Universal units and fundamental constants

The sizes of the meter, kilogram, and second were originally selected for convenience. They bear no relationship to things which most physicists would regard as universal or fundamental. In fact, most of the 42 fundamental constants listed in the CODATA adjustment are neither fundamental nor universal. We assert that there are only three truly universal constants:

c — the speed of light
$\hbar$ — the quantum of action
G — the gravitational constant

These quantities involve relativity, the quantum, and gravitation. Universal units are defined by setting c, $\hbar$, and G all equal to 1. By taking suitable combinations of these fundamental constants, this defines units for M, L, and T. Together they set what is called the **Planck scale**—the scale at which relativistic quantum gravity effects are expected to become significant. In SI units, the Planck mass is 5.4×10^{-8} kg, the Planck length is 4×10^{-35} m, and the Planck time is 1.3×10^{-43} s.

Next, we come to atomic constants, whose magnitude is determined by the size of the quantized matter which we find all around us. Clearly the most fundamental of these is

e — the quantum of charge

because it is the same (except for the sign) for all particles. Even though the existence of quantized charge seems independent of the physics which underlies the construction of universal units, this is probably not the case because charge does not have independent units. In fact, e appears in the *dimensionless* quantity

$$\alpha = \frac{e^2}{\hbar c} = (137.035\,999\,074\ldots)^{-1}$$

which is called the **Fine structure constant**, and is (I think) the only fundamental constant truly worthy of that name. When we really understand E&M, quantum mechanics, and the origins of quantized matter, we should be able predict it. The fact that we have to measure it is a sign of our ignorance, but the good agreement of the many seemingly independent ways of measuring it shows that we are beginning to understand some things.

Other atomic constants like masses (m_e, m_p, m_n, ...) and magnetic moments (μ_e, μ_p, μ_n, ...) seem to be rather arbitrary at our current level of physical knowledge, except that certain relationships are given by QED ($\mu_e = g_s e\hbar/2m_e c$, with g_s a little more than 2) and the quark model of the nucleons.

C. Atomic units

The system of atomic units is defined according to the scale of quantities in a typical atom, usually hydrogen. It is obtained by setting $\hbar = m = e = 1$, where m is the electron mass. Units for other physical quantities are formed by dimensionally suitable combinations of these units. In this book, expressions will be given in atomic units to facilitate interpretation; numerical evaluations should be done in SI units. In atomic units, the units of length and energy are the most important—they are called the Bohr ($a_\circ$) and Hartree (H), respectively.* Expressions for length and energy can generally be expressed in terms of $a_\circ$ or H, and powers of the dimensionless fine structure constant α introduced earlier. Most of the important atomic units (along with their SI values) are listed in the table below.

Physical quantity	Name	Atomic unit	SI value
Charge	Electron charge	e	1.602×10^{-19} C
Angular momentum	h-bar	$\hbar$	1.055×10^{-34} J s
Mass	Electron mass	m	9.110×10^{-31} kg
Length	Bohr	$a_\circ = \dfrac{\hbar^2}{me^2}$	5.292×10^{-11} m
Energy	Hartree	$H = \dfrac{me^4}{\hbar^2}$ (= 2 Ry)	4.360×10^{-18} J (27.2 eV)
Velocity		$\dfrac{e^2}{\hbar} = \alpha c$	2.180×10^6 m/s
Magnetic moment	Bohr magneton	$\mu_B = \dfrac{e\hbar}{2mc}$	1.400×10^4 MHz/T
Electric Field		$\dfrac{e}{a_\circ^2}$	5.142×10^{11} V/m

*One Hartree is two times the more familiar Rydberg.

1. Fine structure constant

The fine structure constant that we saw earlier (and reproduced below) is ubiquitous in atomic physics.

$$\alpha = \frac{e^2}{\hbar c} = [\,137.035\,999\,074(44)\,]^{-1}$$

where the numerical value is the 2010 CODATA recommended value. The number in round brackets is the uncertainty in the last digit, representing a relative uncertainty of 3.2×10^{-10}, and is the standard way of expressing a quantity with error.

The name fine structure reflects the appearance of this quantity (squared) in the ratio of the hydrogenic fine structure splitting to the Rydberg.

$$\frac{\Delta(\text{fine structure})}{\text{Rydberg}} = \alpha^2 \frac{Z^4}{n^3 \ell(\ell+1)}$$

The fine structure constant will often crop up as the ratio between different physical quantities having the same dimensions. An impressive example of this is length, as seen in the table below.

Physical quantity	**Expression**	**SI value**
Bohr (radius of electron orbit in Bohr's model of hydrogen)	$a_\circ = \dfrac{\hbar^2}{me^2}$	5.29×10^{-11} m
Reduced Compton wavelength ($\lambda_c/2$). λ_c is the change in wavelength of a photon scattered at 90° by a stationary electron due to recoil of the electron.	$\lambda\!\!\!^{-}_c = \alpha a_\circ = \dfrac{\hbar}{mc}$	3.86×10^{-13} m
Classical radius of the electron (size for which electrostatic self energy equals rest mass energy).	$r_\circ = \alpha^2 a_\circ = \alpha \lambda\!\!\!^{-}_c = \dfrac{e^2}{mc^2}$	2.82×10^{-15} m

D. Problems

1. Atomic units for E and B fields

The atomic unit of electric field is $\mathcal{E}_{\mathrm{au}} \equiv e/a_\circ^2$, the field of the electron at the proton in hydrogen.

(a) Find the magnetic field B_N of the electron at the proton.

(b) Find the magnetic field B_H which has one Hartree interaction with a Bohr magneton.

(c) Express these fields in terms of $\mathcal{E}_{\mathrm{au}}$ and α.

(d) Are there strong reasons to prefer B_N or B_H as the atomic unit of magnetic field?

Solution

Recall the definitions of the atomic constants

$$\alpha = \frac{e^2}{\hbar c} \qquad \text{and} \qquad a_\circ = \frac{\hbar^2}{me^2}$$

(a) The magnetic field at the center of a loop of radius a and carrying current I is

$$B = \frac{2\pi I}{ca}$$

For an electron in a circular orbit of radius $a_\circ$, the current is

$$I = \frac{e}{\tau} = \frac{e}{2\pi a_\circ / v} = \frac{e\alpha c}{2\pi a_\circ}$$

Therefore the magnetic field of the electron at the proton is

$$B_N = \frac{2\pi}{ca_\circ}\frac{e\alpha c}{2\pi a_\circ} = \frac{e\alpha}{a_\circ^2}$$

(b) Using $\mu_B = e\hbar/(2mc)$ and one Hartree is $e^2/a_\circ$, the required magnetic field is

$$B_H \mu_B = \frac{e^2}{a_\circ} \qquad \Longrightarrow \qquad B_H = \frac{2mce}{a_\circ \hbar}$$

(c) In terms of $\mathcal{E}_{\mathrm{au}}$ and α, the fields are

$$B_N = \mathcal{E}_{\mathrm{au}}\alpha \qquad \text{and} \qquad B_H = \frac{2\mathcal{E}_{\mathrm{au}}}{\alpha}$$

(d) No preference. Though the choice is finally a matter of convenience, it should be noted that B_N is nearly 10 000 times larger.

2. Unit of charge

(a) How is the unit of charge defined, and what are its dimensions, in the SI system? (Give the most pertinent equations.)

(b) Same question for Gaussian units.

(c) Would it be possible to define a unit of mass *à la* the Gaussian definition of charge, thereby eliminating the artifact mass standard?

Solution

(a) The SI unit of charge "coulomb" is defined through current.

$$I = \frac{dq}{dt} \qquad \text{and} \qquad \frac{dF}{d\ell} = 2\mu_\circ \frac{I_1 I_2}{4\pi r}$$

The ampere is defined as that current which when flowing through two infinitely long conductors placed one meter apart (cross-section ~ 0), produces a force/length of 2×10^{-7} N/m. Coulomb is then defined as the current flowing per unit time.

Its dimensions are IT^{-1}.

(b) Gaussian unit of charge "esu" is defined through the Coulomb force

$$F = \frac{q_1 q_2}{r^2}$$

Two unit charges placed 1 cm apart produce a force of 1 dyne.

Its dimensions are $M^{1/2}L^{3/2}T^{-1}$.

(c) There are two issues with using the Coulomb force analogy to define mass.

(i) The Coulomb force already has a mass term in it because the force is mass times the acceleration (from Newton's second law of motion). This mass is called the "inertial" mass. The mass term that appears in the gravitational force equation is the "gravitational" mass, and should be correctly called the "gravitational charge." The fact that these two masses are equal is a fundamental result from what is called the "equivalence principle." The equality is actually an empirical fact (subject to experimental verification), which was used by Einstein in formulating the general theory of relativity. If we accept this as a true fact, then mass can indeed be defined using the gravitational force equation

$$F = \frac{m_1 m_2}{r^2}$$

The definition would be that two unit masses placed one meter apart have an acceleration of 1 m/s^2.

(ii) The second issue with this definition is that it is not **practical** because gravity is such a weak force that it would be overwhelmed by the electrostatic force between the bodies, which cannot be shielded perfectly.

3. Universal units (Planck units)

Universal units are defined by setting c, $\hbar$, and G equal to 1, which defines units for M, L, and T. Find expressions for these units and calculate their magnitude in SI units. These units are often called the Planck mass (or energy since $c = 1$), Planck length, and Planck time.

Solution

Let the universal units be M_u, L_u and T_u

$$c = 1 \quad \Longrightarrow \quad 2.998 \times 10^8 \text{ m/s} = 1\,\frac{L_u}{T_u}$$

$$\hbar = 1 \quad \Longrightarrow \quad 1.055 \times 10^{-34} \text{ kg m}^2\text{/s} = 1\,\frac{M_u L_u^2}{T_u}$$

$$G = 1 \quad \Longrightarrow \quad 6.674 \times 10^{-11} \text{ m}^3\text{/(s}^2\text{ kg)} = 1\,\frac{L_u^3}{T_u^2 M_u}$$

This yields

$$L_u^2 = \frac{M_u L_u^2}{T_u} \frac{L_u^3}{T_u^2 M_u} \frac{T_u^3}{L_u^3} = \frac{\hbar G}{c^3}$$

$$L_u = \sqrt{\frac{\hbar G}{c^3}} \quad \Longrightarrow \quad L_u = 1.616 \times 10^{-35} \text{ m}$$

and

$$M_u = \frac{M_u L_u^2}{T_u} \frac{T_u}{L_u} \frac{1}{L_u} = \frac{\hbar}{c L_u} \quad \Longrightarrow \quad M_u = 2.177 \times 10^{-8} \text{ kg}$$

$$T_u = \frac{T_u}{L_u} L_u = \frac{L_u}{c} \quad \Longrightarrow \quad T_u = 5.391 \times 10^{-44} \text{ s}$$

Chapter 2

Preliminaries

UNDERSTANDING the harmonic oscillator is the key to understanding physics. In fact, it is no exaggeration to say that you understand 90% of physics if you understand the classical harmonic oscillator, and 99% of physics if you understand the quantum harmonic oscillator. This is because nature abounds in such examples—from the simple pendulum to LCR circuits. Collective excitations like normal modes or phonons in a solid obey the harmonic oscillator equation of motion. Perhaps the most important reason is that it is the first step in understanding light. A light wave is like a harmonic oscillator, except that instead of position and velocity oscillating (out of phase) as in a normal oscillator, it is the electric and magnetic fields that oscillate in a light wave. It is called a wave only because the disturbance propagates in some direction, but orthogonal to this direction the fields just oscillate in time. This chapter, therefore, deals with the harmonic oscillator and radiation, both classical and quantized. The radiation part will be of use in Chapter 6, "Interaction."

A. Classical harmonic oscillator

We first consider the behavior of the one-dimensional harmonic oscillator in the classical regime. In the presence of damping and driving, it obeys the standard equation of motion

$$\ddot{x} + \Gamma\dot{x} + \omega_\circ^2 x = \frac{1}{m}F(t)$$

where Γ is the damping, $\omega_\circ$ is the frequency in the absence of damping (also called the **natural frequency** of the oscillator), m is the mass of the particle, and $F(t)$ is the driving force.

1. Not driven

When the oscillator is **not driven**, $x(t)$ will decay to zero. The equation of motion simplifies to

$$\ddot{x} + \Gamma\dot{x} + \omega_\circ^2 x = 0$$

and is called **homogeneous**, because if $x(t)$ is a solution, then so is $A \times x(t)$. The solution is obtained by substituting $x(t) = e^{\alpha t}$, which yields the characteristic equation

$$\alpha^2 + \Gamma\alpha + \omega_\circ^2 = 0$$

with roots

$$\alpha_\pm = -\Gamma/2 \pm \sqrt{(\Gamma/2)^2 - \omega_\circ^2}$$

Depending on the relative sizes of the damping and the frequency, we define the following three regimes:

(i) **Strong damping** for $\Gamma > 2\omega_\circ$. The solution is

$$x(t) = Be^{\alpha_+ t} + Ce^{\alpha_- t}$$

which does not oscillate.

(ii) **Critical damping** for $\Gamma = 2\omega_\circ$. This is a transition regime, with the solution

$$x(t) = (B + Ct)e^{-\Gamma t/2}$$

which is solved in a slightly different manner because the two roots of α become equal.

(iii) **Weak damping** for $\Gamma < 2\omega_\circ$. One defines $\omega_m = \sqrt{\omega_\circ^2 - (\Gamma/2)^2}$ (the **pulled** frequency), and the solution becomes

$$\begin{aligned} x(t) &= e^{-\Gamma t/2}(B\cos\omega_m t + C\sin\omega_m t) \\ &= Ae^{-\Gamma t/2}\cos(\omega_m t + \phi) \end{aligned}$$

where $A^2 = B^2 + C^2$ and $\phi = \tan^{-1}(C/B)$.
The **quality factor**—usually just called Q—is defined as

$$Q = \omega_\circ/\Gamma$$

and is the number of radians of oscillation it takes to reduce the energy by e. The energy is proportional to $x^2(t)$ and therefore damps like $e^{-\Gamma t}$. In most cases, we will be dealing with a **high Q oscillator**, which means that the change in amplitude per cycle is negligibly small. Then the Q can also be interpreted as the ratio of the energy stored to the energy lost per cycle.

The behavior for three cases of damping considered above are shown graphically in Fig. 2.1, for the case $\omega_\circ = 10$ rad/s.

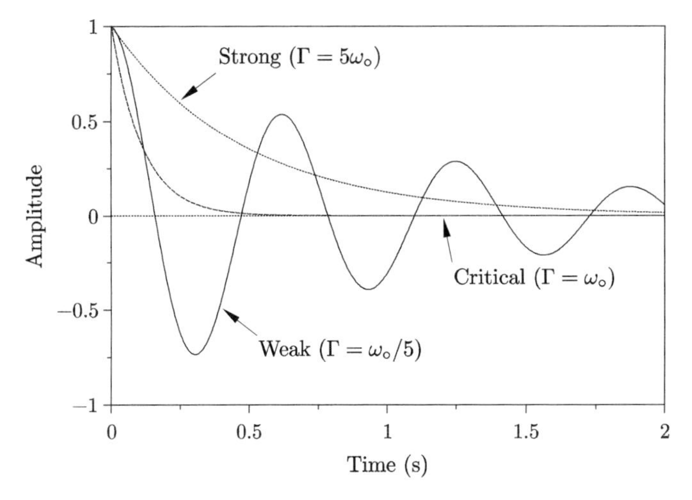

Figure 2.1: Behavior of an undriven harmonic oscillator for the three cases of damping—strong, critical, and weak.

2. Driven with weak damping

The **driven damped harmonic oscillator** is one which is driven by a sinusoidally varying applied force with amplitude F_d. It has a **steady-state** response at the driving frequency ω. It also has the transient response of an undriven damped harmonic oscillator whose motion adds on to the steady-state solution to meet the initial conditions. The oscillator is assumed to be high Q, i.e. one where the damping is weak $\Gamma \ll \omega_\circ$. The amplitude and phase vary with the detuning from resonance $\omega - \omega_\circ$. The resulting equation of motion

$$\ddot{x} + \Gamma\dot{x} + \omega_\circ^2 x = \frac{1}{m} F_d \cos\omega t$$

is typically solved by the **complex exponential** method, where the complex equation of motion is found by changing $x \to z$ and $\cos\omega t \to e^{i\omega t}$. x is the real part of the complex solution

$$x = \text{Re}\{z\} = \text{Re}\left\{z_\circ e^{i\omega t}\right\}$$

Therefore, the solution to the equation of motion is

$$x(\omega, t) = \text{Re}\left\{\frac{F_d e^{i\omega t}/m}{(\omega_\circ^2 - \omega^2) + i\Gamma\omega}\right\} = A(\omega)\cos(\omega t + \phi)$$

(i) The **amplitude** of the response is

$$A(\omega) = \frac{F_d/m}{\sqrt{(\omega_\circ^2 - \omega^2)^2 + \Gamma^2\omega^2}} \approx \frac{F_d/(2m\omega_\circ)}{\sqrt{(\omega - \omega_\circ)^2 + (\Gamma/2)^2}} \tag{2.1}$$

The approximate result (here and below) is valid near resonance ($\omega \approx \omega_\circ$), which is the only place where the response is non-negligible for high Q oscillators.

(ii) The **phase** of the response is

$$\phi(\omega) \equiv \arg(z_\circ) = \tan^{-1}\frac{\Gamma\omega}{\omega^2 - \omega_\circ^2} \approx \tan^{-1}\frac{\Gamma}{2(\omega - \omega_\circ)}$$

and is always $0 \geq \phi \geq -\pi$. It represents motion that always lags behind the drive: by a little when $\omega \ll \omega_\circ$ (inertia has negligible effect) and
by approximately π when $\omega \gg \omega_\circ$ (inertia dominates).

(iii) The **power** averaged over a cycle supplied by the force (and dissipated by the damping) is

$$P(\omega) = \langle F\dot{x}(\omega)\rangle = F_d\langle\cos\omega t\,\text{Re}\left\{i\omega z_\circ e^{i\omega t}\right\}\rangle = \frac{\omega F_d}{2}\,\text{Im}\{-z_\circ\}$$

Therefore

$$P(\omega) = \frac{\Gamma F_d^2}{2m} \frac{\omega^2}{(\omega_\circ^2 - \omega^2)^2 + \Gamma^2\omega^2} \approx \frac{F_d^2}{8m} \frac{\Gamma}{(\omega - \omega_\circ)^2 + (\Gamma/2)^2}$$

where the approximate result is written so that it shows explicitly that the lineshape is **Lorentzian.**

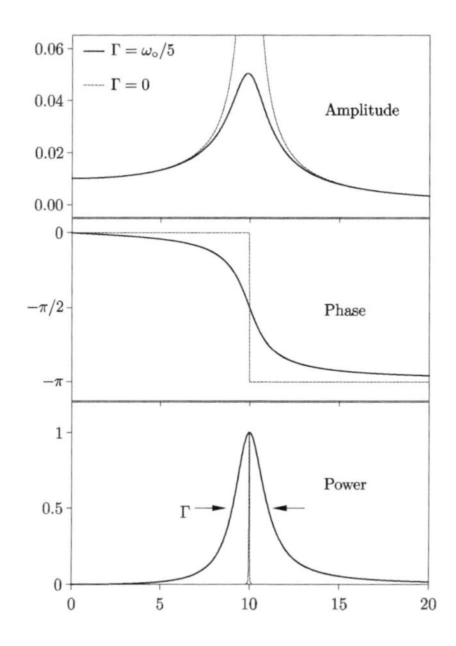

Figure 2.2: Amplitude, phase, and (normalized) power dissipated per cycle of a driven harmonic oscillator. The solid curves are for the weak damping case ($\Gamma = \omega_\circ/5$), shown in Fig. 2.1. The power dissipated has a Lorentzian lineshape with linewidth Γ. The dotted curves are for the case of no damping.

In Fig. 2.2, we plot the amplitude, phase, and power dissipated of the driven oscillator discussed above, comparing the cases of weak damping ($\Gamma = \omega_\circ/5$) with no damping. It is important to note that the amplitude remains finite throughout, except on resonance in the no-damping case. The lineshape is that given by Eq. (2.1), which is not a Lorentzian. On the other hand, the average power dissipated as already mentioned has a Lorentzian lineshape near resonance—this curve is a universal feature of all resonance phenomena. Its linewidth Γ, defined as the full width at half maximum (FWHM), is equal to $\Delta\omega$, so that the Q can also be defined as

$$Q = \frac{\omega_\circ}{\Delta\omega}$$

When there is **no damping**, the power dissipated is zero away from resonance. This is because whatever energy is supplied by the drive in one half of the cycle is recovered in the second half, and the net energy transferred *per cycle* is zero. On resonance, the power becomes infinite because the amplitude blows up. The lineshape becomes that of a Dirac delta function.

3. Harmonically bound electron

Consider a harmonically bound electron of charge e and mass m, with its natural frequency equal to the absorbed or emitted frequency when the atom makes a transition between two states. To start with, let it be undamped and driven by a field $\mathcal{E}\cos\omega t$, so that

$$F_d = e\,\mathcal{E}\cos\omega t$$

The resulting motion can be regarded as an oscillating dipole moment. From the amplitude response in Eq. (2.1), we get

$$d = eA(\omega) = \frac{e^2}{m}\frac{1}{\omega_\circ^2 - \omega^2}\,\mathcal{E}\cos\omega t$$

This expression will be useful when dealing with **oscillator strengths** for describing atomic polarizability, which we will encounter in Chapter 3, "Atoms."

Even in the absence of other kinds of damping, the motion will be damped because of what is called **radiation damping**. From classical electrodynamics, we know that any accelerated charge will emit radiation. For example, an electron in a circular orbit around a proton is constantly being accelerated, and therefore loses energy. This is why such a "planetary" model for the hydrogen atom failed until Bohr postulated stationary orbits where the electron does not lose energy. The total power radiated by an accelerated electron in the full solid angle of 4π is

$$P = \frac{2}{3}\frac{e^2}{c^3}\,|\dot{v}|^2 \qquad \dot{v} \text{ is acceleration}$$

For an oscillating electron, $\dot{v} = -\omega_\circ^2 x$, hence the total energy lost per cycle is

$$E_{\text{lost}} = \frac{2}{3}\frac{e^2}{c^3}\frac{\omega_\circ^3 x_\circ^2}{2} \qquad x_\circ \text{ is amplitude of the motion}$$

As mentioned earlier, the Q for such a high Q oscillator is the ratio of the energy stored ($= \frac{1}{2}m\omega_\circ^2 x_\circ^2$) to the energy lost. Hence the damping term becomes

$$\Gamma_{\text{radiation}} = \frac{2}{3}\frac{e^2\omega_\circ^2}{mc^3}$$

4. Coupled oscillators

The simplest coupled oscillator problem consists of two identical masses on springs so that they have identical natural frequencies (degenerate), which are coupled together using a third spring with a different spring constant, as shown in Fig. 2.3(a). Due to the coupling, the equations of motion for x_1 and x_2 are coupled and cannot be solved independently. But there are linear superpositions of these displacements, called **normal modes**, which do obey the harmonic oscillator equation and oscillate at some characteristic frequency.

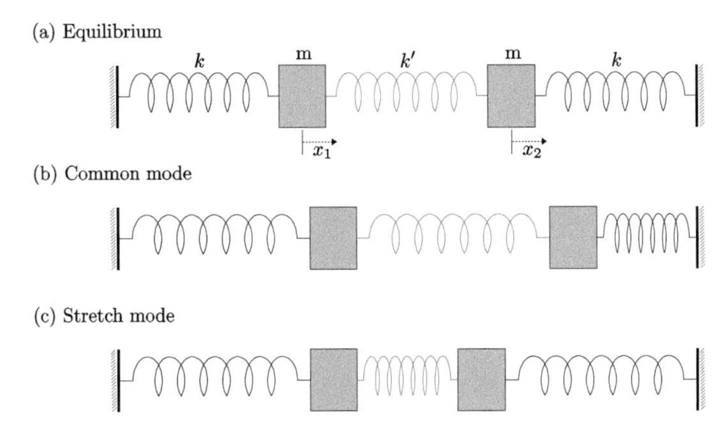

Figure 2.3: Coupled oscillator system. (a) At equilibrium. (b) When a normal mode called common mode—corresponding to in-phase motion of the two masses—is excited. (c) When a normal mode called stretch mode—corresponding to out-of-phase motion of the two masses—is excited.

The problem of finding these normal modes is a standard problem in classical mechanics. It consists of diagonalizing the following matrix

$$A = \begin{bmatrix} k+k' & -k' \\ -k' & k+k' \end{bmatrix}$$

Defining $\omega_\circ = \sqrt{k/m}$ and $\omega' = \sqrt{k'/m}$, this yields the normal modes as

$$\eta_c = (x_1 + x_2)/\sqrt{2} \qquad \omega_c = \omega_\circ \qquad \textbf{common mode}$$

$$\eta_s = (x_1 - x_2)/\sqrt{2} \qquad \omega_s = \sqrt{\omega_\circ^2 + 2\omega'^2} \qquad \textbf{stretch mode}$$

which are depicted in Fig. 2.3(b) and (c). What this means is that if one of the normal modes is excited, then the energy stays in that mode forever. On the other hand, if x_1 is given some initial amplitude, then both the normal modes are excited because

$$x_1(0) = 1 \text{ and } x_2(0) = 0 \qquad \Longrightarrow \qquad \eta_c(0) = 1/\sqrt{2} \text{ and } \eta_s(0) = 1/\sqrt{2}$$

The time dependence of each normal mode is just $\eta(t) = \eta(0)\cos\omega t$. Projecting back to x_1 and x_2, one gets

$$x_1(t) = \frac{1}{2}\left(\cos\omega_c t + \cos\omega_s t\right) = \cos\left(\frac{\omega_c + \omega_s}{2}t\right)\cos\left(\frac{\omega_s - \omega_c}{2}t\right)$$

$$x_2(t) = \frac{1}{2}\left(\cos\omega_c t - \cos\omega_s t\right) = \sin\left(\frac{\omega_c + \omega_s}{2}t\right)\sin\left(\frac{\omega_s - \omega_c}{2}t\right)$$

These two motions are shown in Fig. 2.4. One sees that the amplitude swaps between x_1 and x_2 at a frequency equal to half the difference in frequencies between the two normal modes. In the language of atomic physics, this is called **Rabi flopping**—each motion is a rapid oscillation that is amplitude modulated at the **Rabi frequency**. If we turn on the coupling for a time equal to exactly one swap, i.e. for a complete transfer from x_1 to x_2, it would be called a "pi pulse" in NMR.

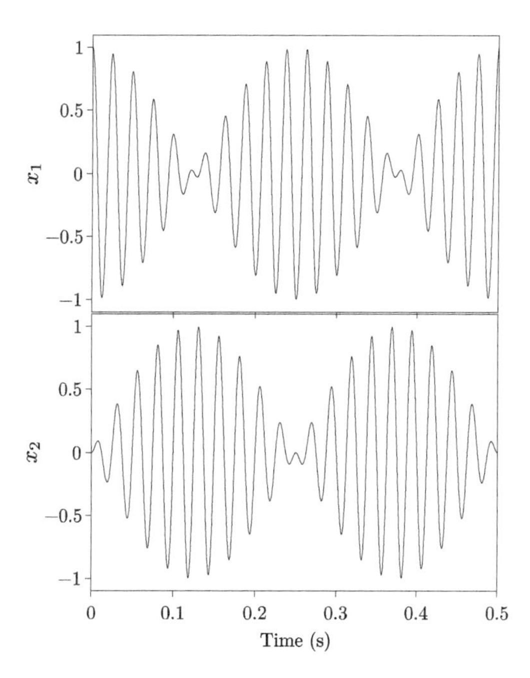

Figure 2.4: Rabi oscillations in the two coupled oscillator system shown in Fig. 2.3, when x_1 is started with an initial amplitude 1. Upper curve is $x_1(t)$ and lower curve is $x_2(t)$, showing out-of-phase amplitude modulation at the Rabi frequency. Parameters used for the simulation: $\omega_c/2\pi = 38$ Hz and $\omega_s/2\pi = 42$ Hz.

The underlying reason for the difference in frequencies of the two normal modes is the difference in symmetries in terms of the original variables—the common mode is the symmetric combination, while the stretch mode is the antisymmetric combination. The symmetric combination always has lower frequency (hence lower energy) compared to the antisymmetric one. In chemistry, if two atoms come close enough to interact and form a molecule, then the symmetric combination of their wavefunctions leads to the lower-energy **bonding** orbital, whereas the antisymmetric combination leads to the higher-energy **antibonding** orbital. In a crystalline solid, the orbitals extend into a band because of the large number of atoms (on the order of $N_A \approx 10^{23}$) involved. The bonding orbital then becomes the **valence band**, while the antibonding orbital becomes the **conduction band**. The energy gap between the two—**bandgap**—is a measure of the coupling between the atoms. This coupling (and hence the bandgap) can be changed by either heating the solid or applying pressure.

B. Quantum harmonic oscillator

We briefly review the standard approach to quantization of the harmonic oscillator. The first step is to replace the classical variables x and p with the corresponding quantum mechanical operators X and P. Then the quantum mechanical Hamiltonian becomes

$$H = \frac{P^2}{2m} + \frac{1}{2}m\omega^2 X^2 \qquad \text{with } [X, P] = i\hbar$$

We now define *dimensionless* **quadrature operators** as

$$\widehat{X} = \sqrt{\frac{m\omega}{2\hbar}}X$$
$$\widehat{Y} = \frac{1}{\sqrt{2m\hbar\omega}}P$$

The Hamiltonian then simplifies to

$$H = \hbar\omega(\widehat{X}^2 + \widehat{Y}^2) \qquad \text{with } [\widehat{X}, \widehat{Y}] = i/2$$

Define new operators

Raising (Creation) operator: $a^\dagger \equiv \widehat{X} - i\widehat{Y}$

Lowering (Annihilation) operator: $a \equiv \widehat{X} + i\widehat{Y}$

Then

$$\widehat{X} = (a^\dagger + a)/2$$
$$\widehat{Y} = i(a^\dagger - a)/2 \tag{2.2}$$

and the Hamiltonian becomes

$$H = \hbar\omega(a^\dagger a + 1/2) = \hbar\omega(N + 1/2) \qquad \text{with } [a, a^\dagger] = 1 \tag{2.3}$$

where $N \equiv a^\dagger a$ is the **number operator**.

1. Energy eigenstates

In the time-independent Schrödinger picture, one finds the **stationary** states $|n\rangle$ of the Hamiltonian by solving the eigenvalue equation

$$H|n\rangle = E_n|n\rangle$$

From Eq. (2.3) we can see that $|n\rangle$ is also an eigenstate of N

$$N|n\rangle = n|n\rangle$$

which is why N is called the number operator. Therefore

$$E_0 = \frac{1}{2}\hbar\omega \qquad \text{and} \qquad E_n = \left(n + \frac{1}{2}\right)\hbar\omega$$

The operator a is not Hermitian and therefore **not an observable** because

$$\langle (n-1)|a|n\rangle = \sqrt{n} \qquad \text{but} \qquad \langle n|a|(n-1)\rangle = 0$$

Some important properties of the eigenstates $|n\rangle$ are as follows.

(i) $|n\rangle$ is not degenerate.

(ii) $|n\rangle$ is orthogonal to $|n'\rangle$.

(iii) $|n\rangle$ has definite parity. Under $x \to -x$, $|n\rangle \to (-1)^n|n\rangle$.

(iv) Action of operators

$$a|n\rangle = \sqrt{n}\,|(n-1)\rangle \qquad \leftarrow \textbf{Lowering}$$

$$a^\dagger|n\rangle = \sqrt{n+1}\,|(n+1)\rangle \qquad \leftarrow \textbf{Raising}$$

(v) Wavefunctions

$$\phi_0(x) \sim \exp\left(-\frac{1}{2}\frac{m\omega}{\hbar}x^2\right) \qquad \leftarrow \textbf{Ground}$$

$$\phi_n(x) \sim (n^{\text{th}} \text{ order Hermite polynomial}) \times \exp\left(-\frac{1}{2}\frac{m\omega}{\hbar}x^2\right)$$

Since the state $|n\rangle$ is obtained by n operations of the raising operator $a^\dagger$ on the state $|0\rangle$, the normalized **number (Fock) state** is

$$|n\rangle = \frac{1}{\sqrt{n!}}(a^\dagger)^n|0\rangle$$

2. Time dependence

The foregoing treatment is in the time-independent Schrödinger picture. In general the time dependence is found by expanding the initial state in the basis of energy eigenstates*

$$|\psi(0)\rangle = \sum_n c_n |n\rangle$$

so that the the time evolution operator

$$U(t) = e^{-iHt/\hbar}$$

can be applied easily. Then each term in the expansion picks up a phase corresponding to its energy, so that the time dependent wavefunction becomes

$$|\psi(t)\rangle = e^{-iHt/\hbar} |\psi(0)\rangle = \sum_n c_n e^{-iE_n t/\hbar} |n\rangle$$

In this picture, the operators $\widehat{X}$, $\widehat{Y}$, $a^\dagger$, etc., are also time independent, but their expectation values do evolve in time, e.g.

$$\begin{aligned}
\frac{d}{dt}\langle a\rangle_t &\equiv \frac{d}{dt}\langle\psi(t)|a|\psi(t)\rangle \\
&= \frac{1}{i\hbar}[-\langle\psi(t)|Ha|\psi(t)\rangle + \langle\psi(t)|aH|\psi(t)\rangle] \\
&= \frac{1}{i\hbar}\langle\psi(t)|[a,H]|\psi(t)\rangle \\
&= \frac{1}{i\hbar}\langle\psi(t)|\hbar\omega a|\psi(t)\rangle \\
&= -i\omega\langle a\rangle_t \\
\implies \langle a\rangle_t &= \langle a\rangle_0 e^{-i\omega t}
\end{aligned} \tag{2.4}$$

*This procedure is similar to how the initial displacement was expanded in terms of the normal modes for the classical coupled oscillator case in Section A4.

3. Quantum uncertainties

The 1/2 in the Hamiltonian [Eq. (2.3)], and the consequential ground state energy being $\hbar\omega/2$, is an entirely quantum phenomenon. The lowest energy state of a *classical* HO is at rest, i.e. both its position and momentum are zero. But, in quantum mechanics, this would violate the Heisenberg uncertainty principle: $\Delta X\, \Delta P \geq \hbar/2$. Hence, the lowest energy state allowed in quantum mechanics—the $|0\rangle$ state—is one with *minimum uncertainty*, i.e. $\Delta X\, \Delta P = \hbar/2$. In terms of the quadrature operators $\widehat{X}$ and $\widehat{Y}$ introduced earlier, the uncertainty relationship is $\Delta\widehat{X}\, \Delta\widehat{Y} \geq 1/4$, and the $|0\rangle$ state has $\Delta\widehat{X}\, \Delta\widehat{Y} = 1/4$. The energy of this state of $\hbar\omega/2$ is called the *zero-point* energy.

In general, the uncertainties are found from the standard deviation of the relevant quantity in a state, e.g.

$$\Delta\widehat{X} \equiv \sqrt{\langle(\Delta\widehat{X})^2\rangle} = \sqrt{\langle(\widehat{X} - \langle\widehat{X}\rangle)^2\rangle} = \sqrt{\langle\widehat{X}^2\rangle - \langle\widehat{X}\rangle^2}$$

For number states $|n\rangle$ this may easily be evaluated by noting that $\langle\widehat{X}\rangle = 0$. Therefore

$$\begin{aligned}
(\Delta\widehat{X})^2 &= \langle\widehat{X}^2\rangle \\
&= \langle n|[(a + a^\dagger)(a + a^\dagger)]|n\rangle /4 \\
&= \langle n|[aa^\dagger + a^\dagger a]|n\rangle /4 \\
&= \langle n|2a^\dagger a + 1|n\rangle /4 \\
&= (2n + 1)/4
\end{aligned}$$

$$\Longrightarrow \Delta\widehat{X} = \sqrt{(2n + 1)/4}$$

Similarly we can show that $\Delta\widehat{Y} = \sqrt{(2n + 1)/4}$. Thus we see that in general number states are not minimum uncertainty states. But the ground state $|0\rangle$ is indeed a **minimum uncertainty** state with

$$\Delta\widehat{X} = \Delta\widehat{Y} = 1/2$$

Recall that its wave function (in the x coordinate) is pure Gaussian with width of $\sqrt{\hbar/m\omega}$. Therefore, the $\widehat{X}$ uncertainty is as big as the whole wave function, which is characteristic of a thermal distribution. There might be a deeper significance to this in the sense that the uncertainty principle has a thermal origin, *à la* Einstein's statement "God does not play dice with the Universe."

Because the two uncertainties are equal, the distribution in the $\widehat{X}\widehat{Y}$ plane is circularly symmetric. As seen in Fig. 2.5, the ground state is like a "fuzz ball" at the origin, with the points distributed *radially* in a Gaussian manner. Each point should be thought of as the values of $\widehat{X}$ and $\widehat{Y}$ obtained from a single measurement. After a large number of measurements, one gets the mean and standard deviation for each variable, which are 0 and 1/2 respectively.

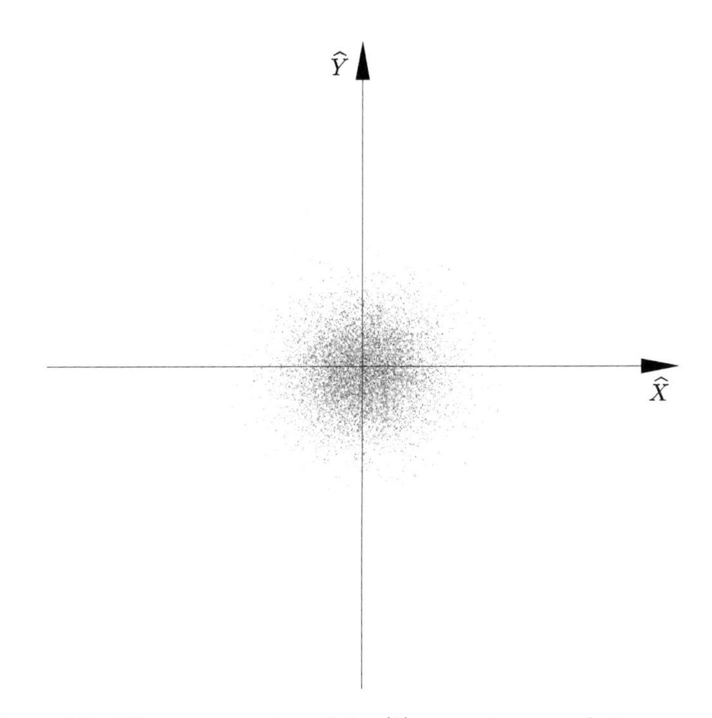

Figure 2.5: Phase space plot of the $|0\rangle$ ground state of the quantum harmonic oscillator, showing the distribution in the quadrature variables $\widehat{X}$ and $\widehat{Y}$ due to the uncertainty principle. Each point is the result of a single shot of measurement of the two variables. The distribution is a circularly symmetric **fuzz ball** at the origin.

C. Coherent states

The number states of a harmonic oscillator are stationary states of the system—the dynamical variables do not oscillate, in fact $\langle \widehat{X} \rangle$ and $\langle \widehat{Y} \rangle$ are identically 0. In order to get non-zero expectation values for $\widehat{X}$ or $\widehat{Y}$ (which oscillate at frequency ω and behave like the classical variables, as a consequence of Ehrenfest's theorem), we need linear superpositions of the $|n\rangle$. As we will see below, coherent states satisfy this requirement; in fact they are the most classical-like states that quantum mechanics will allow. They are important in modern spectroscopy because they represent the state coming out of a laser.

1. Definition and properties

Coherent states are defined to be eigenstates of the annihilation operator a, i.e.

$$a\,|\alpha\rangle = \alpha\,|\alpha\rangle \qquad (\alpha \text{ is complex because } a \text{ is not an observable})$$
$$\Longrightarrow \langle\alpha|\, a^\dagger = \alpha^*\,\langle\alpha|$$

A little bit of algebra shows that its expansion in number states is

$$|\alpha\rangle = e^{-\frac{|\alpha|^2}{2}} \sum_n \frac{\alpha^n}{\sqrt{n!}}\,|n\rangle$$

Some important properties of coherent states are as follows.

(i) Average number of quanta, $\langle n\rangle_\alpha \equiv \langle\alpha|N|\alpha\rangle = |\alpha|^2$.

(ii) Average of n^2, $\langle n^2\rangle_\alpha \equiv \langle\alpha|N\cdot N|\alpha\rangle = |\alpha|^4 + |\alpha|^2$.

(iii) Probability of finding n quanta $\equiv |\,\langle n|\alpha\rangle\,|^2 = e^{-\langle n\rangle_c} \dfrac{(\langle n\rangle_c)^n}{n!}$, which is **Poissonian.**

(iv) $|\alpha\rangle$ and $|\beta\rangle$ are not orthogonal, because $\langle\alpha|\beta\rangle = e^{[|\alpha|^2/2+|\beta|^2/2]} \neq 0$, which is to be expected because a is not an observable.

(v) Coherent states do not form a basis, which is again expected because a is not an observable. In fact, they form an **overcomplete** basis, $\int |\alpha\rangle\langle\alpha|\, d^2\alpha = \pi \neq 1$.

2. Time evolution

The time evolution of a coherent state results in another coherent state. To see this, we evaluate the formal expression for time evolution

$$\begin{aligned}
|\alpha(t)\rangle &= e^{-iHt/\hbar}\,|\alpha\rangle \\
&= e^{-iHt/\hbar}e^{-|\alpha|^2/2}\sum_n \frac{\alpha^n\,|n\rangle}{\sqrt{n!}} \\
&= e^{-|\alpha|^2/2}\sum_n \frac{\alpha^n e^{-i\omega(n+\frac{1}{2})t}\,|n\rangle}{\sqrt{n!}} \\
&= e^{-i\omega t/2}e^{-|\alpha|^2/2}\sum_n \frac{(\alpha e^{-i\omega t})^n\,|n\rangle}{\sqrt{n!}} \\
&= e^{-i\omega t/2}\,|\alpha e^{-i\omega t}\rangle
\end{aligned}$$

The last line shows that the time evolved coherent state is just the original state with a time varying argument whose phase evolves like the classical phase.

From Eq. (2.4), we know that $\langle a\rangle_t = \langle a\rangle_0\, e^{-i\omega t}$. For a coherent state $\langle a\rangle_0 = \alpha$. Therefore

$$\langle a\rangle_t = \alpha e^{-i\omega t} \qquad \Longrightarrow \qquad \langle a^\dagger\rangle_t = \alpha^* e^{i\omega t}$$

Using this we can show that $\langle\widehat{X}\rangle$ and $\langle\widehat{Y}\rangle$ evolve like the corresponding classical variables.

$$\begin{aligned}
\langle\widehat{X}\rangle_t &= \langle\alpha(t)|(a^\dagger + a)|\alpha(t)\rangle\,/2 \\
&= [\langle a^\dagger\rangle_t + \langle a\rangle_t]/2 \\
&= [\alpha^* e^{i\omega t} + \alpha e^{-i\omega t}]/2 \\
&= \mathrm{Re}\{\alpha e^{-i\omega t}\}
\end{aligned}$$

$$\begin{aligned}
\langle\widehat{Y}\rangle_t &= i[\alpha^* e^{i\omega t} - \alpha e^{-i\omega t}]/2 \\
&= \mathrm{Im}\{\alpha e^{-i\omega t}\}
\end{aligned}$$

Thus both $\langle\widehat{X}\rangle_t$ and $\langle\widehat{Y}\rangle_t$ are *real* oscillating functions. The amplitude of $\langle\widehat{X}\rangle_t$ is $|\alpha|$ and its phase is $-$Phase(α). Clearly α is the quantum analog of the complex amplitude $z_\circ$ in the classical oscillator. The quantity $z_\circ e^{-i\omega t}$ is called the **phasor** in classical mechanics; its analog here is $\alpha e^{-i\omega t}$.

3. Coherent states have minimum uncertainty

As we shall now show, coherent states are **minimum uncertainty** states for all values of α. We note that

$$\begin{aligned}\langle\alpha|\widehat{X}^2|\alpha\rangle &= \langle\alpha|(a^\dagger + a)(a^\dagger + a)|\alpha\rangle/4 \\ &= \langle\alpha|a^\dagger a^\dagger + a^\dagger a + aa^\dagger + aa|\alpha\rangle/4 \\ &= (\alpha^{*2} + 2|\alpha|^2 + 1 + \alpha^2)/4 \\ \langle\alpha|\widehat{X}|\alpha\rangle &= (\alpha^* + \alpha)/2\end{aligned}$$

Therefore

$$(\Delta\widehat{X}_\alpha)^2 = [(\alpha^{*2} + 2|\alpha|^2 + 1 + \alpha^2) - (\alpha^* + \alpha)^2]/4 = 1/4$$

Similarly we can show that

$$(\Delta\widehat{Y}_\alpha)^2 = 1/4$$

Hence the coherent state is a minimum uncertainty state with $\Delta\widehat{X}_\alpha = \Delta\widehat{Y}_\alpha = 1/2$.

4. Phase space behavior

The previous discussion shows that the coherent state is exactly like the ground state of the harmonic oscillator in terms of the uncertainty distribution; the difference is that the distribution is centered not at the origin but at α in phase space. Graphically, this means that the fuzz ball at the origin in the $\widehat{X}\widehat{Y}$ plane that we saw in Fig. 2.5 now gets displaced to α. This is shown in Fig. 2.6. At $t = 0$, the center of the fuzz ball is at $\langle\widehat{X}\rangle_0 = \mathrm{Re}\{\alpha\}$ and $\langle\widehat{Y}\rangle_0 = \mathrm{Im}\{\alpha\}$. At later times, the ball rotates counterclockwise along the circular trajectory, which is the orbit followed by the equivalent classical HO. As expected from Ehrenfest's theorem, the center of the fuzz ball is always at the expectation values of the two variables, i.e. at $\langle\widehat{X}\rangle_t = \mathrm{Re}\{\alpha e^{-i\omega t}\}$ and $\langle\widehat{Y}\rangle_t = \mathrm{Im}\{\alpha e^{-i\omega t}\}$ at t.

If we define the displacement operator as

$$D(\alpha) \equiv \exp\left[\alpha a^\dagger - \alpha^* a\right]$$

then the fact that $|\alpha\rangle = D(\alpha)\,|0\rangle$ shows that a coherent state is simply the ground state displaced by α in phase space. Noting that $\langle\widehat{X}\rangle$ and $\langle\widehat{Y}\rangle$ oscillate, one sees that the wave function for a coherent state is an oscillating Gaussian—it's just the wave function $\phi_0(x)$ oscillating back and forth in the potential (and correspondingly in momentum space).

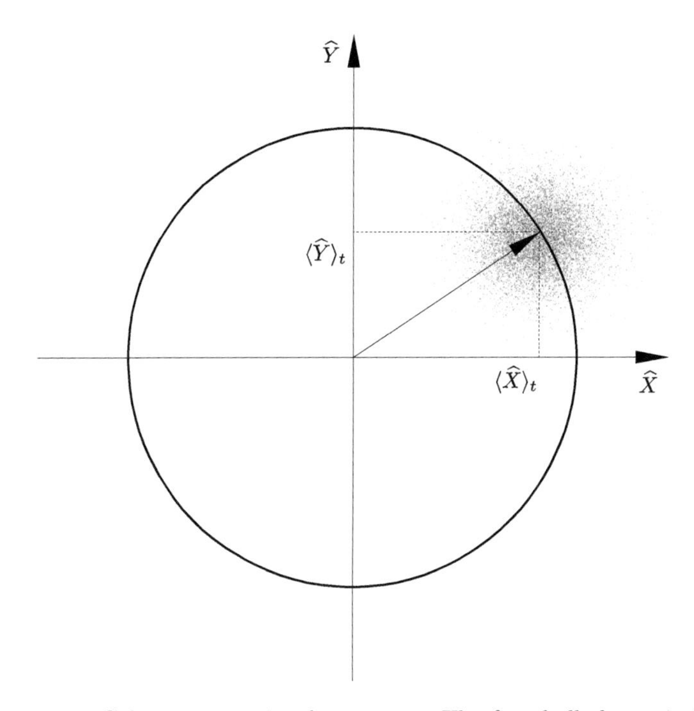

Figure 2.6: Coherent state in phase space. The fuzz ball shown in Fig. 2.5 (representing the minimum uncertainty ground state in the $\widehat{X}$ and $\widehat{Y}$ variables) gets displaced so that its center is located at $\langle\widehat{X}\rangle_t$ and $\langle\widehat{Y}\rangle_t$. They are given by the real and imaginary parts of the complex number $\alpha e^{-i\omega t}$, respectively. α evolves along the circle, which is the trajectory followed by the equivalent classical HO.

D. Squeezed states

Squeezed states are important for precision measurements because they have the potential to get precision better than that set by the uncertainty principle. They are defined as quantum states which have **imbalanced** uncertainties between two conjugate variables, such as P and X in a harmonic oscillator. They must, of course, satisfy the uncertainty principle $\Delta P \Delta X \geq \hbar/2$. But if the signal is encoded in the X variable, one gains in signal-to-noise ratio (SNR) by having a smaller ΔX (if one is limited by quantum noise), while the increased uncertainty in P does not affect the measurement.

A **balanced** distribution of uncertainties means that the quantities ΔP and ΔX remain constant in time. In a harmonic oscillator this occurs when

$$\Delta P = \omega \Delta X$$

If the oscillator is prepared with a different distribution, then ΔP and ΔX will **oscillate**. For example, if ΔP is small and ΔX is big, the potential will cause the portions of the wavepacket at large X to accelerate so that ΔP will increase dramatically during the next quarter period. If one thinks *classically* of an ensemble of oscillators prepared at the bottom of the potential with a statistical distribution having a narrow ΔX and large ΔP, it is clear that they will rapidly spread due to their high initial velocity distribution. A quarter period later they will all have stopped, so ΔP will be a minimum; however, ΔX will be a maximum. This out-of-phase oscillation of uncertainties for conjugate variables is a hallmark of squeezed states; **note that it occurs at** 2ω.

1. Hyperbolic transform of the HO Hamiltonian

We now present a mathematical treatment of squeezing in a harmonic oscillator. The basic idea is to make a hyperbolic rotation of the operators a and $a^\dagger$ to b and $b^\dagger$, a transformation which distorts the $\widehat{X}\widehat{Y}$ phase space while preserving the commutator. A new Hamiltonian with the same form as the old one is obtained, which has number states which are called **generalized number** states. Investigation of the generalized vacuum* will show that this new state has squeezed uncertainties in the old quadrature variables, $\widehat{X}$ and $\widehat{Y}$.

The transformation to the generalized raising and lowering operators is

$$b \equiv a \cosh r - a^\dagger e^{2i\theta} \sinh r$$
$$b^\dagger \equiv a^\dagger \cosh r - a e^{-2i\theta} \sinh r$$

*The ground state is often called the vacuum state even for non-electromagnetic harmonic oscillators.

where r is the distortion parameter, stretching by a factor e^r along an axis at angle θ to the $\widehat{X}$ axis, and shrinking by a factor of e^{-r} perpendicular to this axis, while preserving the area. This is called a Bogoliubov transformation and corresponds closely to a rotation with hyperbolic trigonometric functions. It can be verified that the transformation preserves the commutator, $[b, b^\dagger] = 1$. Defining the unitary squeezing operator $S(r, \theta)$ which does this transformation, i.e. $SaS^{-1} = b$ or $Sa^\dagger S^{-1} = b^\dagger$, one finds

$$S = \exp\left[\frac{r}{2}\left(e^{-2i\theta}a^2 - e^{2i\theta}a^{\dagger 2}\right)\right]$$

Hence the transformed Hamiltonian $H' = SHS^{-1}$ is

$$H' = \hbar\omega(b^\dagger b + 1/2)$$

which has the same form as the original Hamiltonian H. The corresponding quanta are not eigensolutions of H, but correspond to a superposition state that reforms itself after one period, and which has squeezed variances that oscillate twice each period.

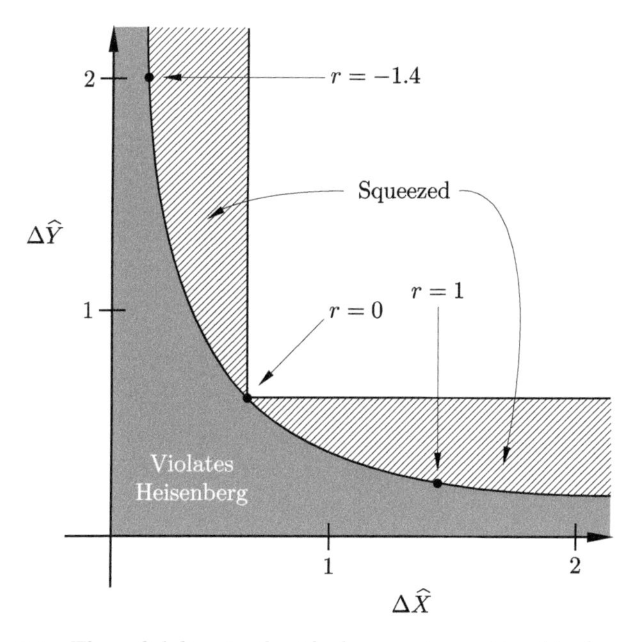

Figure 2.7: The solid line is the ideal squeezing achievable for various values of the squeeze parameter r (with $\theta = 0$). The shaded regions have significant but not ideal squeezing. The region with $\Delta\widehat{X} > 1/2$ and $\Delta\widehat{Y} > 1/2$ is unsqueezed.

In the following we specialize to $\theta = 0$ for clarity. The generalized quadrature operators are defined analogously to $\widehat{X}$ and $\widehat{Y}$

$$\widehat{X}_b = (b^\dagger + b)/2 = \widehat{X}e^{-r}$$
$$\widehat{Y}_b = i(b^\dagger + b)/2 = \widehat{Y}e^{r}$$

showing that the stretching occurs along the $\widehat{X}$ and $\widehat{Y}$ axes. The Hamiltonian then becomes

$$H' = \hbar\omega(\widehat{X}_b^2 + \widehat{Y}_b^2) = \hbar\omega(\widehat{X}^2 e^{-2r} + \widehat{Y}^2 e^{2r})$$

which may be interpreted in two ways.

(i) It may be written using the original position and momentum operators as

$$H' = \frac{P^2}{2me^{-2r}} + \frac{1}{2}me^{-2r}\omega^2 X^2$$

which is an oscillator with the same frequency but mass smaller by e^{2r}.

(ii) Alternatively, one can write

$$H' = \hbar\omega\left[e^{-2r}(\widehat{X}^2 + \widehat{Y}^2) + 2\widehat{Y}^2 \sinh 2r\right]$$

which is the HO Hamiltonian plus a nonlinear perturbation in $\widehat{Y}$.

2. Squeezed vacuum

Since both the commutator and the transformed Hamiltonian, when expressed in terms of b and $b^\dagger$, have exactly the same mathematical form as the HO Hamiltonian when a and $a^\dagger$ were used, the eigensolutions are the same. So we can immediately write down $\Delta\widehat{X}_b = \Delta\widehat{Y}_b = 1/2$. Hence

$$\langle 0_b|\Delta\widehat{X}|0_b\rangle = e^r/2$$
$$\langle 0_b|\Delta\widehat{Y}|0_b\rangle = e^{-r}/2$$

showing squeezing **but maintaining the uncertainty product**. Therefore the generalized vacuum state $|0_b\rangle$ represents a squeezed state with respect to the original oscillator.

How do we express the squeezed vacuum $|0_b\rangle$ in terms of the old states $|n\rangle$? The usual approach is to use the unitary squeezing operator S defined earlier so that

$$|0_b\rangle = S\,|0\rangle = A\left\{|0\rangle + \sum_{n=1}^{\infty}(-1)^n \tanh^{2n} r \left[\frac{(2n-1)(2n-3)\cdots 1}{2n(2n-2)\cdots 2}\right]^{1/2} |2n\rangle\right\}$$

where A is a normalization constant. Note that only even number states are involved. This reflects the appearance of a^2 and $a^{\dagger 2}$ in S, and also means that $|0_b\rangle$ has even parity. For this reason, squeezed states are sometimes referred to as two-photon states. $|0_b\rangle$ is the time-independent state; in $|0_b(t)\rangle$ each $|n\rangle$ evolves with a prefactor $e^{-i(n+1/2)\omega t}$. Since $|0_b\rangle$ involves only even n, it will return to itself (within a phase) in half the period.

In Fig. 2.8, we compare the effect of squeezing on the vacuum state in phase space. The circular fuzz ball for the normal vacuum state that we saw earlier (Fig. 2.5) becomes elliptical for the squeezed vacuum. The squeezing parameters are $r = 0.7$ and $\theta = 0$ (stretched along the $\widehat{X}$ axis).

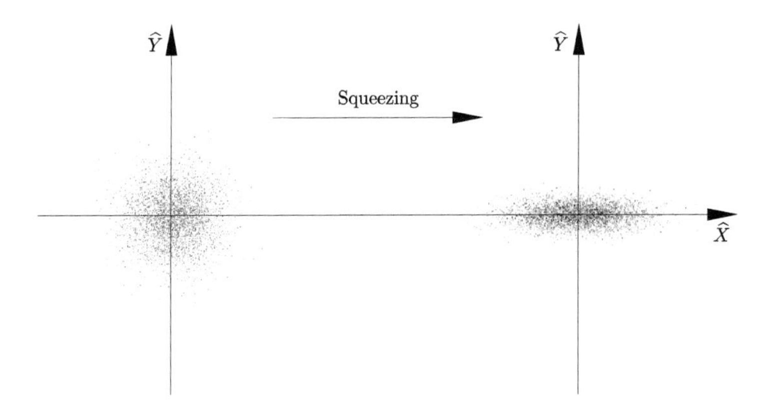

Figure 2.8: Phase space plot showing how the circular distribution for the normal vacuum (left) gets transformed into an elliptic distribution for the squeezed vacuum (right). The squeezing parameters are $r = 0.7$ and $\theta = 0$.

3. Classical squeezing by FM at $2\omega_\circ$

Anyone who has played on a swing as a child knows that one can *self* amplify the motion by modulating the length at twice the oscillation frequency, a process called **parametric amplification**. One learns instinctively to crouch at the bottom (increasing the length of the "pendulum") and stretch at the ends (decreasing the length), thus changing the length twice every period. It is called **parametric** because what is being modulated is a parameter that determines the frequency. Here, we will see that such frequency modulation results in squeezing of the distribution of the quadrature components of motion of an ensemble of (classical) harmonic oscillators in thermal equilibrium.

Consider a mass m which moves in the following potential

$$U(t) = \frac{1}{2}kx^2 + \frac{\varepsilon}{2}kx^2 \sin 2\omega_\circ t \qquad \text{with } \omega_\circ = \sqrt{k/m}$$

This is a harmonic oscillator potential which is modulated in frequency by a small amount (i.e. $\varepsilon \ll 1$) at twice its natural frequency. The corresponding equation of motion is

$$\ddot{x} + \omega_\circ^2 x = -\omega_\circ^2 x \varepsilon \sin 2\omega_\circ t \tag{2.5}$$

Since ε is small, we guess the solution

$$x(t) = B(t)\cos\omega_\circ t + C(t)\sin\omega_\circ t$$

This is the usual solution for an undamped harmonic oscillator *except* that B and C are not constants but functions of time. But the time variation is taken to be very slow, slow enough that second derivatives can be neglected. Therefore

$$\ddot{x}(t) \approx -\omega_\circ^2 x(t) - \omega_\circ \dot{B}\sin\omega_\circ t + \omega_\circ \dot{C}\cos\omega_\circ t$$

and substitution in the equation of motion [Eq. (2.5)] yields

$$-\omega_\circ \dot{B}\sin\omega_\circ t + \omega_\circ \dot{C}\cos\omega_\circ t = \omega_\circ^2 \varepsilon [B\cos\omega_\circ t + C\sin\omega_\circ t]\sin 2\omega_\circ t$$

Using trigonometric identities and averaging away the terms rapidly oscillating at $3\omega_\circ$ gives

$$-\dot{B}\sin\omega_\circ t + \dot{C}\cos\omega_\circ t = \frac{-\varepsilon\omega_\circ}{2}[B\sin\omega_\circ t + C\cos\omega_\circ t]$$

The coefficients of $\sin\omega_\circ t$ and $\cos\omega_\circ t$ must be separately equal, hence

$$\dot{B} = +\frac{\varepsilon\omega_\circ}{2}B \qquad \dot{C} = -\frac{\varepsilon\omega_\circ}{2}C$$

Therefore the coefficients $B(t)$ and $C(t)$ are

$$B(t) = B_\circ e^{+\varepsilon\omega_\circ t/2} \qquad C(t) = C_\circ e^{-\varepsilon\omega_\circ t/2}$$

showing that one quadrature component increases exponentially with time at the expense of the other. Note that the parametric drive increases the total energy of the oscillator, the signature of any amplification process.

If we consider an ensemble of identical harmonic oscillators in equilibrium with a thermal bath, then both quadrature components of the motion will have a Gaussian distribution. By applying the FM drive discussed above, one quadrature component increases and the other decreases. Thus the statistical distribution of the ensemble gets progressively squeezed. Consistent with Liouville's theorem, the coherent nature of the drive means that the area in phase space (proportional to the entropy) cannot change, but the shape can become elongated. This statistical squeezing of an ensemble of classical oscillators is entirely analogous to the quantum mechanical squeezing of the uncertainty principle distribution of a state. The squeezing parameter is $r = \varepsilon\omega_\circ t/2$, showing that the degree of squeezing increases with time.

4. Generating squeezed light

The classical calculation of squeezing by quadrature-sensitive parametric amplification using a **pump** at $2\omega_\circ$ and a coupling potential which contains a nonlinear coupling term x^2 provides suggestions as to how squeezing might be accomplished in a quantum mechanical system. First of all, a nonlinear coupling to the quantum oscillator to be squeezed is needed—i.e. the perturbation Hamiltonian must contains a term like

$$\widehat{X}^2 \sim (a^\dagger + a)(a^\dagger + a) = a^{\dagger 2} + (a^\dagger a + aa^\dagger) + a^2$$

The $(a^\dagger a + aa^\dagger)$ portion can be absorbed into the HO Hamiltonian, leaving

$$\Delta H = Ka^{\dagger 2} + K^* a^2$$

a form which is Hermitian and which is (reassuringly) like the exponent of the squeezing operator.

The first two ways to generate squeezing which were experimentally demonstrated involved parametric down conversion and four-wave mixing.

(i) In parametric down conversion, the pump is a strong coherent wave at frequency 2ω and

$$\Delta H = \chi^{(2)}[a_{2\omega} a^{\dagger 2} + a_{2\omega}^\dagger a^2]$$

where $a_{2\omega}$ is the destruction operator for pump photons. For a coherent pump in a state $|\alpha_{2\omega}\rangle$ the expectation value of $a_{2\omega}$ is

$$\langle a_{2\omega}\rangle = \text{Re}\{\alpha_{2\omega}e^{-2\omega_\circ t}\}$$

so $a_{2\omega}$ acts like a classical pump at frequency $2\omega_\circ$. The term $a_{2\omega}a^{\dagger 2}$ destroys one pump photon at $2\omega_\circ$ and generates two photons with definite relative phase at $\omega_\circ$, conserving energy.

(ii) In four-wave mixing

$$\Delta H = \chi^{(3)}[a'^2(a^\dagger)^2 + (a'^\dagger)^2a^2]$$

where a' and $a'^\dagger$ represent annihilation and creation operators at frequency $\omega_\circ$, but for a mode field different from a and $a^\dagger$ (e.g. laser beams going in different directions).

The factors $\chi^{(n)}$ in the foregoing perturbation interactions are termed n^{th} order nonlinear susceptibilities, which will be discussed in a later chapter.

Why/how do perturbation Hamiltonians like those above generate squeezed light? The most straightforward way to see this is to recall the time evolution operator $U(t)$ in the Schrödinger picture, where $H_\circ = \hbar\omega(a^\dagger a + 1/2)$ and ΔH is given above. This shows that the time evolution of wave function is given by

$$\psi(t) = \psi(0)e^{-i(H_\circ+\Delta H)t/\hbar}$$

Thus (after a disentangling theorem is used)

$$\psi(t) = e^{-H_\circ t/\hbar}\psi(0)e^{-i\Delta Ht/\hbar}$$

Since ΔH is the squeezing operator, an initially unsqueezed state will become progressively squeezed. The amount of squeezing is proportional to the time, just as it was in the classical case.

5. Balanced homodyne detector

A balanced homodyne detector, shown schematically in Fig. 2.9, is the optical analog of the ubiquitous lock-in amplifier found in experimental labs. It allows detection of a weak signal in a *phase sensitive* manner with noise below the vacuum fluctuations.

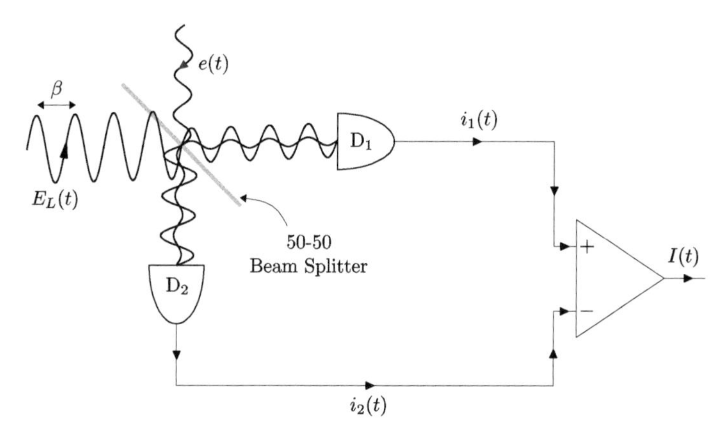

Figure 2.9: Schematic of an optical balanced homodyne detector.

Balanced refers to subtracting the outputs of the two detectors to get the output,

$$I(t) = i_1(t) - i_2(t)$$

Homodyne means the local oscillator has the same frequency as the signal.

$$\begin{aligned} e(t) &= \widehat{Y}\cos\omega t - \widehat{X}\sin\omega t && \text{(weak signal)} \\ E_L(t) &= E_L\cos(\omega t + \beta) && \text{(local oscillator signal)} \\ E_{LN}(t) &= \text{noise in local oscillator} && \text{(at least vacuum fluctuations)} \end{aligned}$$

β is an experimenter-adjustable phase that allows detection of any phase of the signal with respect to the local oscillator. For example, many applications may require the *in-phase* component to be detected, in which case β will be set to 0.

The photodetector currents are

$$\begin{aligned} i_1(t) &\propto [E_L(t) + E_{LN}(t) + e(t)]^2 && \text{(}\textbf{square law}\text{ detector)} \\ &\simeq E_L^2(t) + 2E_L(t)E_{LN}(t) + 2E_L(t)e(t) && \text{(neglect } e^2 \text{ and } eE_{LN}\text{)} \end{aligned}$$

$$i_2(t) \propto E_L^2(t) + 2E_L(t)E_{LN}(t) - 2E_L(t)e(t)$$

The $-$ve sign arises at the beamsplitter; the interference term between E_L and e must have opposite signs at the two detectors to make the power out equal to the power in.

Thus the final detected current (after using a low pass filter to remove 2ω components) is

$$I(t) \propto 4E_L(t)e(t) = 2E_L[\widehat{Y}(t)\cos\beta + \widehat{X}(t)\sin\beta]$$

This shows the insensitivity of a balanced homodyne detector to technical and vacuum noise in the local oscillator.

If we are interested in measuring a weak signal like that from a gravitational wave, then we will gain in sensitivity by using a squeezed state for $e(t)$. The signature of such a state will show up in the statistics of $I(t)$—it will be Poissonian for a normal coherent state and **sub-Poissonian** for the squeezed state.

E. Radiation

We start by describing radiation in a classical electromagnetic field, and take up quantization* later. This approach of treating the radiation classically while treating the atom quantum mechanically is called **semiclassical**. Although fundamentally inconsistent, it provides a natural and intuitive approach to the study of interaction of EM radiation with atomic systems. Furthermore, it is completely justified in cases where the radiation fields are large, i.e. there are many photons in each mode, as for example in the case of microwave or laser spectroscopy. The chief disadvantage is that this treatment does not predict **spontaneous emission**, thus forcing us to use complex eigenvalues to mimic the presence of this decay in excited states. Nevertheless, phenomenological properties such as selection rules, radiation rates and cross sections, can be developed naturally with this approach.

Classical electrodynamics is based on the following set of four Maxwell's equations

$$\nabla \cdot \vec{E} = 4\pi\rho$$

$$\nabla \times \vec{E} = -\frac{1}{c}\frac{\partial \vec{B}}{\partial t}$$

$$\nabla \cdot \vec{B} = 0$$

$$\nabla \times \vec{B} = \frac{4\pi \vec{j}}{c} + \frac{1}{c}\frac{\partial \vec{E}}{\partial t}$$

These equations are for the $\vec{E}$ and $\vec{B}$ fields; their form allows us to define potentials—a scalar potential ϕ and a vector potential $\vec{A}$. In terms of the potentials, the two fields are

$$\vec{E} = -\nabla\phi - \frac{1}{c}\frac{\partial \vec{A}}{\partial t} \qquad \text{and} \qquad \vec{B} = \nabla \times \vec{A}$$

In quantum mechanics, the potentials are considered more fundamental than the fields, as exemplified by the **Aharonov–Bohm effect** where the electron wavefunction picks up a phase in a region where $\vec{B}$ is zero but $\vec{A}$ is non-zero. As we will see below, even the quantization of the radiation field relies on defining operators for $\vec{A}$.

Since the curl of gradient of any function is zero, we can always add the gradient of any function to the vector potential without changing $\vec{B}$—this is called **gauge invariance**. Of course, to keep $\vec{E}$ unchanged, ϕ must also be changed appropriately. Hence, the complete gauge transformation which keeps the fields unchanged is

$$\vec{A} \to \vec{A} + \nabla\chi \qquad \text{and} \qquad \phi \to \phi - \frac{1}{c}\frac{\partial \chi}{\partial t}$$

*My personal view on the nature of a photon is detailed in Appendix B, "What Is a Photon?"

If we use the following **Lorenz gauge**

$$\nabla \cdot \vec{A} + \frac{1}{c}\frac{\partial \phi}{\partial t} = 0$$

then the Maxwell's equations result in the following **decoupled** wave equations for the potentials

$$\begin{aligned} \nabla^2 \phi - \frac{1}{c^2}\frac{\partial^2 \phi}{\partial t^2} &= -4\pi\rho \\ \nabla^2 \vec{A} - \frac{1}{c^2}\frac{\partial^2 \vec{A}}{\partial t^2} &= -\frac{4\pi\vec{j}}{c} \end{aligned} \tag{2.6}$$

1. Field modes are classical oscillators

Light is just an EM wave propagating in a **source-free** region, i.e. in a region where $\rho = 0$ and $\vec{j} = 0$, which implies that the wave equations in Eq. (2.6) reduce to their homogeneous form. In such a region, we can make a gauge transformation that eliminates ϕ by choosing χ such that*

$$\phi = \frac{1}{c}\frac{\partial \chi}{\partial t}$$

Therefore the Lorenz gauge becomes what is called the **radiation gauge**

$$\nabla \cdot \vec{A} = 0$$

If we divide the vector potential into its longitudinal and transverse components —$\vec{A}_{\|}$ and $\vec{A}_{\perp}$—and note that

$$\nabla \cdot \vec{A}_{\perp} = 0 \qquad \text{and} \qquad \nabla \times \vec{A}_{\|} = 0$$

then the radiation gauge implies that only transverse fields survive in the radiation zone. This is in fact the defining characteristic of radiation—**that the $\vec{E}$ and $\vec{B}$ fields are transverse**. One way to think about this is to say that the gauge freedom allows the longitudinal component of the vector potential to cancel the scalar potential, so that only its transverse component survives. Thus in the radiation zone $\vec{A}$ satisfies the homogeneous wave equation

$$\nabla^2 \vec{A} - \frac{1}{c^2}\frac{\partial^2 \vec{A}}{\partial t^2} = 0 \qquad \text{with } \nabla \cdot \vec{A} = 0 \tag{2.7}$$

Radiation fields are found by solving the above wave equation. Each mode with a particular value of the wave vector $\vec{k}$ and the polarization λ is independent. The polarization, defined as the path traversed by the tip of the

*Important: This can be done only in a source-free region, and is not relativistically covariant.

electric field vector—**linear** if it oscillates along a line, **circular** if it traverses a circle, and **elliptic** in general—can take on two independent values. If we define the plane perpendicular to the direction of propagation as the xy plane and choose suitable unit vectors, then the two sets of orthogonal polarizations are

$$\hat{\varepsilon}_{\text{lin}} = \hat{\varepsilon}_x \pm \hat{\varepsilon}_y \qquad \text{orthogonal linear}$$
$$\hat{\varepsilon}_{\text{circ}} = \hat{\varepsilon}_x \pm i\hat{\varepsilon}_y \qquad \text{right or left circular}$$

The vector potential that solves Eq. (2.7) is (with $\omega_k = c|\vec{k}|$)

$$\vec{A}(\vec{r},t) = \sum_{k,\lambda} \hat{\varepsilon}_{k\lambda}\left[A_{k\lambda}e^{i(\vec{k}\cdot\vec{r}-\omega_k t)} + A^*_{k\lambda}e^{-i(\vec{k}\cdot\vec{r}-\omega_k t)}\right] \qquad \text{with } \hat{\varepsilon}\cdot\vec{k} = 0 \tag{2.8}$$

in which the spatial modes are traveling waves (with periodic boundary conditions). The corresponding fields are (with $\theta \equiv \vec{k}\cdot\vec{r} - \omega_k t$)

$$\vec{E}(\vec{r},t) = -\frac{1}{c}\frac{\partial \vec{A}}{\partial t} = \frac{i}{c}\sum_{k,\lambda}\omega_k\hat{\varepsilon}_{k\lambda}\left[A_{k\lambda}e^{i\theta} - A^*_{k\lambda}e^{-i\theta}\right]$$
$$\vec{B}(\vec{r},t) = \nabla\times\vec{A} = \frac{i}{c}\sum_{k,\lambda}\omega_k(\hat{k}\times\hat{\varepsilon}_{k\lambda})\left[A_{k\lambda}e^{i\theta} - A^*_{k\lambda}e^{-i\theta}\right] = \hat{k}\times\vec{E}(\vec{r},t)$$

which shows that the three vectors $\vec{k}$, $\vec{E}$, and $\vec{B}$ form a mutually orthogonal set, with the fields being transverse to the propagation direction.

Since the energy is stored in the fields, it is given by

$$W = \frac{1}{8\pi}\int_V \left(\vec{E}\cdot\vec{E} + \vec{B}\cdot\vec{B}\right) dV$$

The expressions for $\int \vec{E}\cdot\vec{E}$ and $\int \vec{B}\cdot\vec{B}$ contain several terms, but volume integrating and adding them together gives the simpler expression

$$W = \frac{V}{2\pi}\sum_{k,\lambda}\omega_k^2 A_{k\lambda}A^*_{k\lambda}$$

Before proceeding further, we will make the time variation *implicit*, so that the expressions may be taken straight over to quantum mechanical operator expressions in the Schrödinger picture, where operators (but not their expectation values) are time independent.

Now define new variables:

$$Q_{k\lambda} \equiv \sqrt{V/4\pi}\,[A_{k\lambda} + A^*_{k\lambda}] \qquad \propto \text{real part of } A_{k\lambda}$$
$$P_{k\lambda} \equiv -i\omega_k\sqrt{V/4\pi}\,[A_{k\lambda} - A^*_{k\lambda}] \qquad \propto \text{imaginary part of } A_{k\lambda} \tag{2.9}$$

In terms of these variables

$$W = \frac{1}{2}\sum_{k\lambda}\left(P_{k\lambda}^2 + \omega_k^2 Q_{k\lambda}^2\right) \equiv H_{\text{EM}} \qquad \text{E-M field Hamiltonian}$$

which looks like a harmonic oscillator Hamiltonian. Using H_{EM}, one can show that P and Q are **canonical variables**, and hence their time variations are

$$\dot{Q}_{k\lambda} = \frac{\partial H_{\text{EM}}}{\partial P_{k\lambda}} = P_{k\lambda} \qquad \text{and} \qquad \dot{P}_{k\lambda} = -\frac{\partial H_{\text{EM}}}{\partial Q_{k\lambda}} = -\omega_{k\lambda}^2 Q_{k\lambda}$$

Therefore each mode of the radiation field is a **harmonic oscillator**. The energy swaps between the electric and magnetic fields, analogous to the way it swaps between kinetic and potential energies in a mechanical oscillator.

2. Quantization

Comparison of Eq. (2.2) with Eq. (2.9) shows that the role of the lowering operator a in the quantum HO is played by the vector potential $\vec{A}$ in the radiation field. Therefore, quantization of the radiation fields is accomplished simply by replacing the $A_{k\lambda}$ and $A_{k\lambda}^*$ in the classical expression for the modes [Eq. (6.1)] by $a_{k\lambda}$ and $a_{k\lambda}^\dagger$, multiplied by $\sqrt{hc^2/(\omega_k V)}$ to get the normalization right in a box of volume V.* Thus the operator for the vector potential is (here alone we use boldface to indicate that it is a quantum mechanical operator)

$$\vec{\mathbf{A}}_{k\lambda}(\vec{r}) = \sqrt{\frac{hc^2}{\omega_k V}}\,\hat{\varepsilon}_{k\lambda}\left[a_{k\lambda}e^{i\vec{k}\cdot\vec{r}} + a_{k\lambda}^\dagger e^{-i\vec{k}\cdot\vec{r}}\right]$$

The operators for the associated $\vec{E}$ and $\vec{B}$ fields are

$$\vec{\mathbf{E}}_{k\lambda}(\vec{r}) = i\sqrt{\frac{h\omega_k}{V}}\,\hat{\varepsilon}_{k\lambda}\left[a_{k\lambda}e^{i\vec{k}\cdot\vec{r}} - a_{k\lambda}^\dagger e^{-i\vec{k}\cdot\vec{r}}\right]$$

$$\vec{\mathbf{B}}_{k\lambda}(\vec{r}) = i\sqrt{\frac{h\omega_k}{V}}\,(\hat{k}\times\hat{\varepsilon}_{k\lambda})\left[a_{k\lambda}e^{i\vec{k}\cdot\vec{r}} - a_{k\lambda}^\dagger e^{-i\vec{k}\cdot\vec{r}}\right]$$

Note that the operators $\vec{\mathbf{A}}$, $\vec{\mathbf{E}}$, and $\vec{\mathbf{B}}$ are all Hermitian, since they are **observables**.

As an example of how we use this formalism, we calculate the expectation

*This is equivalent to the quantization procedure followed for the harmonic oscillator, but we choose this slightly different approach in order to make the role of $\vec{A}$ explicit.

value of the E field in a coherent state $|\alpha\rangle$

$$\begin{aligned}\langle \vec{E}(\vec{r},t)\rangle_\alpha &\equiv \langle \alpha(t)|\mathbf{E}_{k\lambda}(\vec{r})|\alpha(t)\rangle \\ &= i\sqrt{\frac{\hbar\omega_k}{V}}\hat{\varepsilon}_{k\lambda}\left[\langle \alpha(t)|a_{k\lambda}|\alpha(t)\rangle\, e^{i\vec{k}\cdot\vec{r}} - \text{c.c.}\right] \\ &= i\sqrt{\frac{\hbar\omega_k}{V}}\hat{\varepsilon}_{k\lambda}\left[\alpha e^{i\vec{k}\cdot\vec{r}-\omega t} - \text{c.c.}\right]\end{aligned}$$

Therefore

$$\langle \vec{E}(\vec{r},t)\rangle_\alpha = -2\sqrt{\frac{\hbar\omega_k}{V}}\,\hat{\varepsilon}_{k\lambda}\,\text{Im}\left\{\alpha e^{i(\vec{k}\cdot\vec{r}-\omega t)}\right\}$$

3. Zero-point energy and fields

Our quantization procedure for the radiation field now leads to a puzzling set of infinities associated with the zero-point. This arises because of the **zero-point energy** for the vacuum state of each of the infinitely many modes of the field. The energy may be found using the operator for the energy of a volume V filled with EM radiation, which is

$$\mathbf{W} = \frac{1}{2}\sum_{k,\lambda}\hbar\omega_k\left(a_{k\lambda}a^\dagger_{k\lambda} + a^\dagger_{k\lambda}a_{k\lambda}\right)$$

The zero-point energy is the expectation value of W in the vacuum state

$$\langle W\rangle_0 = \frac{1}{2}\sum_{k,\lambda}\hbar\omega_k$$

which (not surprisingly) is the same as the zero-point energy for the quantized harmonic oscillator in the $|0\rangle$ state. The total energy is **infinite** because there are infinite modes that are allowed. But we can get a **finite** value for this energy by defining a cut-off frequency ω_c. Noting that the k index is continuous and using the density of states $\rho(\omega)$, we have

$$\sum_k \to \int_0^{\omega_c}\rho(\omega_k)d\omega_k \qquad \text{and} \qquad \sum_\lambda = 2$$

Using $\rho(\omega) = V\omega^2/(\pi^2c^3)$, we get

$$W(\omega_c) = \frac{\hbar\omega_c^4}{4\pi^2c^3}V$$

In the vacuum state the eigenvalue of the number operator is known (i.e. $n_{k\lambda} = 0$) so the fields must have non-zero variances to satisfy the uncertainty principle (the number operator does not commute with the operators for the $\vec{E}$ and $\vec{B}$ fields). These are called the **vacuum fluctuations** of the field. Since $\langle \vec{E} \rangle = 0$,

$$(\Delta \vec{E})^2 = \langle \vec{E}^2 \rangle = \frac{h\omega_k}{V}$$

For most purposes the zero-point energy and zero-point field fluctuations can be neglected, as they almost invariably are. But we know that for a single harmonic oscillator, e.g. a diatomic molecule, the zero-point energy is real—it is responsible for the isotope shift in the binding energy of molecules (the lowest vibrational state of the heavier isotopes nestles down a little bit further into the potential). There are also other pieces of evidence for the reality of the vacuum fluctuations of the fields. For example, we have already seen that it is possible not only to detect the field fluctuations using an optical homodyne detector, but to reduce them by replacing the normal vacuum with squeezed light. In addition, our present "model" for spontaneous emission from the excited state of an atom is really *stimulated* emission, but one where the stimulation is by the vacuum fields. That is why spontaneous emission is not present in the semiclassical model where we do not quantize the radiation fields. One consequence of this model is that the spontaneous emission rate can be reduced, e.g. by suppressing the vacuum modes by placing the atom in a photonic bandgap material, or by using a control laser on an auxiliary transition, both of which have been demonstrated.

Another example of the reality of the vacuum field fluctuations is Unruh radiation—a mirror accelerating at a scatters the vacuum fluctuations producing radiation with a characteristic temperature of $\hbar a/2\pi c k_B$. Future experiments may be able to observe this in spite of its small size. For a mirror accelerating from zero to c in 3 ps, $a = 10^{13}$ m/s^2 and the temperature is 0.04 K; obviously a non-physical mirror such as the edge of a plasma breakdown must be used. Incidentally, an analogous process in which a moving object scatters zero-point acoustic fluctuations is one answer to the Zen question "What is the sound of one hand clapping?"!

However real the zero-point energy may seem in certain cases, there is good justification for ignoring it. This is provided by the lack of observed gravitational effect—the zero-point energy (divided by c^2) around the Sun would act as a spherically symmetric mass density and would cause a deviation from the GM/r gravitational potential, something which is accurately inferred from planetary motion.

4. Casimir effect

Though the previous discussion raises some questions about the reality of the zero-point energy and fields, changes in these quantities are definitely observable. One such manifestation is the **Casimir force**—a long range force among any pair of atoms, electrons, or walls (conducting or dielectric). In the case of two conducting walls, the exclusion of some of the vacuum modes in the region between the two planes leads to a position-dependent change in the vacuum energy and hence to a force.

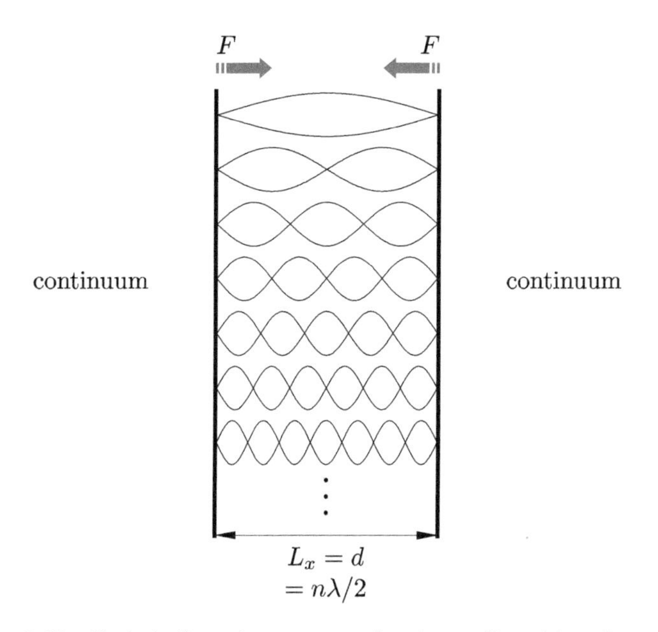

Figure 2.10: Casimir force between conducting walls arising due to exclusion of vacuum modes. The boundary condition on the walls implies that only standing waves of the form shown are allowed, while outside all modes are allowed.

To see this quantitatively, we assume that the two conducting walls have size $L_y = L_z = L$, and separation $L_x = d$, as shown in Fig. 2.10. The boundary condition on the walls implies that the electric field is zero there; therefore the only allowed waves between the walls are standing waves of the form (similar to the modes in a string held between two rigid points)

$$\sin k_n x \qquad \text{with } k_n = n\pi/d$$

which forms a discrete set. Outside the walls all modes are allowed. This difference of vacuum modes leads to the Casimir force. Even though the total energy on both sides is infinite, the difference is finite and can be derived as

$$\Delta W = -\frac{\pi^2 \hbar c}{720\, d^3} L^2$$

This implies that the force *per unit area* is

$$F_{\text{Casimir}} = -\frac{1}{L^2}\frac{\partial \Delta W}{\partial d} = \frac{\pi^2}{240}\frac{\hbar c}{d^4}$$

which shows that the force scales as d^{-4}. To give an idea of its size, it is 1.3×10^{-3} N/m^2 at a distance of 1 µm.

F. Problems

1. An electron on a spring

Consider a model of an atom as an electron on a spring.

(a) Use Larmor's expression for the power radiated ($P_L = 2a^2/3c^3$) to find the energy damping rate for this oscillator Γ.

(b) Find the Q and express it in atomic units.

(c) For $\lambda = 589$ nm (Na D lines), find Q and the corresponding decay time of the atom.

Solution

(a) The position of the electron is given by

$$x(t) = A \sin \omega t$$

where A is the amplitude of the motion. This gives

$$\dot{x} = A\omega \cos \omega t$$
$$\ddot{x} = -A\omega^2 \sin \omega t$$
$$\ddot{x} = a \qquad \Longrightarrow \qquad a^2 = A^2\omega^4 \sin^2 \omega t$$

If $Q \gg 1$, the above expressions are true even in the presence of radiation damping, because the amplitude does not change significantly in one cycle.

Average of $\sin^2 \theta$ over a cycle is

$$\frac{1}{2\pi}\int_0^{2\pi} \sin^2 \theta \, d\theta = \frac{1}{2\pi}\int_0^{2\pi} \frac{1-\cos 2\theta}{2} \, d\theta = \frac{1}{2\pi}\left[\frac{\theta}{2} - \frac{\sin 2\theta}{4}\right]_0^{2\pi} = \frac{1}{2}$$

Therefore

$$\langle a^2 \rangle = \frac{1}{2}A^2\omega^4 \qquad \Longrightarrow \qquad \langle P_L \rangle = \frac{2e^2}{3c^3}\frac{A^2\omega^4}{2}$$

Energy stored in the oscillator is $m\omega^2A^2/2$. Therefore the damping rate is

$$\Gamma = \frac{\text{Power lost}}{\text{Energy stored}} = \frac{2\omega^2e^2}{3c^3m}$$

(b) The Q of the oscillator is

$$Q = \frac{\omega}{\Gamma} = \frac{3c^3 m}{2\omega e^2} = \frac{3}{4\pi\alpha^2}\left(\frac{\lambda}{a_\circ}\right) \qquad \leftarrow \text{in atomic units}$$

where the term in brackets is the wavelength of the radiation in units of the Bohr radius.

(c) For $\lambda = 589$ nm $= 11130.5\, a_\circ$,

$$\omega = 3.19 \times 10^{15} \text{ rad/s}$$

$$Q = 4.99 \times 10^{7}$$

$$\tau = \frac{1}{\Gamma} = \frac{Q}{\omega} = 1.563 \times 10^{-8} \text{ s}$$

2. Quantum harmonic oscillator

Consider a one-dimensional harmonic oscillator of mass m and frequency ω which is in number state $|n\rangle$.

(a) Find both the average and rms position and momentum (four things in all).

(b) Check your results using energy and the virial theorem.

(c) Sketch the wavefunction $\psi_n(x)$ for $n = 0$ and 1.

(d) If a Na atom is confined in the $|0,0,0\rangle$ state of a trap with oscillation frequency $\omega = 2\pi \times 10^2$ rad/s, what is its rms size and velocity?

Solution

(a) The operators for position and momentum are

$$X = \sqrt{\frac{\hbar}{2m\omega}}\left(a^\dagger + a\right)/2$$

$$P = i\sqrt{\frac{m\hbar\omega}{2}}\left(a^\dagger - a\right)/2$$

This shows that $\langle n|X|n\rangle = \langle n|P|n\rangle = 0$, i.e.

$$\text{average position} = 0$$

$$\text{average momentum} = 0$$

In order to find the rms values of position and momentum, we note that

$$\langle(\Delta X)^2\rangle = \langle n|X^2|n\rangle - (\langle n|X|n\rangle)^2 = \langle n|X^2|n\rangle$$
$$\langle(\Delta P)^2\rangle = \langle n|P^2|n\rangle - (\langle n|P|n\rangle)^2 = \langle n|P^2|n\rangle$$

Using

$$X^2 = \frac{\hbar}{2m\omega}\left(a^{\dagger 2} + aa^\dagger + a^\dagger a + a^2\right)$$
$$P^2 = -\frac{m\hbar\omega}{2}\left(a^{\dagger 2} - aa^\dagger - a^\dagger a + a^2\right)$$

and with

$$\langle n|aa^\dagger + a^\dagger a|n\rangle = \langle n|2a^\dagger a + 1|n\rangle = 2n + 1$$

we get the rms values as

$$x_{\text{rms}} \equiv \langle\Delta X\rangle = \sqrt{\frac{\hbar}{m\omega}\left(n + \frac{1}{2}\right)}$$
$$p_{\text{rms}} \equiv \langle\Delta P\rangle = \sqrt{m\hbar\omega\left(n + \frac{1}{2}\right)}$$

(b) According to virial theorem for a potential $V \sim x^2$, we have $\langle V\rangle = \langle K\rangle = E/2$, where

$$\langle V\rangle = \frac{1}{2}m\omega^2\langle X^2\rangle \qquad \text{and} \qquad \langle K\rangle = \frac{1}{2m}\langle P^2\rangle$$

From part (a) we see that

$$\langle V\rangle_n = \left(n + \frac{1}{2}\right)\frac{\hbar\omega}{2} = \frac{E_n}{2}$$
$$\langle K\rangle_n = \left(n + \frac{1}{2}\right)\frac{\hbar\omega}{2} = \frac{E_n}{2}$$

which shows that the previous results are consistent with the virial theorem.

(c) Defining $\sigma = \sqrt{\hbar/(m\omega)}$, we have

$$\psi_0(x) = \left(\frac{1}{\sigma\sqrt{\pi}}\right)^{1/2}\exp\left(-\frac{x^2}{2\sigma^2}\right)$$
$$\psi_1(x) = \left(\frac{2}{\sigma\sqrt{\pi}}\right)^{1/2}\frac{x}{\sigma}\exp\left(-\frac{x^2}{2\sigma^2}\right)$$

These two wavefunctions are sketched in the figure below.

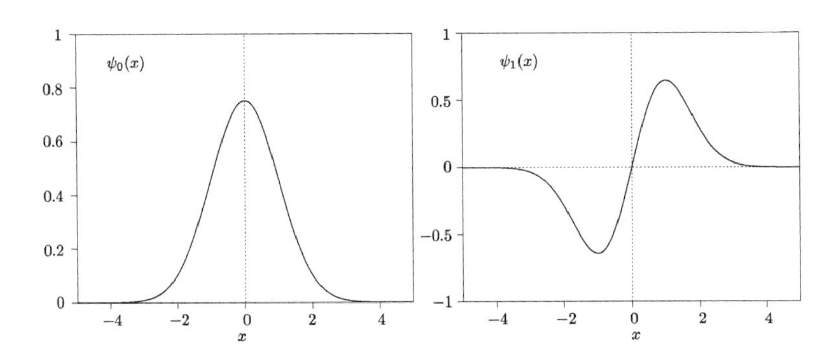

(d) For $n = 0$, the results of part (a) show that

$$\text{rms size} \equiv q_{\text{rms}} = \sqrt{\frac{\hbar}{2m\omega}}$$

$$\text{rms velocity} \equiv \frac{p_{\text{rms}}}{m} = \sqrt{\frac{\hbar\omega}{2m}}$$

For a particle in a state with $n_x = n_y = n_z = 0$

$$\text{rms size} = \sqrt{x_{\text{rms}}^2 + y_{\text{rms}}^2 + z_{\text{rms}}^2} = \sqrt{3}\,q_{\text{rms}} = \sqrt{\frac{3\hbar}{2m\omega}}$$

$$\text{rms velocity} = \sqrt{v_{x,\text{rms}}^2 + v_{y,\text{rms}}^2 + v_{z,\text{rms}}^2} = \sqrt{3}\,\frac{p_{\text{rms}}}{m} = \sqrt{\frac{3\hbar\omega}{2m}}$$

Na has mass of 23 amu $= 3.8 \times 10^{-26}$ kg, so with $\omega = 2\pi \times 10^2$ rad/s the values are

$$\text{rms size} = 2.6 \times 10^{-6}\ \text{m}$$

$$\text{rms velocity} = 1.6 \times 10^{-3}\ \text{m/s}$$

3. Damping in a driven harmonic oscillator

Explain qualitatively and physically. When driven far from resonance the power dissipated in a classical damped oscillator increases linearly with the damping Γ, but on resonance it varies as Γ^{-1}. Why does reducing the damping increase the power dissipated on resonance?

Solution

The amplitude of the response of a driven oscillator is

$$A(\omega) = \frac{F_d/m}{\sqrt{(\omega_\circ^2 - \omega^2)^2 + \Gamma^2\omega^2}} \approx \frac{F_d/(2m\omega_\circ)}{\sqrt{(\omega - \omega_\circ)^2 + (\Gamma/2)^2}}$$

with the approximate result obtained by expansion near resonance ($\omega \approx \omega_\circ$), valid for a high Q oscillator.

The power dissipated is given by

$$P_{\text{dis}} = \langle F_{\text{damping}}\, v\rangle$$

The damping force is

$$F_{\text{damping}} = 2m\Gamma\omega_\circ\, v$$

therefore the power dissipated becomes

$$P_{\text{dis}} = \langle 2m\Gamma\omega_\circ\, \dot{x}\, \dot{x}\rangle \sim A^2\,\Gamma$$

Far from resonance $|\omega - \omega_\circ| \gg \Gamma/2$ the amplitude is independent of Γ, and the power dissipated varies as Γ because

$$P_{\text{dis}} \sim A^2\,\Gamma \sim \Gamma$$

But on resonance when $\omega - \omega_\circ = 0$, the amplitude varies as $2/\Gamma$ and the power dissipated varies as $1/\Gamma$ because

$$P_{\text{dis}} \sim A^2\,\Gamma \sim \frac{1}{\Gamma^2}\Gamma \sim \frac{1}{\Gamma}$$

Qualitatively, this happens because the drive causes the response on resonance to increase to the point where it compensates for the damping. Therefore reducing the damping increases the response which then causes the power dissipated to increase. In the limit of no damping, the power dissipated diverges to infinity whereas off-resonance it goes to zero—i.e. becomes a Dirac delta function.

4. Squeezing operators

The generalized raising and lowering operators for producing squeezed states are

$$b = a \cosh r - a^{\dagger} e^{2i\theta} \sinh r$$
$$b^{\dagger} = a^{\dagger} \cosh r - a e^{-2i\theta} \sinh r$$

Show that they satisfy the commutator relation $\left[b, b^{\dagger}\right] = 1$.

Solution

The commutator of the squeezing operator is calculated as

$$\begin{aligned}
\left[b, b^{\dagger}\right] &\equiv bb^{\dagger} - b^{\dagger} b \\
&= aa^{\dagger} \cosh^2 r - a^2 e^{-2i\theta} \cosh r \sinh r \\
&\qquad - a^{\dagger 2} e^{2i\theta} \cosh r \sinh r + a^{\dagger} a \sinh^2 r \\
&\qquad - a^{\dagger} a \cosh^2 r + a^{\dagger 2} e^{2i\theta} \cosh r \sinh r \\
&\qquad + a^2 c^{-2i\theta} \cosh r \sinh r - aa^{\dagger} \sinh^2 r \\
&= aa^{\dagger} \left(\cosh^2 r - \sinh^2 r\right) - a^{\dagger} a \left(\cosh^2 r - \sinh^2 r\right) \\
&= aa^{\dagger} - a^{\dagger} a \\
&= \left[a, a^{\dagger}\right] \\
&= 1
\end{aligned}$$

5. Squeezed states

(a) Find the expected number of photons in the squeezed vacuum state as a function of r and θ.

(b) Find the expected numbers of photons in a coherently displaced squeezed state with α, r, and θ.

Solution

(a) Since we are working with eigenstates of the Hamiltonian expressed in terms of b, we want to express the number operator N in terms of b.

The definitions of b and $b^\dagger$ are

$$\begin{aligned} b &= a\cosh r - a^\dagger e^{2i\theta}\sinh r \\ b^\dagger &= a^\dagger \cosh r - ae^{-2i\theta}\sinh r \end{aligned}$$

Multiplying these two equations by the appropriate factors and adding gives

$$b\cosh r + b^\dagger e^{2i\theta}\sinh r = a\left(\cosh^2 r - \sinh^2 r\right) = a$$

Thus a and $a^\dagger$ satisfy

$$\begin{aligned} a &= b\cosh r + b^\dagger e^{2i\theta}\sinh r \\ a^\dagger &= b^\dagger \cosh r + be^{-2i\theta}\sinh r \end{aligned}$$

and the number operator is

$$\begin{aligned} N &\equiv a^\dagger a \\ &= b^\dagger b\cosh^2 r + b^2 e^{-2i\theta}\cosh r\sinh r \\ &\quad + b^{\dagger 2}e^{2i\theta}\cosh r\sinh r + bb^\dagger \sinh^2 r \end{aligned}$$

The only term that will give a non-zero result is the $bb^\dagger$ term. Thus the number of photons in the squeezed vacuum is

$$\langle 0_b|N|0_b\rangle = \langle 0_b|bb^\dagger \sinh^2 r|0_b\rangle = \sinh^2 r$$

The result is independent of θ because the number of photons in the vacuum state does not depend on the angle of squeezing.

(b) The coherently displaced squeezed vacuum state is

$$|\alpha, 0_b\rangle = D(\alpha)\,|0_b\rangle$$

where

$$\begin{aligned}
D(\alpha) &= \exp\left(\alpha a^\dagger - \alpha^* a\right) \\
&= \exp\left(\alpha b^\dagger \cosh r + \alpha b e^{-2i\theta} \sinh r - \alpha^* b \cosh r - \alpha^* b^\dagger e^{2i\theta} \sinh r\right) \\
&= \exp\left[b^\dagger \left(\alpha \cosh r - \alpha^* e^{2i\theta} \sinh r\right) - b\left(-\alpha e^{-2i\theta} \sinh r + \alpha^* \cosh r\right)\right] \\
&= \exp\left(\beta b^\dagger - \beta^* b\right)
\end{aligned}$$

with β defined as

$$\beta \equiv \alpha \cosh r - \alpha^* e^{2i\theta} \sinh r$$

This shows that

$$D(\alpha)\left|0_b\right\rangle = D'(\beta)\left|0_b\right\rangle \equiv \left|\beta_b\right\rangle$$

Hence the $|\beta_b\rangle$ states are analogous to the $|\alpha\rangle$ states, i.e. they are eigenstates of b with eigenvalue β.

The average number of photons in this state is

$$\langle n\rangle_\beta = \langle \beta_b | a^\dagger a | \beta_b\rangle$$

From part (a) and using the commutator relation for b, we have

$$\begin{aligned}
a^\dagger a = b^\dagger b \cosh^2 r + b^2 e^{-e^2 i\theta} \cosh r \sinh r \\
+ b^{\dagger 2} e^{2i\theta} \cosh r \sinh r + \left(1 + b^\dagger b\right) \sinh^2 r
\end{aligned}$$

Therefore

$$\begin{aligned}
\langle n\rangle_\beta = |\beta|^2 \cosh^2 r + \beta^2 e^{-2i\theta} \cosh r \sinh r \\
+ \left(\beta^*\right)^2 e^{2i\theta} \cosh r \sinh r + \sinh^2 r + |\beta|^2 \sinh^2 r
\end{aligned}$$

From the definition of β we have

$$\begin{aligned}
|\beta|^2 &= \alpha\alpha^* \cosh^2 r + \alpha\alpha^* \sinh^2 r \\
&\quad - \alpha^2 e^{-2i\theta} \sinh r \cosh r - (\alpha^*)^2 e^{2i\theta} \sinh r \cosh r \\
\beta^2 &= \alpha^2 \cosh^2 r + (\alpha^*)^2 e^{2i\theta} \sinh^2 r - 2\alpha\alpha^* e^{2i\theta} \sinh r \cosh r
\end{aligned}$$

Substituting into the equation for $\langle n\rangle$, we get

$$\begin{aligned}
\langle n\rangle_\beta &= \sinh^2 r + \alpha\alpha^* \left[\left(\cosh^2 r + \sinh^2 r\right)^2 - 4\cosh^2 r \sinh^2 r\right] \\
&= \sinh^2 r + \alpha\alpha^* \left(\cosh^2 r - \sinh^2 r\right)^2 \\
&= \sinh^2 r + |\alpha|^2
\end{aligned}$$

which as expected reduces to the vacuum value when $\alpha = 0$.

6. Size of zero-point energy

(a) Calculate the zero-point energy in a cubic cavity of length L on each side if you stop the integral (arbitrarily) at wave vector k_c.

(b) Find a numerical result for $k_c = 2\pi \times 20\,000\ \mathrm{cm}^{-1}$.

(c) What would be the electric field $\mathcal{E}$ associated with this energy density?

(d) Approximately what would be the force on one of the walls of a 1 m^3 cube if radiation with this energy density is present inside the cube?

(Hint: How are pressure and energy related for a photon gas?)

Solution

(a) Zero-point energy for each mode of the radiation field is

$$E_0 = \frac{1}{2}\hbar\omega = \frac{1}{2}\hbar ck$$

Density of states in k space is

$$\Omega(k) = \left(\frac{L}{2\pi}\right)^3 4\pi k^2 \times 2$$

$\uparrow$ for spin

$$= \frac{L^3}{\pi^2}k^2$$

Therefore the energy up to a cut-off wavevector k_c is

$$E = \int_0^{k_c} E_0(k)\,\Omega(k)\,dk$$

$$= \int_0^{k_c} \frac{1}{2}\hbar ck\,\frac{L^3}{\pi^2}k^2\,dk = \frac{\hbar c}{2\pi^2}L^3\frac{k^4}{4}\bigg|_0^{k_c} = \frac{\hbar c}{8\pi^2}L^3k_c^4$$

and the energy density is

$$\frac{E}{V} = \frac{E}{L^3} = \frac{\hbar c}{8\pi^2}k_c^4$$

(b) For $k_c = 2\pi \times 20\,000\ \mathrm{cm}^{-1}$ we have

$$\frac{E}{V} = 9.989\ \mathrm{J/m^3}$$

(c) Equating the above energy density to that in an electrostatic field we get

$$U = \frac{\mathcal{E}^2}{8\pi} = 9.989 \qquad \Longrightarrow \qquad \mathcal{E} = 1585 \text{ V/cm}$$

(d) Radiation pressure due to a photon gas is $E/(3V)$, therefore

$$F = p \times A = \frac{E}{V}\frac{L^2}{3} = 3.33 \text{ N}$$

Chapter 3

Atoms

ATOMS are the building blocks of matter. But they serve an even more important role in atomic physics since they are not only objects to study and understand as deeply as possible, but also in the words of I. I. Rabi, "Nature's laboratory," where you can study the interaction of electrons and nucleus, the interaction of matter and radiation, or very weak interactions (e.g. parity or time-reversal symmetry violation) with exquisite precision. In this chapter, we shall use the phrase "one-electron atom" to include not only atoms which are iso-electronic with hydrogen (such as He^{+}, Li^{++}, etc.), but also atoms with one-electron which is far less weakly bound than all the others so that the inner electrons may be considered collectively as a core whose interaction with the active electron may be adequately described by parameters such as scattering length, polarizability, etc.

We will study the structure—energy levels and matrix elements—of atoms, both in zero field and in static electric and magnetic fields separately. The emphasis will be on providing a brief catalog of the physics of atomic structure and the field-induced changes in this structure. The dynamics of transitions induced by a time-varying electric field will be covered in later chapters, but we shall discuss the energy levels of an atom in an oscillating field here—**the dressed atom**—for use later on.

A. Spectroscopic notation

Neutral atoms consist of a heavy nucleus with positive charge Z, surrounded by Z electrons with negative charge. Positively charged atomic ions generally have structure similar to the neutral atom with the same number of electrons except for a scale factor; negatively charged ions lack the attractive Coulomb interaction at large electron-core separation and typically have only one bound state. Thus the essential feature of an atom is its number of electrons, and their mutual arrangement as expressed in the quantum numbers.

An isolated atom has two good angular momentum quantum numbers, J and m_j.* In **zero external field**, the atomic Hamiltonian possesses rotational invariance, which implies that each J level is degenerate with respect to the $2J+1$ states for the different values of m_j — traditional spectroscopists refer to these states as "sublevels." (This also makes it clear that the word "level" is used for an energy eigenstate which has possible degeneracy because of unspecified quantum numbers, whereas the word "state" is used to refer to an eigenstate with distinct quantum numbers which are different from others.) For each J and m_j, an atom will typically have a large number of discrete energy levels (plus a continuum) which may be labeled by other quantum numbers. If Russel–Saunders coupling (LS coupling) is a good description of the atom, then the quantum numbers for the following vector operators in an N electron atom

$$\vec{L} \equiv \sum_{i=1}^{N} \vec{L}_i \qquad \text{and} \qquad \vec{S} \equiv \sum_{i=1}^{N} \vec{S}_i$$

are nearly good quantum numbers and may be used to distinguish the levels. In this case the level is designated by a

Term symbol $\quad {}^{2S+1}L_J$

where $2S+1$ and J are written numerically, and L is designated with the following letter code

L :	0	1	2	3	4	...
Letter :	S	P	D	F	G	...

The first four letters stand for Sharp, Principal, Diffuse, and Fundamental—adjectives that apply to the spectral lines of one-electron atoms as recorded by a spectrograph. The term symbol is frequently preceded by the n value of the outermost electron. In addition, when dealing with one-electron atoms, it is customary to omit the $2S+1$ superscript, with the understanding that

*This is strictly true only for atoms whose nuclei have spin $I = 0$. However, J is never significantly destroyed by coupling to I in ground state atoms.

it is equal to 2. Thus the term symbol for the ground state of Na is written as $3\,\mathrm{S}_{1/2}$, where the subscript 1/2 tells us that it is a one-electron atom.

The preceding discussion of the term symbol was based on an external view of the atom. Alternatively, one may have or assume knowledge of the internal structure—the quantum numbers of each electron. These are specified as the

Configuration $1s^2 2s^2 2p^2 \ldots$

i.e. a series of symbols of the form $n\ell^m$, which represents m electrons in the orbital $n\ell$. n is the principal quantum number and characterizes the radial motion, which has the largest influence on the energy. n and m are written numerically, but the $s, p, d, f, \ldots$ coding is used for ℓ (lowercase for the lowercase ℓ). As an example, the configuration for Ca is $1s^2 2s^2 2p^3 3s 3d$, which is frequently abbreviated as $\ldots 3d$. In general, each configuration leads to several terms which may be split apart by several eV—e.g. the above Ca configuration gives rise to terms $^1\mathrm{D}_2$ and $^3\mathrm{D}_{1,2,3}$.

In classifying levels, the term is generally of more importance than the configuration because it determines the behavior of the atom when it interacts with external electric or magnetic fields.* Furthermore the configuration may not be pure—if two configurations can give rise to the same term (and have the same parity) then intra-atomic electrostatic interactions will mix them together. This process, called **configuration interaction**, results in shifts in the level positions and intensities of spectral lines from terms which interact, and also in correlations in the motions of the electrons within the atoms.

*By contrast, selection rules, as we will see in Chapter 6, "Interaction," only deal with changes in internal states, such as ΔJ.

B. Energy levels of one-electron atoms

In trying to understand some new phenomenon for the first time, it is common sense and good science to study it in the simplest situation where it is manifest—alas, this ideal situation is frequently discernible only by intuition until the phenomenon is understood! With atoms it is evident that **hydrogen** is of paramount simplicity and historically most *fundamental* physics which has been discovered in atoms has been discovered in hydrogen, though Na, Rb, and other alkali atoms have taken over since tunable lasers arrived.

1. Bohr atom

Balmer's empirical formula of 1885 had reproduced Angstrom's observations of spectral lines in hydrogen to 0.1 Å accuracy, but it was not until 1913 that Bohr gave an explanation for this based on a quantized mechanical model of the atom. This model is historically important because it is simple enough to be taught in high school, and provided the major impetus for developing quantum mechanics. It involves the following **Postulates of Bohr Atom:**

(i) Electron and proton are point charges whose interaction is Coulombic at all distances.

(ii) Electron moves in circular orbits about the center of mass in **stationary states** with orbital angular momentum $L = n\hbar$.

These two postulates give the energy levels (with −ve sign showing that these are bound states)

$$E_n = -\frac{1}{2}\left[\frac{me^4}{\hbar^2}\frac{m_p}{m+m_p}\right]\frac{1}{n^2} = -\frac{R_H}{n^2}$$

where the factor involving m_p comes because of the reduced mass $\mu \equiv \dfrac{m\,m_p}{m+m_p}$

R_H is the Rydberg in hydrogen $= \dfrac{m_p}{m+m_p} R_\infty$

(iii) One quantum of radiation is emitted when the system changes between two of these energy levels.

(iv) Its wavenumber is given by the Einstein frequency criterion

$$\nu_{n\to m} = \frac{E_n - E_m}{hc}$$

Note that the wavenumber is the number of wavelengths per cm: $\nu = \omega/(2\pi c) = k/2\pi$, where k is the usual wavevector.

The mechanical spirit of the Bohr atom was extended by Sommerfeld in 1916 using the **Wilson–Sommerfeld quantization rule**

$$\oint p_i dq_i = n_i h$$

where q_i and p_i are conjugate coordinate and momentum pairs for each degree of freedom of the system. This extension yielded elliptical orbits which were found to have energy nearly degenerate with respect to the orbital angular momentum for a particular value of the principal quantum number n. The dependency was lifted only by a relativistic correction, and the splitting was in agreement with the observed fine structure of hydrogen. Although triumphant in hydrogen, simple mechanical models of two electron atoms which reproduced reality could not be developed during the following decade and further progress in understanding atoms required the development of quantum mechanics, as discussed in the next section.

2. Radial Schrödinger equation for central potentials

The quantum mechanical problem in an atom consists of solving the following **Time dependent Schrödinger equation**

$$i\hbar \frac{\partial \psi(\vec{r}, t)}{\partial t} = H(\vec{r})\psi(\vec{r}, t)$$

Stationary solutions are obtained by the substitution

$$\psi(\vec{r}, t) = e^{-iE_n t/\hbar}\psi_n(\vec{r})$$

where n stands for all quantum numbers necessary to label the state. This leads to the **Time independent Schrödinger equation**

$$[H(\vec{r}) - E_n]\psi(\vec{r}) = 0$$

The most important and pervasive application of this equation in atomic physics is to the case of a one particle system of mass μ in a spherically symmetric (or isotropic) potential. Such potentials are called **central potentials** because they only depend on the scalar distance from a *center of force*, as typified by the Coulomb force in an atom. In mathematical terms, $V(\vec{r}) = V(r)$. In such a case, the Hamiltonian is

$$\begin{aligned}
H &\equiv \text{Kinetic Energy} + \text{Potential Energy} \\
&= \frac{P^2}{2\mu} + V(r) \\
&= -\frac{\hbar^2 \nabla^2}{2\mu} + V(r) \\
&= -\frac{\hbar^2}{2\mu}\left[\frac{1}{r^2}\frac{\partial}{\partial r}\left(r^2 \frac{\partial}{\partial r}\right) + \frac{1}{r^2 \sin\theta}\frac{\partial}{\partial \theta}\left(\sin\theta \frac{\partial}{\partial \theta}\right) + \frac{1}{r^2 \sin^2\theta}\frac{\partial^2}{\partial \phi^2}\right] + V(r)
\end{aligned}$$

where in the last line the kinetic energy operator ∇^2 has been written in spherical coordinates. Since V is spherically symmetric, the angular dependence of the solution is characteristic of spherically symmetric systems in general and may be factored out

$$\psi_{n\ell m}(\vec{r}) = R_{n\ell}(r)\, Y_{\ell m}(\theta, \phi)$$

where $Y_{\ell m}$'s are the **Spherical Harmonics**, with ℓ being the eigenvalue of the vector operator for the orbital angular momentum $\vec{L}$

$$\vec{L}^2\, Y_{\ell m} = \ell(\ell+1)\hbar^2\, Y_{\ell m}$$

and m being the eigenvalue of the projection of $\vec{L}$ on the quantization axis (which may be chosen at will)

$$L_z\, Y_{\ell m} = m\hbar\, Y_{\ell m}$$

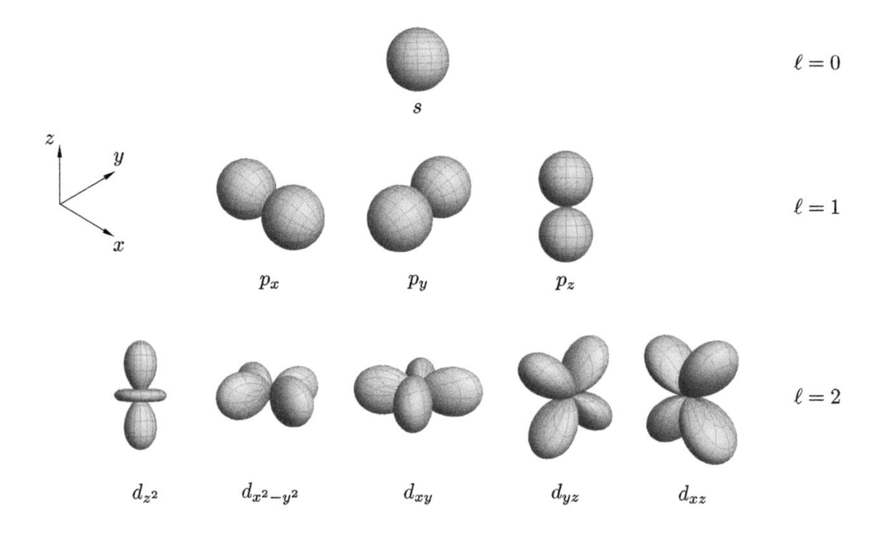

Figure 3.1: Shapes of the first three hydrogen orbitals in $\theta\phi$ space as determined by the spherical harmonics $Y_{\ell m}$'s. Each one has parity $(-1)^\ell$, $2\ell+1$ components, and a node at the origin except for the $\ell = 0$ case.

The shapes of the first three orbitals—corresponding to $\ell = 0, 1, 2$—are shown in Fig. 3.1.

Substitution of $Y_{\ell m}$ into $\psi_{n\ell m}(\vec{r})$ leads to the **Time independent radial Schrödinger equation** (remember $R_{n\ell}$ is only a function of r)

$$\frac{1}{r^2}\frac{d}{dr}\left(r^2\frac{dR_{n\ell}}{dr}\right) + \frac{2\mu}{\hbar^2}\left[E_{n\ell} - V(r)\right]R_{n\ell} - \frac{\ell(\ell+1)}{r^2}R_{n\ell} = 0$$

This is the equation which is customarily solved for the hydrogen atom's radial wave functions.

For many applications (e.g. scattering by a central potential, diatomic molecules) it is convenient to make a further substitution

$$R_{n\ell}(r) = \frac{y_{n\ell}(r)}{r}$$

which leads to

$$\frac{d^2 y_{n\ell}(r)}{dr^2} + \frac{2\mu}{\hbar^2}\left[E_{n\ell} - V(r) - \frac{\hbar^2 \ell(\ell+1)}{2\mu r^2}\right] y_{n\ell}(r) = 0 \tag{3.1}$$

with the boundary condition $y_{n,\ell}(0) = 0$.

This equation is identical with the time independent Schrödinger equation for a particle of mass μ in an effective one-dimensional potential

$$V_{\text{eff}}(r) = V(r) + \frac{\vec{L}^2}{2\mu r^2} = V(r) + \frac{\hbar^2 \ell(\ell+1)}{2\mu r^2}$$

The additional term is called the **Centrifugal barrier** and is a pseudo-potential that arises because we are in a non-inertial rotating frame.

3. Radial equation for hydrogen

For hydrogen, Eq. (3.1) becomes

$$\frac{d^2 y_{n\ell}(r)}{dr^2} + \frac{2\mu}{\hbar^2}\left[E_n + \frac{e^2}{r}\right] y_{n\ell}(r) - \frac{\ell(\ell+1)}{r^2} y_{n\ell}(r) = 0 \tag{3.2}$$

We first consider the limit $r \to 0$. Then the dominant terms lead to the simplified equation

$$\frac{d^2 y_{n\ell}}{dr^2} - \frac{\ell(\ell+1)}{r^2} y_{n\ell} = 0$$

for any value of E_n. It is easily verified that the two independent solutions are $y_{n\ell} \sim r^{\ell+1}$ and $y_{n\ell} \sim r^{-\ell}$. For $\ell \geq 1$, the only physically acceptable (i.e. normalizable) solution is

$$y_{n\ell} \sim r^{\ell+1} \tag{3.3}$$

Next, we look at the limit $r \to \infty$. We may investigate the simpler equation if the potential falls to zero sufficiently rapidly, i.e. $\lim_{r\to\infty} V(r) = 0$. Then, for large r

$$\frac{d^2 y_{n\ell}}{dr^2} + \frac{2\mu E_n}{\hbar^2} y_{n\ell} = 0$$

For $E_n > 0$, this equation has oscillating solutions corresponding to a free particle, which are important in scattering problems.

For $E_n < 0$, the equation has exponential solutions, but only the decaying exponential of the following form is normalizable

$$R_{n\ell}(r) = \frac{y_{n\ell}(r)}{r} = \frac{1}{r} e^{-r\sqrt{-2\mu E_n/\hbar^2}} \tag{3.4}$$

When $E_n < 0$, it is possible to obtain physically reasonable solutions to Eq. (3.1)—or indeed any **bound state** problem—only for certain discrete values of E_n called **eigenvalues**. This situation arises from the requirement that the solution be square integrable (i.e. $\int_0^\infty y_{n\ell}^2(r)dr$ is finite); obviously Eq. (3.1) is a prescription for generating a function $y_{n\ell}(r)$ for arbitrary $E_n < 0$ given $y_{n\ell}$ and $dy_{n\ell}/dr$ at any point. To obtain a normalizable solution one can proceed as follows:

(i) Starting at large r, a "solution" of the form of (3.4) is selected and extended to some intermediate value of r, say r_m, using a trial value of $E_n < 0$.

(ii) At the origin one must select the solution of the form $y_{n\ell} \propto r^{\ell+1}$ from (3.3); this "solution" is then extended out to r_m.

(iii) The two "solutions" may be made to have the same value at r_m by multiplying one by a constant; however, the resulting function is a valid solution only if the first derivative is continuous at r_m, and this occurs only for a discrete set of values of $E_{n\ell}$.

The procedure described above is, in fact, the standard **Numerov–Cooley** technique for finding bound states with digital computers; its most elegant feature is a rapidly converging procedure for adjusting the trial eigenenergy using the discontinuity in the derivative.

For the hydrogen atom, the eigenvalues can be determined analytically because the substitution

$$y_{n\ell}(r) = r^{\ell+1} e^{-r\sqrt{-2\mu E_n/\hbar^2}} v_\ell(r)$$

leads to a particularly simple equation if one also changes the variable from r to x

$$x = 2r\sqrt{-2\mu E_n/\hbar^2}$$

so the exponential becomes $e^{-x/2}$. Defining the constant

$$\nu = \frac{\hbar}{a_\circ\sqrt{-2\mu E_n}} = \alpha\sqrt{\frac{\mu c^2}{-2E_n}} \tag{3.5}$$

(with α and $a_\circ$ as defined earlier), Eq. (3.2) in terms of x becomes

$$\left[x\frac{d^2}{dx^2} + (2\ell + 2 - x)\frac{d}{dx} - (\ell + 1 - \nu)\right] v_\ell = 0 \tag{3.6}$$

This is a Laplace equation, and its solution is a confluent hypergeometric series. To find the eigenvalues one now tries a Taylor series expansion for v_ℓ

$$v_\ell(x) = 1 + a_1 x + a_2 x^2 + \dots$$

This satisfies Eq. (3.6) only if the coefficients of each power of x are satisfied, i.e.

$$\begin{aligned} x^0 : &\quad (2\ell + 2)a_1 - (\ell + 1 - \nu) = 0 \\ x^1 : &\quad 2(2\ell + 3)a_2 - (\ell + 2 - \nu)a_1 = 0 \\ x^{p-1} : &\quad p(2\ell + 1 + p)a_p - (\ell + p - \nu)a_{p-1} = 0 \end{aligned}$$

The first line fixes a_1, the second then determines a_2, and last line gives the general coefficient as

$$a_p = \frac{(\ell + p - \nu)}{p(2\ell + 1 + p)} a_{p-1}$$

The above expression will give a p^{+n} coefficient on the order of $1/p!$ so that

$$v_\ell(x) \sim \sum_{p=0}^{\infty} \frac{x^p}{p!} = e^x$$

This spells disaster because it means that $y_{n\ell} = r^{\ell+1} e^{-x/2} v_\ell(r)$ diverges. The only way in which this can be avoided is if the series truncates, i.e. if ν is an integer

$$\nu = n = n' + \ell + 1 \qquad n' = 0, 1, 2, \dots$$

so that all the coefficients a_p for $p \geq (n' + 1)$ will be zero. Since $n' \geq 0$ it is clear that you must look at an energy level $n \geq \ell + 1$ to find a state with angular momentum ℓ (e.g. the $2d$ configuration does not exist).

This gives the eigenvalues of hydrogen from Eq. (3.5) as

$$\begin{aligned} E_n &= -\frac{1}{2}\alpha^2 \mu c^2 \frac{1}{(n' + \ell + 1)^2} \\ &= -\frac{1}{2}\frac{\mu e^4}{\hbar^2}\frac{1}{n^2} \\ &= -\frac{R_H}{n^2} \end{aligned}$$

which agrees with the Bohr formula.

4. One-electron atoms with cores — Quantum defect

(i) Phenomenology

It is observed that the eigenvalues of atoms which have one valence electron have the same density as those of hydrogen, but not the same positions. If one defines an effective quantum number

$$n^* = \sqrt{R/E_n}$$

such that the Bohr formula will reproduce the energy levels of a particular term, it is found that the n^* values for adjacent levels differ by almost exactly 1.000, especially after the first few terms. Thus the **quantum defect** is defined as

$$\delta_\ell \equiv n - n^*$$

where n is the principal quantum number of the valence electron for that term.* It remains very constant with respect to n, but decreases rapidly with respect to ℓ.

A more accurate empirical formula for the term values of a series is the Balmer–Ritz formula

$$T_n = \frac{Z^2 R}{(n - \delta_\ell - \beta_\ell/n^2)^2} \tag{3.7}$$

When comparing x-ray spectra of isoelectronic ions, a useful empirical formula was suggested by Moseley

$$T_n = \frac{(Z - \delta Z)^2 R}{n^2}$$

in which the charge is adjusted, rather than n. δZ may be regarded as the amount of charge shielded by the core.

(ii) Explanation

It must always be kept in mind that the quantum defect is a phenomenological result. To explain how such a simple result arises is obviously an interesting challenge, but it is not to be expected that the solution of this problem will lead to great new physical insight. The only new results from understanding quantum defects in one-electron systems are the connection between the quantum defect and the electron scattering length for the same system which may be used to predict low-energy electron scattering

*If you do not know n, the use of the next larger integer than n^* still leads to useful results.

cross-sections, and the simple expressions relating δ_ℓ to the polarizability of the core for larger δ_ℓ. The principal use of the quantum defect is to predict the positions of higher terms in a series for which δ_ℓ is known.

Explanations of the quantum defect range from the elaborate fully quantal explanation of Seaton, to the extremely simple treatment of Parsons and Weisskopf. P&W assume that the electron cannot penetrate inside the core at all, but use the boundary condition $R(r_c) = 0$ which requires relabeling the lowest ns state to $1s$ since it has no nodes outside the core. This viewpoint has a lot of merit because the exclusion principle and the large kinetic energy of the electron inside the core combine to reduce the amount of time it spends in the core. This is reflected in the true wavefunction which has n nodes in the core and therefore never has a chance to reach a large amplitude in this region.

To show the physics without much math (or rigor) we turn to the JWKB solution to the radial Schrödinger equation. Defining the wavenumber as

$$k_\ell(r) = \frac{\sqrt{2m[E - V_{\text{eff}}(r)]}}{\hbar} \qquad \text{(remember } V_{\text{eff}} \text{ depends on } \ell\text{)}$$

the phase accumulated in the classically allowed region is

$$\phi_\ell(E) = \int_{r_i}^{r_o} k_\ell(r)dr$$

where r_i and r_o are the inner and outer turning points.

Bound state eigenvalues are found by setting*

$$[\phi_\ell(E) - \pi/2] = n\pi \qquad \Longrightarrow \qquad \phi_\ell(E) = (n + 1/2)\pi \tag{3.8}$$

The Bohr formula permits us to evaluate the accumulated phase for hydrogen

$$\phi_{\ell H}(E) = \pi\sqrt{R_H/E} + \pi/2 \qquad \text{independent of } \ell$$

In the spirit of the JWKB approximation this is regarded as a continuous function of E.

Now consider a one-electron atom with a core of inner electrons which lies entirely within r_c. Since it has a hydrogenic potential outside of r_c, its phase can be written as

$$\begin{aligned}
\phi_\ell(E) &= \int_{r_i}^{r_{oH}} k_\ell(r)\, dr \\
&= \int_{r_i}^{r_c} k_\ell(r)\, dr + \int_{r_c}^{r_{oH}} k_{\ell H}(r)\, dr \\
&= \int_{r_i}^{r_c} k_\ell(r)\, dr - \int_{r_{iH}}^{r_c} k_{\ell H}(r)\, dr + \int_{r_{iH}}^{r_{oH}} k_{\ell H}(r)\, dr
\end{aligned}$$

*The $\pi/2$ comes from the connection formulae and would be $1/4$ for $\ell = 0$ states where $r_i = 0$. Fortunately it cancels out.

where r_{iH} and r_{oH} are the corresponding turning points in hydrogen.

The last integral is the phase accumulated by an equivalent hydrogen atom, which is by definition $(n^* + 1/2)\pi$. If we designate the sum of the first two integrals as $\Delta\phi$, then Eq. (3.8) allows us to rewrite the above equation as

$$(n + 1/2)\pi = \Delta\phi + (n^* + 1/2)\pi \qquad \Longrightarrow \qquad \Delta\phi = \pi\delta_\ell(E)$$

Hence the total phase accumulated is related to the quantum defect as follows

$$\phi_\ell(E) = \pi\delta_\ell(E) + \phi_{\ell H}(E)$$

The defect δ_ℓ may be expanded in a power series in E

$$\delta_\ell(E) = \delta_\ell^{(0)} + \delta_\ell^{(1)} E + \ldots$$

noting that the turning points are determined by $E + V_{\text{eff}}(r) = 0$ so that δ_ℓ approaches a constant as $E \to 0$.

Now we can find the bound state energies for the atom with a core. Starting with Eq. (3.8)

$$\begin{aligned} n\pi &= \phi_\ell(E) - \pi/2 \\ &= \pi\delta_\ell(E) + \pi\sqrt{R_H/E} \end{aligned}$$

we get

$$E = \frac{R_H}{[n - \delta_\ell(E)]^2} \approx \frac{R_H}{[n - \delta_\ell^{(0)} - \delta_\ell^{(1)} R_H/n^2]^2}$$

Thus we have explained the Balmer–Ritz formula of Eq. (3.7).

If we look at the radial Schrödinger equation for the electron-ion core system in the region where $E > 0$ we are dealing with the scattering of an electron by a modified Coulomb potential. Intuitively one would expect that there would be an intimate connection between the bound state eigenvalue problem described earlier and this scattering problem, especially in the limit $E \to 0$ (from above and below). Since the quantum defects characterize the bound state problem accurately in this limit one would expect that they should be directly useful in the scattering phase shifts $\sigma_\ell(k)$ (k is the momentum of the **free** particle) which obey

$$\lim_{k\to 0} \cot[\sigma_\ell(k)] = \pi\delta_\ell^0 \tag{3.9}$$

This has great intuitive appeal: $\pi\delta_\ell^0$ as discussed above is precisely the phase shift of the wave function with the core present relative to the one with $V = e^2/r$. On second thought Eq. (3.9) might appear puzzling since the scattering phase shift is customarily defined as the shift relative to the one with $V = 0$. The resolution of this paradox lies in the long range nature of the Coulomb interaction; it forces one to redefine the scattering phase shift to be the shift relative to a pure Coulomb potential.

C. Interaction with magnetic fields

In this section we study the interaction of the electron's orbital and spin angular momentum with an external static magnetic field. In addition we consider the **spin-orbit interaction**: the coupling of electron spin to the magnetic field generated by the nucleus (which appears to move about the electron in the electron's rest frame). The spin-orbit interaction causes the orbital and spin angular momenta of the electron to couple together to produce a total angular momentum which then couples to the external field; the magnitude of this coupling is calculated for weak external fields.

1. Magnetic moment of circulating charge

The energy of interaction of a magnetic moment $\vec{\mu}$ with a magnetic field $\vec{B}$ is

$$U = -\vec{\mu} \cdot \vec{B}$$

indicating that the torque tends to align the moment along the field. In classical electrodynamics, the magnetic moment of a moving point particle about some point in space is independent of the path which it takes, but depends only on the product of ratio of its charge to mass and angular momentum. This result follows from the definitions of

$$\text{Angular Momentum} \qquad \vec{L} \equiv \vec{r} \times \vec{p} = m[\vec{r} \times \vec{v}]$$

$$\text{Magnetic Moment} \qquad \vec{\mu} \equiv \frac{1}{2c}\vec{r} \times \vec{i} = \frac{q}{2c}[\vec{r} \times \vec{v}]$$

where $\vec{v}$ is the velocity and $\vec{i}$ is the current. The equality of the bracketed terms implies

$$\vec{\mu} = \frac{q}{2mc}\vec{L} \equiv \gamma \vec{L} \qquad (3.10)$$

where γ is referred to as the **gyromagnetic ratio**.

For an electron with orbital angular momentum $\vec{L}$

$$\vec{\mu}_\ell = \frac{-e}{2mc}\vec{L} \equiv -\frac{\mu_B}{\hbar}\vec{L}$$

where μ_B is the

$$\textbf{Bohr magneton} \qquad \mu_B = \frac{e\hbar}{2mc} = 1.39983 \text{ MHz/G}$$

Note that the electron moment is negative, which is indicated by putting a − sign in front.

2. Intrinsic electron spin and moment

When Uhlenbeck and Goudsmit suggested that the electron had an intrinsic spin $s = 1/2$, it soon became apparent that it had a magnetic moment twice as large as would be expected on the basis of Eq. (3.10).* This is accounted for by writing

$$\text{Intrinsic Electron Moment} \qquad \vec{\mu}_s = -g_s \mu_B \vec{S}/\hbar$$

where the quantity g_s is called the electron g factor or the **spin gyromagnetic ratio**, and the minus sign again showing that the moment is negative with a positive g. The value of g_s was predicted to be 2 by the Dirac theory, probably its greatest triumph.

Later experiments by Kusch, by Crane et al., and by Dehmelt and coworkers have shown that (for both electrons and positrons) the value is slightly larger, the result being called the **anomalous magnetic moment** of the electron

$$g_s/2 = 1.\underbrace{0011}_{\text{Kusch}}\overbrace{59}^{\text{Crane}}\;\underbrace{620}_{\text{Dehmelt}}\;\ldots$$

This result must be explained using quantum electrodynamics (QED), which gives

$$g_s/2 = 1 + 1/2\left[\frac{\alpha}{\pi}\right] - 0.3258\left[\frac{\alpha}{\pi}\right]^2 + 0.13\left[\frac{\alpha}{\pi}\right]^3 + \ldots$$

Note that each successive term has a higher power of the fine structure constant α, and is therefore roughly 100 times smaller. This shows both the importance of α in QED, and that it is a perturbative theory with progressively smaller corrections.

*This implies that the electron cannot be made out of material with a uniform ratio of charge to mass.

3. Spin-orbit interaction

The existence of intrinsic magnetic moment for the electron implies that it will interact with any magnetic field present in an atom. One such field arises from the motion of the electron through the Coulomb field of the nucleus.

$$\vec{B}_{\text{mot}} = \frac{1}{c}\vec{E}\times\vec{v} \qquad \text{Electrodynamics}$$

$$= \frac{1}{c}\frac{Ze}{r^3}\vec{r}\times\vec{v} = \frac{1}{c}\frac{Ze}{mr^3}(\vec{r}\times\vec{p}) \qquad \vec{E}\text{ field of nucleus}$$

$$= \frac{Ze}{mc}\frac{1}{r^3}\vec{L} \qquad \text{Definition of } \vec{L}$$

However, there is another contribution to the magnetic field arising from the Thomas precession . The relativistic transformation of a vector between two moving coordinate systems with a relative acceleration $\vec{a}$ between them involves not only a dilation but also a rotation. The rotation is called Thomas precession, and its rate is given by

$$\vec{\Omega}_{\text{Thomas}} = \frac{1}{2}\frac{\vec{a}\times\vec{v}}{c^2}$$

Thus the precession vanishes for co-linear acceleration. However, it is non-zero for a vector moving around a circle, as in the case of the spin vector of the electron as it circles the proton. From the point of view of an observer fixed to the nucleus, the precession of the electron is identical to the effect of a magnetic field

$$\vec{B}_{\text{Thomas}} = \frac{\hbar}{g_s\mu_B}\vec{\Omega}_{\text{Thomas}} = -\frac{1}{2}\frac{Ze}{mc}\frac{1}{r^3}\vec{L}$$

where we have used that the acceleration of the electron is $-Ze^2\vec{r}/mr^3$. Hence the total effective magnetic field is

$$\vec{B}_{\text{eff}} = \frac{Ze}{2mc}\frac{1}{r^3}\vec{L}$$

which results in a spin-orbit coupling of

$$E_{\text{so}} = -\vec{\mu}_s\cdot\vec{B}_{\text{eff}} = -\frac{-e}{mc}\frac{Ze}{2mc}\frac{1}{r^3}\vec{S}\cdot\vec{L} = \frac{Ze^2}{2m^2c^2}\frac{1}{r^3}\vec{S}\cdot\vec{L}$$

For hydrogen-like atoms, the above expression can be evaluated exactly, and gives the following result for the fine structure (or spin-orbit) splitting

$$\Delta E^{\text{fs}} = \alpha^2 R_\infty \frac{Z^4}{n^3\ell(\ell+1)}$$

The strong dependence on Z results from both the reduction in size of the electron wavefunction for larger Z and its higher velocity (remember this is a relativistic effect).

For one-electron atoms with cores, it is possible to get estimates of ΔE^{fs} good to $\sim 15\%$ by considering that the electron sees a charge $Z_\circ = Z - Z_{\mathrm{core}}$ outside the core, and Z inside. Then

$$\Delta E^{\mathrm{fs}} = \alpha^2 R_\infty \frac{Z_\circ^2 Z^2}{n^3 \ell(\ell+1)} \tag{3.11}$$

Since the spin-orbit interaction is a relativistic effect, it is inconsistent to consider it in the absence of other relativistic energy corrections. When this is done for hydrogen atom one obtains

$$\begin{aligned} E_{nj} &= E_{\mathrm{electrostatic}} + E_{\mathrm{so}} + E_{\mathrm{relativistic}} \\ &= -\frac{Z^2 R_\infty}{n^2}\left[1 + \frac{Z^2\alpha^2}{n}\left(\frac{1}{J+1/2} - \frac{3}{4n}\right)\right] \end{aligned}$$

which predicts that only n and J are necessary to determine the energy of a level. Thus **the spin-orbit splitting remains unchanged by the other relativistic effects.**

4. The Landé vector model of g_j — weak field

In a weak magnetic field, the spin-orbit interaction couples $\vec{S}$ and $\vec{L}$ together to form

$$\vec{J} = \vec{L} + \vec{S}$$

This resultant angular momentum interacts with the applied magnetic field with an energy

$$U = -g_j \mu_B \vec{B} \cdot \vec{J}/\hbar$$

The interaction of the field is actually with $\vec{\mu}_s$ and $\vec{\mu}_\ell$; however g_j is not simply related to these quantities because $\vec{\mu}_s$ and $\vec{\mu}_\ell$ precess about $\vec{J}$ instead of the field. As Landé showed in investigations of angular momentum coupling of different electrons, it is a simple matter to find g_j by calculating the sum of the projections of $\vec{\mu}_s$ and $\vec{\mu}_\ell$ onto $\vec{J}$.

Projection of $\vec{\mu}_\ell$ on $\vec{J}$ $\quad \mu_{\ell j} = -\mu_B \dfrac{|\vec{L}|}{\hbar} \dfrac{\vec{L}\cdot\vec{J}}{|\vec{L}||\vec{J}|}$

Projection of $\vec{\mu}_s$ on $\vec{J}$ $\quad \mu_{sj} = -g_s \mu_B \dfrac{|\vec{S}|}{\hbar} \dfrac{\vec{S}\cdot\vec{J}}{|\vec{S}||\vec{J}|}$

This yields

$$
\begin{aligned}
g_j &= -\frac{\mu_{\ell j} + \mu_{sj}}{\mu_B |\vec{J}|/\hbar} && \text{Definition of } g_j \\
&= \frac{\vec{L} \cdot (\vec{L} + \vec{S}) + 2\vec{S} \cdot (\vec{L} + \vec{S})}{|\vec{J}|^2} && \text{Taking } g_s = 2 \\
&= 1 + \frac{\vec{S} \cdot (\vec{L} + \vec{S})}{|\vec{J}|^2}
\end{aligned}
$$

Using the relation

$$
\begin{aligned}
&|\vec{J}|^2 = |(\vec{L} + \vec{S})|^2 = |\vec{L}|^2 + |\vec{S}|^2 + 2\vec{L} \cdot \vec{S} \\
&\implies \vec{L} \cdot \vec{S} = \frac{1}{2}\left(|\vec{J}|^2 - |\vec{L}|^2 - |\vec{S}|^2\right)
\end{aligned}
$$

one gets

$$
g_j = 1 + \frac{J(J+1) + s(s+1) - \ell(\ell+1)}{2J(J+1)}
$$

where ℓ, s, and J are the respective quantum numbers of the state in which g_j is being evaluated, with the condition that $J \neq 0$. If $J = 0$, then $g_j = 0$.

If a transition from a level with angular momentum J to a level with angular momentum J' takes place in the presence of a magnetic field, the resulting spectral line will be split into three or more components—a phenomenon known as the **Zeeman effect**. These transitions are called **electric dipole** allowed transitions and the relevant selection rules (which we will see in Chapter 6, "Interaction") require that the only values allowed for Δm are $-1, 0, +1$. The components will hence have shifts

$$
\begin{aligned}
\Delta E_{\Delta m=-1} &= [g_j m - g_{j'}(m-1)]\mu_B B = [(g_j - g_{j'})m + g_{j'}]\mu_B B \\
\Delta E_{\Delta m=0} &= [g_j m - g_{j'} m]\mu_B B = [(g_j - g_{j'})m]\mu_B B \\
\Delta E_{\Delta m=+1} &= [g_j m - g_{j'}(m+1)]\mu_B B = [(g_j - g_{j'})m - g_{j'}]\mu_B B
\end{aligned}
$$

If $g_j = g_{j'}$ (e.g. for a $^1\mathrm{P}_1 \to {}^1\mathrm{D}_2$ transition, where $s = 0$ and $J = \ell$ so that $g_j = g_{j'} = 1$), then ΔE will not depend on m. Alternately, if either J or J' is 0 so that the corresponding g_j is 0 (e.g. for a $^1\mathrm{S}_0 \to {}^1\mathrm{P}_1$ transition), then there will be only one shift for each Δm. Thus in both cases there will be three components of the line corresponding to the three shifts listed above; this is called the **normal** Zeeman splitting. If neither of these conditions holds (e.g. for a $^2\mathrm{S}_{1/2} \to {}^2\mathrm{P}_{3/2}$ transition), the line will be split into more than three components and the Zeeman structure is termed **anomalous**—it cannot be explained with classical atomic models because it involves the spin of the electron.

D. Atoms in static electric fields — Stark effect

This section deals with the behavior of the energy levels of an atom that is placed in a **static** electric field. The basic treatment involves Rayleigh–Schrödinger stationary perturbation theory, because the perturbing field is assumed to be weak. The treatment will cover two important topics: the restrictions placed on the qualitative nature of the interaction by symmetry considerations, and the concept of oscillator strength and its use to simplify the dipole matrix elements of atoms.

1. Restrictions due to parity

The operation of inverting the coordinates is called **parity reversal**. Its effect on the coordinates is:

$$\begin{array}{ll} \text{Cartesian} & (x, y, z) \rightarrow (-x, -y, -z) \\ \text{Spherical} & (r, \theta, \phi) \rightarrow (r, \pi - \theta, \phi + \pi) \end{array}$$

Polar vectors (e.g. the position $\vec{r}$ or momentum $\vec{p}$) change sign under parity reversal. **Axial vectors** (e.g. angular momentum $\vec{L} \equiv \vec{r} \times \vec{p}$ or torque $\vec{N} \equiv \vec{r} \times \vec{F}$) do not change sign. Axial vectors are (cross) products of two polar vectors, and are actually second rank tensors with the requirement of being antisymmetric; only in three dimensions an antisymmetric second rank tensor has three components, which can be written conveniently as the components of a vector. Axial vectors are called that because they require an axis and sign convention (e.g. right hand rule) for their definition. The electric field vector $\vec{E}$ is polar, while the magnetic field vector $\vec{B}$ (remember it can be written as $\nabla \times \vec{A}$) is axial.

Electromagnetic interactions at low energies (so we can neglect the weak interactions, which are known to violate parity) are invariant under parity reversal. Therefore the Hamiltonian for an isolated atom $H_\circ$ may be taken to be invariant under parity reversal.* Symbolically

$$I^{-1} H_\circ I = H_\circ \qquad \Longrightarrow \qquad [H_\circ, I] = 0$$

where I is the inversion operator (and not the identity operator $\mathbb{1}$). This means that each eigenstate of $H_\circ$ must have a definite parity. In case there are degenerate states with the same energy, it must be possible to find linear combinations of them which have definite parity (although these are not the best basis states for all problems).

The parity operator has the property (shared with operators for time reversal, charge conjugation, reflection, etc.) that $I^2 = \mathbb{1}$: two inversions bring

*If we want to discuss parity non-conservation measurements in atoms, we have to add a specifically parity violating term to the Hamiltonian.

you back to where you started. Thus for any state $|E_n, i\rangle$, with energy E_n and parity i,

$$I\,|E_n, i\rangle = i\,|E_n, i\rangle$$

Then $i^2 = 1$ which implies $i = \pm 1$. The plus and minus signs are called even and odd parity states, respectively.

Consider the interaction of a static external electric field of strength $\mathcal{E}$ along the $+z$ axis with a single electron. Choosing z as the operator for the electron position, and since the field $\vec{E} = \mathcal{E}\hat{z}$ implies $\phi = -\mathcal{E}z$, the interaction Hamiltonian is

$$H' = -e\phi = e\mathcal{E}z$$

This interaction can be considered as a perturbation if the field is weak ($\mathcal{E}$ is small) compared to the other fields present in the atom, like for example the Coulomb field of the nucleus. The perturbation has **negative** parity because $I^{-1}zI = -z$ which implies

$$I^{-1}H'I = -H'$$

Noting that $II = I^{-1}I^{-1} = \mathbb{1}$ we can write the perturbation matrix element as

$$\begin{aligned}\langle E_{n'}, i'|H'|E_{n''}, i''\rangle &= \langle E_{n'}, i'|I^{-1}I^{-1}H'II|E_{n''}, i''\rangle \\ &= -i'i''\,\langle E_{n'}, i'|H'|E_{n''}, i''\rangle\end{aligned}$$

This forces the matrix element to be zero unless i' and i'' have opposite signs. Thus H' has matrix elements **only between states of opposite parity**. Matrix elements of H' between states of the same parity vanish. In particular,

$$\langle E_n, i|H'|E_n, i\rangle = 0$$

Hence there is no first order perturbation (i.e. linear Stark effect) if $|E_n, i\rangle$ is a non-degenerate state. Any energy shift resulting from the interaction of an atomic system with a static electric field must be quadratic in $\mathcal{E}$.

In electrostatics the potential energy of a system at $\vec{r}$ in an electric field is

$$U(\vec{r}) = q\phi(\vec{r}) - \vec{d}\cdot\vec{E}(\vec{r}) - \alpha\vec{E}(\vec{r})\cdot\vec{E}(\vec{r}) - \frac{1}{6}\sum_{i,j} Q_{ij}\frac{\partial E_j(\vec{r})}{\partial x_i} \tag{3.12}$$

$$\begin{aligned} q &= \text{charge} \\ \vec{d} &= \text{permanent electric dipole moment} \\ \alpha &= \text{isotropic polarizability} \\ Q_{ij} &= \text{quadrupole moment}\end{aligned}$$

Thus an atomic system which has no linear Stark shift is said to have no permanent electric dipole moment (EDM). The existence of an atomic EDM forms an important class of experiments, and will be discussed in the next section.

The preceding argument does not apply if the system has two degenerate energy levels of opposite parity, since then one can form superposition states which do not have a definite parity. This happens, for example, in polar molecules which have degenerate levels of different orientations called **isomers**. As a consequence, molecules can have permanent electric dipole moments whereas atoms do not.

A system can have a dipole moment even if the two energy states of opposite parity, $|E_{n+}\rangle$ and $|E_{n+}\rangle$, are not exactly degenerate, but only **nearly** so. Because if they satisfy the condition

$$\langle E_{n+}|e\mathcal{E}z|E_{n-}\rangle > (E_{n+} - E_{n-}) \tag{3.13}$$

then perturbation theory is not valid. If the two-state secular equation is solved exactly it will lead to eigenstates of

$$E^{\pm}_{\text{field}} = \pm e\mathcal{E}\,\langle E_{n+}|z|E_{n-}\rangle$$

so that the system will appear to have a dipole moment equal to

$$e\,\langle E_{n+}|z|E_{n-}\rangle$$

but only for field strengths for which (3.13) holds. The $2S_{1/2}$ state of H is generally said to have a linear Stark effect because the field for which (3.13) is an equality is around 5 V/cm and it was not possible to observe the Stark splitting at lower fields (where it is quadratic) using classical spectroscopy. Furthermore the energies of the $2S_{1/2}$ and $2P_{1/2}$ states were thought to be exactly degenerate before the development of quantum electrodynamics.

The discussion above shows that a **macroscopic object** can have a permanent dipole moment, provided that it is in a state that is a superposition of two (or more) nearly degenerate states for which i is a good quantum number, and which therefore do not have dipole moments. If a dipole moment is created by an electric field satisfying (3.13), it will persist for a time $\sim \hbar(E_+ - E_-)^{-1}$. The $E_+ - E_-$ splitting is directly proportional to the tunneling rate of the system between states with the dipole up and down; since this rate decreases **exponentially** with the $\sqrt{\text{length} \times \text{height}}$ of any barrier in the middle, the dipole moments of macroscopic objects persist essentially forever.

2. Stationary perturbation theory

We first recall some results of stationary perturbation theory because they will be used to calculate the DC polarizability and the induced dipole moment.

Assume that the Hamiltonian of a system may be written as the sum of two parts

$$H = H_\circ + H'$$

where $H_\circ$ is the unperturbed Hamiltonian and H' is a weak perturbation.

Furthermore, assume that the problem of obtaining eigenstates and eigenvalues for $H_\circ$ has been accomplished, so that

$$H_\circ \left| n^{(0)} \right\rangle = E_n^{(0)} \left| n^{(0)} \right\rangle$$

Assuming that $H_\circ$ is independent of time, there are basically two types of perturbation theory depending on whether or not H' is time dependent. If H' is **time independent**, the problem is called stationary and the appropriate perturbation theory is the **Rayleigh–Schrödinger stationary perturbation theory**.

Since H' is weak, the eigenvalues and eigenstates change only slightly from their unperturbed values. Hence we can expand them in the form

$$E_n = E_n^{(0)} + E_n^{(1)} + E_n^{(2)} + \dots$$
$$|n\rangle = \left| n^{(0)} \right\rangle + \left| n^{(1)} \right\rangle + \left| n^{(2)} \right\rangle + \dots$$

where we are expressing the ith order perturbation as $E^{(i)}$ and $\left| n^{(i)} \right\rangle$. The matrix elements of H' are normalized by the condition $\langle n|n\rangle = 1$ implying $\langle n^{(i)} | n^{(0)} \rangle = 0$ for $i \neq 0$.

For the energy one obtains

$$E_n^{(i)} = \langle n^{(0)} | H' | n^{(i-1)} \rangle$$

meaning the ith order perturbed energy depends on the $(i-1)^{\text{th}}$ order perturbed wavefunction.

The results are as follows—*all rhs quantities have superscript* 0 *(except H') denoting the unperturbed quantities.*

(i) **First order**

$$E_n^{(1)} = \langle n | H' | n \rangle$$
$$\left| n^{(1)} \right\rangle = \sum_m{}' \frac{\langle m | H' | n \rangle}{E_n - E_m} |m\rangle$$

The $\sum'$ means to omit the term $m = n$ in the summation, and it is understood that the sum extends over continuum states also.

(ii) **Second order**

$$E_n^{(2)} = \sum_m{}' \frac{|\langle m|H'|n\rangle|^2}{E_n - E_m}$$

$$|n^{(2)}\rangle = -\sum_m{}' \frac{\langle m|H'|n\rangle \langle n|H'|n\rangle}{(E_n - E_m)^2} |m\rangle + \sum_{m,p}{}' \frac{\langle m|H'|p\rangle \langle p|H'|n\rangle}{(E_n - E_m)(E_n - E_p)} |m\rangle$$

Note that the effect of a coupling of n and m together by H' is always to push the levels apart (to raise the upper and lower the lower) in second order perturbation theory, independent of the value of H'_{nm}. This leads to repulsion of states coupled by H'.

(iii) **Third order**

$$E_n^{(3)} = -\sum_m{}' \frac{|\langle n|H'|m\rangle|^2 \langle n|H'|n\rangle}{(E_n - E_m)^2} + \sum_{m,p}{}' \frac{\langle n|H'|m\rangle \langle m|H'|p\rangle \langle p|H'|n\rangle}{(E_n - E_m)(E_n - E_p)}$$

Problems arise if the level n is a degenerate level such that $E_m = E_n$ for some m, since then the energy denominators blow up. This problem may be solved by using linear combinations of these degenerate eigenvalues constructed so that $\langle m|H'|n\rangle = 0$.

3. DC polarizability and dipole moment

We now give expressions for the energy and polarizability of an atom placed in a static electric field polarized along the $+z$ direction

$$\vec{E} = \mathcal{E}\,\hat{z}$$

We assume that the field is weak so that we can use the results of stationary perturbation theory. As shown before, the perturbation Hamiltonian is

$$H' = e\mathcal{E}z$$

and parity forces $H'_{mm} = 0$ for all m, so there will be no first order perturbation of the energy. Then, to second order

$$\Delta E_n = -e^2\mathcal{E}^2 \sum_m{}' \frac{|\langle m|z|n\rangle|^2}{E_m - E_n} \tag{3.14}$$

which is the energy shift of the atom due to the Stark effect. This quantity is often called the DC Stark shift, to distinguish it from the AC Stark shift in an oscillating field which we will see later in the chapter.

If we compare the above expression with the potential energy U of a charge distribution interacting with an electric field in Eq. (3.12), one could identify the polarizability with the sum. This identification is incorrect, however: U represents the potential energy of the charge distribution interacting with the external field; ΔE_n also includes the energy required to polarize the atom. Making an analogy with the energy stored in a capacitor shows that the stored energy is $-\frac{1}{2}$ the interaction energy. Thus

$$\Delta E_n = \Delta U + \Delta U_{\text{int}} = -\frac{1}{2}\alpha_n \mathcal{E}^2$$

As a result the DC polarizability of an atom in the state $|n\rangle$ is

$$\alpha_n = 2e^2 \sum_m{}' \frac{|\langle m|z|n\rangle|^2}{E_m - E_n} \tag{3.15}$$

which has the dimensions l^3, i.e. it is a **volume**.

The **induced dipole moment** can be found from electrostatics,

$$\vec{d}_n = \alpha_n \vec{E} = 2e^2\mathcal{E}\hat{z} \sum_m{}' \frac{|\langle m|z|n\rangle|^2}{E_m - E_n} \tag{3.16}$$

An alternative way to find the dipole moment would be to define a dipole operator

$$\vec{P} \equiv -e\vec{r} \tag{3.17}$$

and find $\vec{d}_n$ from the expectation value of this operator using the first order perturbed state vector $|n_1\rangle \equiv |n^{(0)}\rangle + |n^{(1)}\rangle$ while noting that $H' = -\vec{P}\cdot\vec{E}$

$$\begin{aligned}
\vec{d}_n &= \langle n_1|\vec{P}|n_1\rangle \\
&= 2\,\mathrm{Re}\{\langle n^{(0)}|\vec{P}|n^{(1)}\rangle\} \\
&= 2e^2\,\mathrm{Re}\left\{\sum_{r,m}{}' \frac{\langle n^{(0)}|r|m\rangle\,\langle m|\vec{r}\cdot\vec{E}|n^{(0)}\rangle}{E_m - E_n}\,\hat{r}\right\}
\end{aligned}$$

where the sum is over $r = x, y, z$. Only the term $r = z$ will give a non-zero result, and it will yield results in accord with Eq. (3.16).

The polarizability may be approximated using Unsold's approximation, where the energy term E_m in the summation is approximated by an average energy $\bar{E}_m$ and taken out. Using the closure relation $\sum_m |m\rangle\,\langle m| = \mathbb{1}$, one gets*

$$\begin{aligned}
\alpha_n &= \frac{2e^2}{\bar{E}_m - E_n}\sum_m \langle n|z|m\rangle\,\langle m|z|n\rangle \\
&= \frac{2e^2\,\langle n|z^2|n\rangle}{\bar{E}_m - E_n}
\end{aligned}$$

In order to get an estimate of the polarizability in the ground state of hydrogen, we take $\bar{E}_m = 0$ (good enough for H because $|g\rangle$ is far from the other states) and use the result of the virial theorem (again valid for H where $V \sim r^{-1}$) to relate E_g and r

$$\begin{aligned}
E_g &= \langle U\rangle_g + \langle K\rangle_g && U \text{ is potential and } K \text{ is kinetic} \\
&= \frac{1}{2}\langle U\rangle_g && \text{Virial theorem} \\
&= \frac{-e^2}{2}\langle r^{-1}\rangle_g
\end{aligned}$$

Noting that $\langle g|z^2|g\rangle = \frac{1}{3}\langle r^2\rangle_g$ gives the result

$$\alpha_g = \frac{4}{3}\frac{\langle r^2\rangle_g}{\langle r^{-1}\rangle_g}$$

which shows that α_g is closely related to the **volume of the atom.**

*The term $m = n$ does not need to be excluded from the sum because $\langle n|z|n\rangle = 0$.

4. Beyond the quadratic Stark effect

It should be clear from the previous discussion that the Stark effect for a state $|a\rangle$ is quadratic only if the electric field is well below a critical value, given by

$$\mathcal{E}_{\text{crit}} \ll \frac{E_b - E_a}{e\,|\langle b|\vec{r}|a\rangle|}$$

where $|b\rangle$ is the nearest state of opposite parity to $|a\rangle$.

If $|a\rangle$ is the ground state $|g\rangle$, then the critical field is very large. To see this, we can estimate its value in a ground state. The energy difference to the nearest state of opposite parity is

$$\Delta E \equiv E_b - E_a \approx 0.5\,E_g$$

and the relevant matrix element to this state is

$$|\langle r\rangle|^{-1} \approx |\langle r^{-1}\rangle| = 2E_g/e^2 \qquad \text{Virial theorem}$$

Using $E_g \sim 0.3$ Hartree, the critical field can be estimated to be

$$\begin{aligned} \mathcal{E}_{\text{crit}} &= \left(0.3\,\frac{me^4}{\hbar^2}\right)^2 \frac{1}{e^3} \\ &\approx 0.1 \times \left(\frac{e}{a_\circ^2}\right) \qquad \leftarrow \text{atomic unit of E field} \end{aligned}$$

or about 5×10^8 V/cm, which is a field that is three orders of magnitude in excess of what can be produced in a laboratory except in a vanishingly small volume.

If $|a\rangle$ is an excited state, say $|n,\ell\rangle$, then this situation changes *dramatically*, with $\mathcal{E}_{\text{crit}}$ becoming quite small. In general, ΔE to the next state of opposite parity depends on the quantum defect

$$\Delta E \equiv E_{n,\ell+1} - E_{n,\ell} = \frac{-R_H}{(n-\delta_{\ell+1})^2} - \frac{-R_H}{(n-\delta_\ell)^2} \approx \frac{2R_H(\delta_{\ell+1}-\delta_\ell)}{n^{*3}}$$

and the matrix element to that state is

$$\langle n,\ell+1|\vec{r}|n,\ell\rangle \sim n^2 a_\circ$$

Thus the critical field is lowered to

$$\begin{aligned} \mathcal{E}_{\text{crit}} &= \frac{me^4}{\hbar^2}\,\frac{1}{ea_\circ}\,\frac{\delta_{\ell+1}-\delta_\ell}{n^{*5}} \\ &= \left(\frac{e}{a_\circ^2}\right) \times \frac{\delta_{\ell+1}-\delta_\ell}{n^{*5}} \end{aligned}$$

Considering that quantum defects are typically $\leq 10^{-5}$ when $\ell \geq \ell_{\text{core}} + 1$, where ℓ_{core} is the largest ℓ of an electron in the core, it is clear that even fields of value 1 V/cm will exceed $\mathcal{E}_{\text{crit}}$ for higher ℓ states if $n^* > 7$. Large laboratory fields (of order 10^5 V/cm) can exceed $\mathcal{E}_{\text{crit}}$ even for s states if $n^* \geq 5$.

When the electric field exceeds $\mathcal{E}_{\text{crit}}$, states with different ℓ but the same n are degenerate to the extent that their quantum defects are small. Once ℓ exceeds the number of core electrons, these states will easily become completely mixed by the field and they must be diagonalized exactly. The result is eigenstates possessing apparently permanent electric dipoles with a resulting linear Stark shift. As the field increases, these states spread out in energy. First they run into states with the same n but different quantum defects; then the groups of states with different n begin to overlap. At this point a matrix containing all $|n, \ell\rangle$ states with $\ell \geq m_\ell$ must be diagonalized. The only saving grace is that the lowest n states do not partake in this strong mixing; however, the n states near the continuum always do if there is an electric field present.

The situation described above differs qualitatively for hydrogen since it has no quantum defects and the energies are degenerate. In this case the zero field problem may be solved using a basis which diagonalizes the Hamiltonian for both the atom and the presence of an electric field. This approach corresponds to solving the hydrogen atom in parabolic-ellipsoidal coordinates and results in the presence of an integral quantum number which replaces ℓ. The resulting states possess permanent dipole moments which vary with this quantum number, and therefore have linear Stark effects even in infinitesimal fields. Moreover the matrix elements which mix states from different n manifolds vanish at all fields, so the upper energy levels from one manifold cross the lower energy levels from the manifold above without interacting with them.

The dramatic difference between the physical properties of atoms with $n > 10$ and the properties of the same atoms in their ground state, coupled with the fact that these properties are largely independent of the type of atom which is excited, justifies the application of the name **Rydberg atoms** to highly excited atoms in general.

5. Field ionization

If an atom is placed in a sufficiently high electric field it will be ionized, a process called **field ionization**. An excellent order of magnitude estimate of the field $\mathcal{E}_{\text{ion}}$ required to ionize an atom which is initially in a bound state with energy $-E$ can be obtained by the following purely classical argument: the presence of the field adds the term $U(z) = e\mathcal{E}z$ to the potential energy of the atom. This produces a potential with a maximum along the negative z axis with $U_{\text{max}} < 0$, and the atom will ionize if $U_{\text{max}} < -E$.

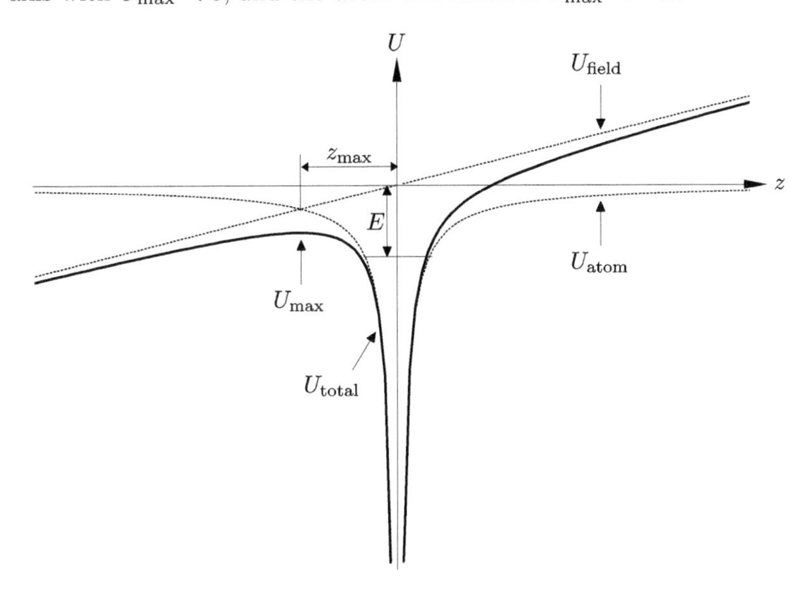

Figure 3.2: Potential energy vs. z for field ionization.

Fig. 3.2 shows U_{atom}, U_{field}, and the total potential

$$U_{\text{total}}(z) = U_{\text{atom}}(z) + U_{\text{field}}(z) = \frac{-e^2}{|z|} + e\mathcal{E}z$$

The maximum as determined by $dU/dz = 0$ occurs at

$$z_{\text{max}} = -\sqrt{\frac{e}{\mathcal{E}}}$$

Equating U_{max} to $-E$ yields

$$\mathcal{E}_{\text{ion}} = \frac{E^2}{4e^3} \qquad \text{for a state with energy } -E$$

For a general level with quantum number n^*

$$E = -\frac{me^4}{\hbar^2}\frac{1}{n^{*2}} = -\frac{e^2}{2a_\circ}\frac{1}{n^{*2}}$$

so the ionizing field becomes

$$\mathcal{E}_{\text{ion}} = \frac{1}{16\,n^{*4}} \times \left(\frac{e}{a_\circ^2}\right) \qquad \leftarrow \text{atomic unit of E field}$$

$$= 3.2 \times 10^8\,(n^*)^{-4}\ \text{V/cm}$$

The predictions of this formula for $\mathcal{E}_{\text{ion}}$ are usually accurate within 20% in spite of its neglect of both quantum tunneling and the change in E produced by the field. Tunneling manifests itself as a finite decay rate for states which classically lie lower than the barrier. The increase of the ionization rate with field is quite dramatic; however, the details of the experiment do not influence the field at which ionization occurs very much. Calculations show the ionization rate increasing from 10^5/s to 10^{10}/s for a 30% increase in the field.

Oddly enough the classical prediction works worse for H than for any other atom. This is a reflection of the fact that certain matrix elements necessary to mix the $|n, \ell\rangle$ states (so the wavefunction samples the region near U_{max}) are rigorously zero in H. Hence the orbital ellipse of the electron does not precess and can remain on the side of the nucleus. There its energy will increase with $\mathcal{E}$, but it will not spill over the lip of the potential and ionize.

E. Permanent atomic electric dipole moment (EDM)

In this section, we discuss when and how a permanent EDM can exist in an atom, and how to look for it experimentally. We first define the three discrete symmetry operators: parity P , time reversal T , and charge conjugation C. There is a theorem—the CPT theorem—which states that the combined operation of the three is conserved in all physical laws. Thus the observation of CP violation in neutral kaon decay is believed to imply T violation.

1. EDM implies P and T violation

In the previous section, we saw that an atom cannot have an EDM unless P is violated. We can also show that the existence of an EDM implies T violation. The simplest way to see this is to refer to Fig. 3.3 below.

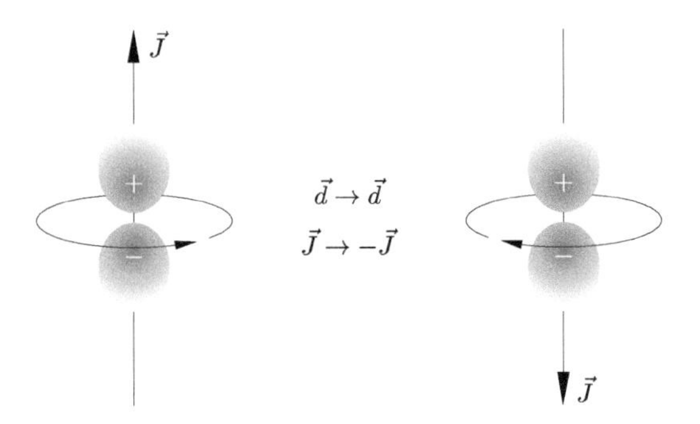

Figure 3.3: Effect of time reversal operation: $\vec{d} \to \vec{d}$ and $\vec{J} \to -\vec{J}$. Since the EDM $\vec{d}$ is always in the direction of $\vec{J}$, it has to be zero unless time reversal symmetry is violated.

From the **Wigner–Eckart theorem**, we know that the angular momentum vector $\vec{J}$ is the only vector in the body-fixed frame—all other vectors must be proportional to it. Applied to the EDM $\vec{d}$, this means that

$$\vec{d} = d\,\frac{\vec{J}}{J}$$

In the time-reversed world, $\vec{J}$ (which recall is $\vec{r} \times \vec{p}$) changes sign, but $\vec{d}$ does not. Thus

$$\vec{d} = -d\,\frac{\vec{J}}{J} \qquad \textbf{after time reversal}$$

which shows either that $\vec{d} = 0$, or that a non-zero $\vec{d}$ implies T violation. The argument for P violation is similar—under P transformation, the polar vector $\vec{d}$ changes sign but the axial vector $\vec{J}$ does not.

The Standard Model (SM) of particle physics accommodates CP violation seen in neutral kaon decay, but predicts EDMs for fundamental particles that are in the range of 10^{-36} e-cm.* This is about 7 to 8 orders-of-magnitude less than current experimental precision. However, theories that go beyond the SM, such as supersymmetry, predict EDMs within experimental range, and are strongly constrained by measured limits on EDMs. Thus EDM searches form an important tool in looking for new physics beyond the SM, though none has been found so far.

In atoms, an EDM can arise due to either (i) an intrinsic electron EDM, (ii) an intrinsic EDM of the neutron or proton, or (iii) a PT-violating nucleon-nucleon or electron-nucleon interaction. Different atoms have different sensitivities to these sources of EDM. In heavy paramagnetic atoms, such as Cs or Tl, the atomic EDM is enhanced by a factor of 100 to 1000 times the intrinsic electron EDM due to relativistic effects. Therefore experiments on such atoms put limits on the existence of an electron EDM. On the other hand, diamagnetic atoms, such as Hg or Yb, are more sensitive to the nuclear Schiff moment and any PT-odd interactions.

2. Experimental method

Atomic EDMs are measured using spin-polarized atoms in the presence of parallel (or anti-parallel) electric and magnetic fields. Recalling that both the electric and magnetic moments are proportional to $\vec{J}$, the total interaction energy is

$$U = -\left(d\vec{E} + \mu\vec{B}\right) \cdot \frac{\vec{J}}{J}$$

The Larmor precession frequency (which we will see more of in Chapter 5, "Resonance") is determined by the interaction of $\vec{\mu}$ with $\vec{B}$, and changes when the direction of the applied $\vec{E}$ field is reversed from being parallel to anti-parallel with respect to $\vec{B}$. The change is given by

$$\Delta\Omega_L = \frac{2dE}{\hbar}$$

Measurement of $\Delta\Omega_L$ therefore constitutes a measurement of the EDM. One experimental advantage of E field reversal is that any imperfections in the B field are inconsequential.

*The standard unit for measuring EDM is e-cm, corresponding to two charges of $\pm e$ separated by 1 cm.

The above analysis shows that, in order to measure the EDM precisely, one needs to (i) measure the Larmor precession frequency very precisely, (ii) have a large E field, and (iii) keep the interaction time with the E field as large as possible. Atomic EDM measurements are usually performed using atomic beams or in vapor cells. With atomic beams, the main limitation is that the interaction time is quite short even if the E-field region is 100 cm long. In vapor-cell experiments, the applied E field is limited by the high pressure to about 10 kV/cm; whereas, the use of an atomic beam in a vacuum system allows the E field to be about 100 kV/cm.

The best current limit (as of 2014) is that the electron EDM is less than $\sim 10^{-27}$ e-cm from a measurement in a thermal beam of Tl atoms; and that the atomic EDM is less than $\sim 10^{-29}$ e-cm from a measurement in a vapor cell containing Hg atoms.

F. Atoms in oscillating electric fields

If an atomic system is placed in an electric field which oscillates at a frequency far from any resonant frequencies, a polarization will be induced in the system which oscillates in phase with the field. This polarization will have two effects: (i) the energy of the system in the field will be different from its field-free value (i.e. its energy levels will be shifted), and (ii) there will be a macroscopic polarization of a gas of such systems causing a departure of the index of refraction from unity.

1. AC polarizability

In this subsection we calculate the response of an atom to the oscillating field

$$\vec{E}(\omega,t) = \mathcal{E}\cos\omega t\,\hat{\varepsilon} \qquad \hat{\varepsilon} \text{ is the polarization of the field}$$

We assume that the field is weak so that we can use the results of first order **time dependent** perturbation theory. As usual, we break the Hamiltonian into two parts

$$H = H_\circ + H'(t) = H_\circ - \frac{1}{2}\mathcal{E}\,(e^{+i\omega t} + e^{-i\omega t})\,\hat{\varepsilon}\cdot\vec{P} \tag{3.18}$$

where $\vec{P}$ is the dipole operator introduced in (3.17).

We express the solution of the time dependent Schrödinger equation as an expansion in the basis of eigenstates $|n\rangle$ of $H_\circ$

$$|\psi(t)\rangle = \sum_n a_n(t)e^{-iE_n t/\hbar}\,|n\rangle$$

Substitution into the Schrödinger equation shows that the a_n must satisfy

$$\dot{a}_k = (i\hbar)^{-1}\sum_n \langle k|H'(t)|n\rangle\, a_n e^{i\omega_{kn}t}$$

where $\omega_{kn} = (E_k - E_n)/\hbar$.

This is an exact equation (it corresponds to the **Interaction picture**), and it must frequently be solved by perturbation theory. This consists of a (hopefully convergent) set of approximations to a_k labeled $a_k^{(i)}(t)$. Starting with

$$a_n^{(0)}(t) = a_n(0)$$

one sets

$$\dot{a}_k^{(i+1)}(t) = (i\hbar)^{-1}\sum_n \langle k|H'(t)|n\rangle\, a_n^{(i)}(t)\, e^{i\omega_{kn}t} \tag{3.19}$$

and solves for the successive approximations by integration.

Let us now turn to the problem of an atom which is in its ground state $|g\rangle$ at $t = 0$, and which is subject to the varying field whose Hamiltonian is given in Eq. (3.18). Substituting into the Eq. (3.19) and integrating from 0 to t gives

$$\begin{aligned} a_i^{(1)}(t) &= (i\hbar)^{-1} \int_0^t \langle i|H'(t')|g\rangle\, e^{i\omega_{ig}t'}\, dt' \\ &= -(i\hbar)^{-1} \langle i|\hat{\varepsilon}\cdot\vec{P}|g\rangle \frac{\mathcal{E}}{2} \int_0^t \left[e^{i(\omega_{ig}+\omega)t'} + e^{i(\omega_{ig}-\omega)t'}\right] dt' \\ &= \frac{\mathcal{E}}{2\hbar} \langle i|\hat{\varepsilon}\cdot\vec{P}|g\rangle \left[\frac{e^{i(\omega_{ig}+\omega)t} - 1}{\omega_{ig}+\omega} + \frac{e^{i(\omega_{ig}-\omega)t} - 1}{\omega_{ig}-\omega}\right] \end{aligned}$$

The -1's come from the persistence of transients generated at $t = 0$ when the field was turned on. If there is damping in the system, and t times the damping rate is $\gg 1$, then these terms may be ignored, as we do in the following.

The term with $\omega_{ig} + \omega$ in the denominator is called the **counter-rotating term**, and may be neglected if one is looking at $\omega \approx \omega_{ig}$ (i.e. near resonance). Ignoring the counter-rotating term is called making the **rotating wave approximation**, which we will see more of later. In order for the co-rotating term not to blow up, a small imaginary term is added to ω_{ig} which takes into account spontaneous emission from state $|i\rangle$.

In the present case (far from resonance) we retain both terms and calculate the expectation value in the ground state of the first order time dependent dipole operator

$$\begin{aligned} \langle\vec{P}(\omega,t)\rangle_g &= 2\,\mathrm{Re}\left\{\langle g|\vec{P}|\sum_i a_i^{(1)}(t)e^{-i\omega_{ig}t}|i\rangle\right\} \\ &= \mathcal{E}\,\mathrm{Re}\left\{\sum_i \frac{\langle g|\vec{P}|i\rangle\,\langle i|\hat{\varepsilon}\cdot\vec{P}|g\rangle}{\hbar}\left[\frac{e^{i\omega t}}{\omega_{ig}+\omega} + \frac{e^{-i\omega t}}{\omega_{ig}-\omega}\right]\right\} \end{aligned}$$

which is by definition the induced dipole moment.

If we now specialize to the case linearly polarized incident light ($\hat{\varepsilon} = \hat{z}$) then

$$\vec{d}_g(\omega,t) = \frac{2e^2}{\hbar}\hat{z}\sum_i \frac{\omega_{ig}|\langle i|z|g\rangle|^2}{\omega_{ig}^2 - \omega^2}\mathcal{E}\cos\omega t$$

This means that the AC polarizability at ω is

$$\alpha_g(\omega) = \frac{2e^2}{\hbar}\sum_i \frac{\omega_{ig}|\langle i|z|g\rangle|^2}{\omega_{ig}^2 - \omega^2} \tag{3.20}$$

which reduces to the DC value in (3.15) when $\omega \to 0$.

2. Oscillator strength expression for polarizabilities

The expressions for the DC and AC polarizabilities shows that the matrix element for the dipole operator $\langle m|\vec{P}|n\rangle$ is by far the most important matrix element for the electrons in an atom because it governs the interaction of electric fields with the atom. The expression bears a close resemblance to the response of a harmonically bound electron driven by an oscillating electric field that we saw in Chapter 2, "Preliminaries." It leads us to express these matrix elements in terms of the oscillator strength f_{mn} for the transition—a dimensionless quantity that relates the strength of this matrix element to the strength of the same matrix element for a harmonically bound electron whose resonant frequency is ω_{mn}.

The oscillator strength is defined for a transition between two levels, and not two states. We highlight this fact by using uppercase letters to label the levels. Thus the oscillator strength for a transition from level $|A\rangle$ to level $|B\rangle$ is

$$f_{AB} \equiv \frac{2}{3}\frac{m}{e^2\hbar}\omega_{BA}\frac{1}{2J_A+1}\sum_{m_A,m_B}|\langle B,m_B|\vec{P}|A,m_A\rangle|^2$$

where $2J_A + 1$ is the multiplicity of $|A\rangle$. We have averaged over all the m states to indicate the rotational symmetry of the definition. For one particular value of m_A, the sum is only over all values of m_B, so that

$$f_{AB} = \frac{2}{3}\frac{m}{e^2\hbar}\omega_{BA}\sum_{m_B}|\langle B,m_B|\vec{P}|A,m_A\rangle|^2$$

If we choose the z axis as the quantization axis, as we have done for the polarizability, and note that z^2 is $r^2/3$, then

$$f_{AB} = 2\frac{m}{\hbar}\omega_{BA}\frac{1}{2J_A+1}\sum_{m_A,m_B}|\langle B,m_B|z|A,m_A\rangle|^2$$

If we try to express the polarizability in terms of the oscillator strength, a problem arises due to the fact that f_{AB} is defined to reflect the rotational invariance of the atomic levels, whereas the polarizability as defined in (3.20) does not

$$\alpha_n(\omega) = \frac{2e^2}{\hbar}\sum_b \frac{\omega_{bn}|\langle b|z|n\rangle|^2}{\omega_{bn}^2 - \omega^2}$$

Therefore, α_n will depend on m_n. This reflects the fact that m_n influences the spatial distribution of the charge and therefore its interaction with a field along z. Thus we must first define an **Isotropic Polarizability** using an average over m

$$\begin{aligned}\alpha_N &= \frac{1}{2J_N+1}\sum_{m_N}\alpha_n(\omega, m_N)\\ &= \frac{2}{3\hbar}\frac{1}{2J_N+1}\sum_{B,m_B,m_N}\frac{\omega_{BN}|\langle B, m_B|\vec{P}|N, m_N\rangle|^2}{\omega_{BN}^2-\omega^2}\end{aligned}$$

The quantity α_N is rotationally invariant—in fact it is the trace of the polarizability tensor. It can be expressed in terms of the oscillator strengths in the particularly simple form

$$\alpha_N = \frac{e^2}{m}\sum_B \frac{f_{NB}}{\omega_{BN}^2-\omega^2}$$

Thus the two isotropic polarizabilities from Eqs. (3.15) and (3.20) are

$$\text{DC} \qquad \alpha_N = \frac{e^2}{m}\sum_M \frac{f_{NM}}{\omega_{MN}^2}$$

$$\text{AC} \qquad \alpha_G = \frac{e^2}{m}\sum_I \frac{f_{GI}}{\omega_{IG}^2-\omega^2}$$

Our definition for oscillator strengths shows that f_{BA} and f_{AB} are related by

$$f_{BA} = -\frac{2J_A+1}{2J_B+1}f_{AB}$$

This relationship stresses two subtleties of oscillator strengths: their magnitude and sign both depend on which way the transition occurs. Their sign is negative for emission from an upper to a lower level, and positive for absorption, which means we have to say "*the oscillator strength for absorption from level* $|A\rangle$ *to level* $|B\rangle$ ..." etc.

One motivation for expressing quantities which involve atomic dipole matrix elements in terms of the oscillator strengths is that the oscillator strengths obey a number of sum rules which place constraints on their values. The most famous sum rule is the Thomas–Reiche–Kuhn sum rule

$$\sum_B f_{GB} = Z_e \qquad |G\rangle \text{ is the ground state}$$

where Z_e is the number of electrons in the atom. The sum above extends over all bound and continuum states involving excitations of any and all of the electrons. If the sum is restricted to configurations with only one excited electron (and includes the continuum only for this electron) then

$$\sum_B f_{GB} \approx 1 \qquad \text{one-electron only}$$

An interesting feature about transitions that comes out from the study of oscillator strengths is that transitions from an initial state $|n, \ell\rangle$ to a final state $|n', \ell'\rangle$ on the average have stronger oscillator strengths for absorption if $\ell' > \ell$, and stronger oscillator strengths for emission if $\ell' < \ell$. In other words, atoms "like" to increase their angular momentum on absorption of a photon, and decrease it on emission.

3. Susceptibility and index of refraction

If a gas of atoms whose density is N is exposed to an oscillating field, the gas will exhibit a polarization with a frequency-dependent susceptibility

$$\vec{P} = \chi(\omega)\vec{E} \qquad \text{with} \qquad \chi(\omega) = N\alpha(\omega)$$

The dielectric constant and index of refraction will be

$$\epsilon = 1 + 4\pi\chi(\omega)$$
$$n = \sqrt{\epsilon} = 1 + 2\pi\chi(\omega) \qquad \text{for low density gases where } \chi \ll 1$$

If Eq. (3.20) is used for $\alpha(\omega)$ then

(i) χ will be real.

(ii) The induced polarization will be in phase with the applied electric field, and the radiation due to $\vec{P}$ will add coherently to $\vec{E}$.

(iii) The only physical effect will be to cause a change in the index of refraction from unity, and hence change the speed of propagation of the wave.

The above points reflect the fact that Eq. (3.20) did not allow for radiative damping of the atomic system—the excited states were assumed to be infinitely sharp. It is also implicitly assumed that the scattered radiation due to $\vec{P}$ interferes destructively in all directions other than that of $\vec{E}$, an assumption which is valid only for a dense spaced array of N atoms (e.g. in glass). Due to the random distribution of atoms in a gas, destructive interference is imperfect to order $N^{1/2}$ and a scattered E field proportional to $N^{1/2}$ is therefore produced, resulting in scattering whose intensity is proportional to N. This is known as Rayleigh scattering (or resonance fluorescence if you are at resonance, $\omega = \omega_{ig}$ for some $|i\rangle$), and is usually computed simply by taking N times the scattering for a single atom. Such a scattering depletes the intensity of the incident field, and is tremendously important; it permits us to see fluorescence radiation from atomic gases without being blinded by the exciting light.

But coming back to radiative damping, we can account for it in a phenomenological way by replacing ω_{ig} with $(\omega_{ig} - i\Gamma_i/2)$ in Eq. (3.20). As a result, α and χ will become complex. In this case, it is traditional to divide χ into real and imaginary parts explicitly

$$\chi = \chi' + i\chi''$$

When a complex susceptibility is used, the wave equation produces solutions of the form (for a wave traveling along z)

$$\vec{E}(\vec{r},t) = \vec{E}_\circ(x,y)\,\mathrm{Re}\left\{e^{i(kz-\omega t)}\right\}$$

with

$$\left[\frac{kc}{\omega}\right]^2 = 1 + 4\pi\chi \tag{3.21}$$

Since c and ω are real, k must be complex

$$k = (\eta + i\kappa)\omega/c$$

Consequently, the propagating wave becomes

$$\vec{E}(\vec{r},t) = \vec{E}_\circ(x,y)e^{-\kappa\omega z/c}\,\mathrm{Re}\left\{e^{i[\eta\omega z/c-\omega t]}\right\}$$

which shows that the wave is attenuated as it propagates.

The index of refraction is now the real part of k, while its imaginary part is called the attenuation factor. They are determined by solving Eq. (3.21), with the results

$$\eta^2 - \kappa^2 = 1 + 4\pi\chi' \qquad \text{and} \qquad 2\eta\kappa = 4\pi\chi''$$

If χ' and χ'' are both $\ll 1$, then the solution is

$$\eta \approx 1 + 2\pi\chi' \qquad \text{and} \qquad \kappa \approx 2\pi\chi''$$

In order to see the lineshapes of these two components as a function of frequency, we go back to the polarizability in Eq. (3.20). For simplicity, we consider only one excited state $|i\rangle$, and subsume all the constants into one K. Then the expression with Γ_i to make α complex becomes

$$\begin{aligned} \alpha_g(\omega) &= K\,\frac{\omega_{ig} - i\Gamma_i/2}{(\omega_{ig} - i\Gamma_i/2)^2 - \omega^2} \\ &\approx \frac{K}{2}\,\frac{(\omega - \omega_{ig}) + i\Gamma_i/2}{(\omega - \omega_{ig})^2 + (\Gamma_i/2)^2} \qquad \text{for } \Gamma_i \ll \omega_{ig} \text{ and } \omega \approx \omega_{ig} \end{aligned} \tag{3.22}$$

The real and imaginary parts of the susceptibility as a function of $\omega - \omega_{ig}$ are plotted in Fig. 3.4. It is clear that κ has a Lorentzian shape near ω_{ig}, while the index of refraction obeys a dispersion curve. This is, in fact, the origin of the name dispersion as applied to this function—dispersion refers to the frequency dependence of the index of refraction in classical optics, which disperses the light into its various components.

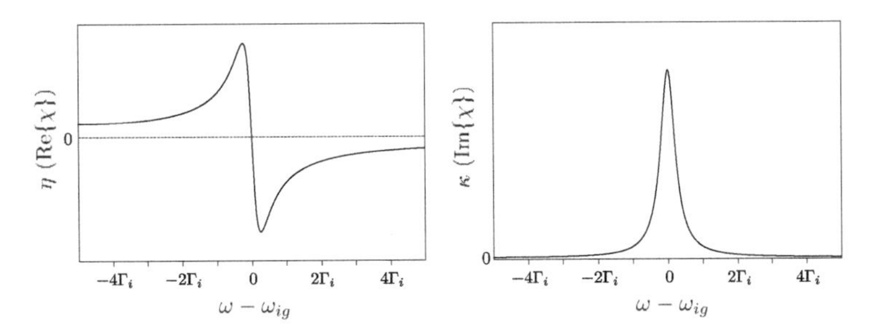

Figure 3.4: Plots of η (corresponding to the real part of χ) and κ (corresponding to the imaginary part of χ) as a function of frequency.

(i) Causality and the Kramers–Kronig relations

It may at first seem that the real and imaginary parts of χ are independent entries—one causing a phase delay and the other corresponding to absorption. This is emphatically not the case: any physical device must satisfy causality, i.e. nothing can come out until after something is put in, and this imposes restriction on the frequency dependence of $\chi'$and χ''.

For example, if χ'' is large and negative except when $|\omega - \omega_\circ| < \Delta\omega$, the gas transmits only if $\omega \approx \omega_\circ$. A δ-function input (at $t = 0$) thus comes out as a nearly monochromatic wave packet which extends for a time $\sim 1/\Delta\omega$ on either side of its maximum. Unless this wave packet is strongly delayed, i.e. $\chi'(\omega \approx \omega_\circ)$ is large, the leading edge will come out before the δ-function goes in!

For the susceptibility of a gas, the causality-imposed restrictions on χ are called the Kramers–Kronig relationships:

$$\chi'(\omega) = \frac{2}{\pi}\mathcal{P}\int_0^\infty \frac{\omega'\chi''(\omega')}{\omega'^2 - \omega^2}\,d\omega' \qquad \text{and} \qquad \chi''(\omega) = -\frac{2\omega}{\pi}\mathcal{P}\int_0^\infty \frac{\chi'(\omega')}{\omega'^2 - \omega^2}\,d\omega'$$

where $\mathcal{P}$ means principal part. The susceptibility obtained from Eq. (3.22) satisfies these relationships.

4. Level shifts — The AC Stark effect

We calculate the shift of an energy level from the isotropic AC polarizability

$$\Delta E_G(\omega) = -\frac{1}{2}\alpha_G(\omega)\mathcal{E}^2(\omega,t) = -\frac{1}{2}\frac{e^2}{m}\sum_I \frac{f_{GI}}{\omega_{IG}^2 - \omega^2}\bar{\mathcal{E}^2} \tag{3.23}$$

which is often called the AC Stark shift, to contrast it with the DC Stark shift that we saw earlier in Eq. (3.14). The two expressions are quite similar, the one above differing only by the addition of the $-\omega^2$ term and the bar denoting the time average over $\mathcal{E}^2$. As a result of the $-\omega^2$ term, the AC shift can be much bigger (if $\omega \to \omega_{ig}$), and it can have either sign.

The averaging of $\mathcal{E}^2$ seems quite reasonable on the grounds that the uncertainty principle prevents one from measuring E_g with an accuracy on the order of the shift in one period of oscillation of the applied field. Nevertheless, it presents a severe philosophical problem because we are now discussing a system whose Hamiltonian has explicit time dependence but we are using the language of stationary perturbation theory. We dodge this issue by insisting on an operational definition of an energy level; if a second weak oscillating field (in addition to the strong one causing the shift) is applied which causes transitions from the state $|g\rangle$ to another state $|k\rangle$ which is not significantly perturbed by the strong field, the resonance will be observed at a frequency $(E_k - E_g - \Delta E_g)/\hbar$.

G. Strong oscillating fields — Dressed atoms

In this section, we solve the problem of an atom with two-states (labeled $|1\rangle$ and $|2\rangle$ and having eigenfrequencies ω_1 and ω_2) in the presence of an arbitrarily strong oscillating electric field of the form $\mathcal{E}\cos\omega t$. This solution is called the **dressed atom**, because we will be looking for eigenstates that are a combination of atom plus field basis states, so in a sense the atom is being "dressed" by the field. The solution is semiclassical because, as mentioned in Chapter 2, "Preliminaries," only the atom, but not the field, is treated quantum mechanically—hence spontaneous decay is not present. It is an exact solution, valid at arbitrary field strength, but limited to a two-state atom. In practice this is not a severe limit if only one intense field is present, and if ω is much closer to ω_{21} than to any other atomic frequencies.

1. The problem

Mathematically we suppose that the matrix representing the Hamiltonian in the $|1\rangle\,,|2\rangle$ basis has the following form

$$H = \frac{\hbar}{2}\begin{bmatrix} -\omega_{21} & \omega_R e^{+i\omega t} \\ \omega_R e^{-i\omega t} & \omega_{21} \end{bmatrix}$$

ω_R is the **Rabi frequency**

Though its exact form will be justified in Chapter 5, "Resonance," the important thing to note for our present purposes is that the off-diagonal coupling element depends on the field strength, i.e. $\omega_R \propto \mathcal{E}$. Thus the weak field limit corresponds to $\omega_R \to 0$.

We begin our search for dressed solutions with the substitution

$$\psi = \begin{bmatrix} a_1\, e^{+i\omega_{21}t/2} \\ a_2\, e^{-i\omega_{21}t/2} \end{bmatrix}$$

The Schrödinger equation then yields the following equations of motion

$$\begin{aligned} \dot{a}_1 &= -\frac{i\omega_R}{2} e^{+i\delta t} a_2 \\ \dot{a}_2 &= -\frac{i\omega_R}{2} e^{-i\delta t} a_1 \end{aligned} \tag{3.24}$$

where $\delta \equiv \omega - \omega_{21}$ is the detuning from resonance.

To find the dressed states we seek a substitution which will yield equations with no explicitly time-dependent terms. The key point to realize is that the atom will exhibit coherences at frequency ω, but not (in steady state) at frequency ω_{21}. This is just like the response of a classical driven harmonic oscillator—it oscillates at the drive frequency ω, not its natural frequency $\omega_\circ$. To achieve this, we substitute

$$a_1 = b_1 e^{+i\delta t/2}$$
$$a_2 = b_2 e^{-i\delta t/2}$$

which yields the coupled equations

$$\dot{b}_1 = \frac{i}{2}(-\delta b_1 - \omega_R b_2)$$
$$\dot{b}_2 = \frac{i}{2}(-\omega_R b_1 + \delta b_2)$$

While the above equations appear more complicated than those in (3.24), they are steady state in the sense that they do not contain time-oscillating terms. They are identical to the Schrödinger equation for a two-state system with

$$H_2 = \frac{\hbar}{2}\begin{bmatrix} \delta & \omega_R \\ \omega_R & -\delta \end{bmatrix}$$

H_2 can be expressed using trigonometric functions as (the use of 2θ instead of θ is for later convenience)

$$H_2 = \frac{\hbar\omega'_R}{2}\begin{bmatrix} \cos 2\theta & \sin 2\theta \\ \sin 2\theta & -\cos 2\theta \end{bmatrix}$$

with

$$\omega'_R = \sqrt{\delta^2 + \omega_R^2} \qquad \text{the } \textbf{effective Rabi frequency}$$
$$\cos 2\theta = \delta/\omega'_R$$
$$\sin 2\theta = \omega_R/\omega'_R$$

Rather than solve for the eigenvalues and eigenvectors of H_2 in the conventional way, we employ a shorter method of solution made possible by the simplicities of a two-state system. This method consists of finding a matrix T which diagonalizes H_2. A suitable choice is

$$T = \begin{bmatrix} \cos\phi & \sin\phi \\ -\sin\phi & \cos\phi \end{bmatrix} \equiv \begin{bmatrix} c & s \\ -s & c \end{bmatrix}$$

T is a general 2×2 unitary matrix—it corresponds to a rotation by ϕ in the plane containing the basis states.

A short calculation will show that the transformation gives the result

$$T^\dagger H_2 T = \frac{\hbar\omega'_R}{2}\begin{bmatrix} D & Q \\ Q & -D \end{bmatrix} \tag{3.25}$$

with

$$D = (c^2 - s^2)\cos 2\theta - 2sc\sin 2\theta$$
$$Q = 2sc\cos 2\theta + (c^2 - s^2)\sin 2\theta$$

for arbitrary ϕ.

To diagonalize H_2, Eq. (3.25) shows that Q must be zero

$$Q = 0 = \sin 2\phi\cos 2\theta + \cos 2\phi\sin 2\theta$$

which implies that $\phi = -\theta$. With this choice of ϕ

$$D = (c^2 - s^2)\cos 2\theta - 2sc\sin 2\theta = 1$$

2. The solution

The eigenvalues of H_2 are [the diagonal elements of $T^\dagger H_2 T$ in Eq. (3.25)]

$$\lambda^\pm = \pm\hbar\omega_R'/2$$

The associated eigenvectors are the columns of T

$$e^+ = \begin{bmatrix} \cos\theta \\ \sin\theta \end{bmatrix} \qquad e^- = \begin{bmatrix} -\sin\theta \\ \cos\theta \end{bmatrix}$$

The resulting eigenfunctions in terms of $|1\rangle$ and $|2\rangle$ are

$$\begin{aligned} |t^+\rangle &= \cos\theta\, e^{-i(2\omega_1+\omega_R'-\delta)t/2}\,|1\rangle + \sin\theta\, e^{-i(2\omega_2+\omega_R'+\delta)t/2}\,|2\rangle \\ &= e^{-i(2\omega_1+\omega_R'-\delta)t/2}\left[\cos\theta\,|1\rangle + e^{-i\omega t}\sin\theta\,|2\rangle\right] \\ |t^-\rangle &= -\sin\theta\, e^{-i(2\omega_1-\omega_R'-\delta)t/2}\,|1\rangle + \cos\theta\, e^{-i(2\omega_2-\omega_R'+\delta)t/2}\,|2\rangle \\ &= e^{-i(2\omega_2-\omega_R'+\delta)t/2}\left[-e^{+i\omega t}\sin\theta\,|1\rangle + \cos\theta\,|2\rangle\right] \end{aligned} \tag{3.26}$$

The second lines of both $|t^+\rangle$ and $|t^-\rangle$ are written such that it is clear that both states are *coherent superpositions* of $|1\rangle$ and $|2\rangle$ in which their relative phase changes at a rate ω (not ω_{21}).

The $\cos\theta$ and $\sin\theta$ factors may be expressed as (for all δ)

$$\cos\theta = \left[\frac{\omega_R' + \delta}{2\omega_R'}\right]^{1/2} = \left[\frac{1}{2}\left(1 + \frac{\delta}{\omega_R'}\right)\right]^{1/2}$$

$$\sin\theta = \left[\frac{\omega_R' - \delta}{2\omega_R'}\right]^{1/2} = \left[\frac{1}{2}\left(1 - \frac{\delta}{\omega_R'}\right)\right]^{1/2}$$

3. Time dependence

Since the eigenfunctions $|t^+\rangle$ and $|t^-\rangle$ diagonalize the atom with the field present, the time evolution of the wave function is expressed easily in terms of these states if we first define the following energy terms

$$
\begin{aligned}
E_1^+ &= \hbar\left(\omega_1 + \frac{\omega_R'}{2} - \frac{\delta}{2}\right)\\
E_1^- &= \hbar\left(\omega_1 - \frac{\omega_R'}{2} - \frac{\delta}{2}\right)\\
E_2^+ &= \hbar\left(\omega_2 + \frac{\omega_R'}{2} + \frac{\delta}{2}\right)\\
E_2^- &= \hbar\left(\omega_2 - \frac{\omega_R'}{2} + \frac{\delta}{2}\right)
\end{aligned}
\tag{3.27}
$$

The energy subscripts 1 and 2 are written to indicate that the components of the dressed states approach the appropriate atomic state in the limit $\omega_R \to 0$ for the two signs of detuning, as will be clear from the expression below. Thus, the time dependent wavefunction is

$$
\begin{aligned}
|\psi(t)\rangle &= a_w\,|t^+\rangle + a_s\,|t^-\rangle\\
&= \left(a_w \cos\theta\, e^{-iE_1^+ t/\hbar} - a_s \sin\theta\, e^{-iE_1^- t/\hbar}\right)\ |1\rangle\\
&\quad + \left(a_w \sin\theta\, e^{-iE_2^+ t/\hbar} - a_s \cos\theta\, e^{-iE_2^- t/\hbar}\right)\ |2\rangle
\end{aligned}
$$

The coefficients a_w and a_s depend on the initial conditions, and are generally time independent.

4. Eigenenergies versus field strength at fixed detuning

We now examine the interaction energy of the dressed atom with the strength of the oscillating field at fixed ω. Recalling from the Hamiltonian that the field strength is proportional to ω_R, this involves the study of the change of the dressed-state eigenvalues $E^{\pm}_{1,2}$ in (3.27) with $\hbar\omega_R$.

Fig. 3.5 below shows the energies of the different branches for the three cases of detuning.

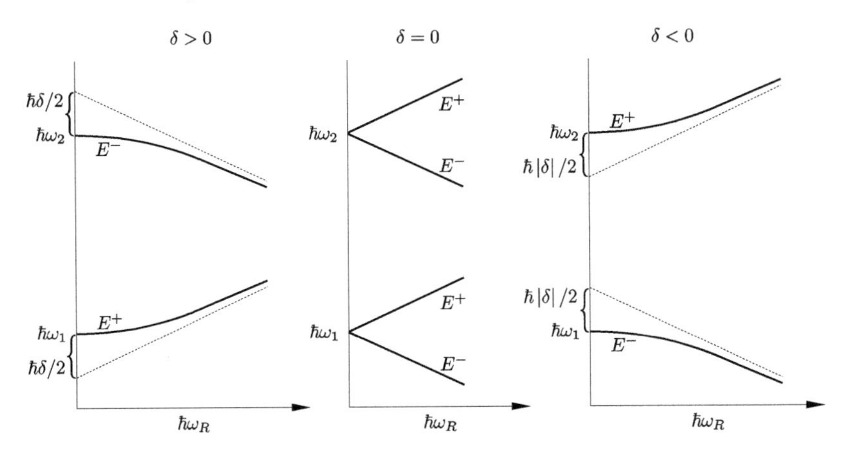

Figure 3.5: Dressed-state energies as a function of field strength for the three cases of detuning. The off-resonant branches approach the atomic states in the weak field limit.

For the off-resonance cases of $\delta \neq 0$, the only branches shown are the ones which approach the atomic states as $\omega_R \to 0$. The limiting values for the parameters are given in the table below

	$\omega_R \to 0$	$\delta < 0$ (ω below ω_{21})	$\delta > 0$ (ω above ω_{21})
$\cos\theta$	$\to$	0	1
$\sin\theta$	$\to$	1	0
θ	$\to$	$\pi/2$	0
$\lvert t^+\rangle$	$\to$	$\lvert 2\rangle$	$\lvert 1\rangle$
$\lvert t^-\rangle$	$\to$	$-\lvert 1\rangle$	$\lvert 2\rangle$

Thus the branches shown in the figure are — for $\delta > 0$, $E^+ = E_1^+$ and $E^- = E_2^-$; while for $\delta < 0$, $E^+ = E_2^+$ and $E^- = E_1^-$. The other components (not shown) are spaced apart by $\pm\hbar\omega$.

For the exact resonance case of $\delta = 0$, it is not possible to identify the eigenvalues E^+ and E^- with either $\hbar\omega_1$ or $\hbar\omega_2$, because even in the limit $\omega_R \to 0$, the eigenfunctions $|t^+\rangle$ and $|t^-\rangle$ are linear combinations of $|1\rangle$ and $|2\rangle$.* In fact, in this limit, $\cos\theta = \sin\theta = 1/\sqrt{2}$; and the two dressed states have equal probabilities of being in either atomic state.

We emphasize that the two eigenstates are coherent superpositions of the two atomic states $|1\rangle$ and $|2\rangle$. They both contain components with two frequencies which differ by ω the driving frequency. Often people refer to the components of these states which is roughly ω from ω_1 or ω_2 (i.e. which is farther from the nearest unperturbed state) as a "virtual level." Do not fall into the trap of thinking that it has a reality of its own; neither its amplitude nor phase, and especially not its population, are distinct and independent—it is only a part of one of the two dressed states. Only the dressed states have a population, independent phase, and amplitude, etc.

The + and − branches are sometimes referred to as

E^+	**weak field seeker**
E^-	**strong field seeker**

This is because the force on the atom will be in the direction of weak (or strong) field in a situation where the field intensity is not constant.

Also, do not be upset by the fact that the E^- energy level appears above E^+ when $\delta > 0$; in the dressed atom picture the number of photons in the field is uncertain (because the electric field is known), and consequently energies of the atom may be adjusted modulo $\hbar\omega$.

*This is similar to the classical coupled oscillators that we saw in Chapter 2, "Preliminaries," where the oscillators were degenerate, and the coupling led to normal modes that were linear combinations of the two displacements.

5. Eigenenergies versus detuning at fixed strength

We now consider the dependence of the eigenenergies on ω at fixed field strength. We shall concentrate on the region near resonance, $\omega \approx \omega_{21}$, going far enough from resonance so that $|\delta| \gg \omega_R$.

When the detuning is many times ω_R, then $|\omega_R/\delta| \ll 1$ and $\omega'_R \approx |\delta|$. As a consequence, the eigenenergies in (3.27) become independent of ω_R and have the following values

	$\lvert\delta\rvert \gg \omega_R$	$\delta < 0$	$\delta > 0$
$E_1^+ = \hbar\left(\omega_1 + \frac{\omega'_R}{2} - \frac{\delta}{2}\right)$	$\rightarrow$	$\hbar(\omega_1 - \delta)$	$\hbar\omega_1$
$E_1^- = \hbar\left(\omega_1 - \frac{\omega'_R}{2} - \frac{\delta}{2}\right)$	$\rightarrow$	$\hbar\omega_1$	$\hbar(\omega_1 - \delta)$
$E_2^+ = \hbar\left(\omega_2 + \frac{\omega'_R}{2} + \frac{\delta}{2}\right)$	$\rightarrow$	$\hbar\omega_2$	$\hbar(\omega_2 + \delta)$
$E_2^- = \hbar\left(\omega_2 - \frac{\omega'_R}{2} + \frac{\delta}{2}\right)$	$\rightarrow$	$\hbar(\omega_2 + \delta)$	$\hbar\omega_2$

We have already mentioned that in the dressed atom picture the eigenvalues are arbitrary up to a multiple of $\hbar\omega$—an ambiguity which arises from the fact that the eigenvalue is for atom plus field, and the number of photons in the field is uncertain when the field strength is known. On this account, we show in Fig. 3.6(a) the behavior of all four energies in (3.27), even though only two of them tend to ω_1 or ω_2 as $\omega_R \to 0$. Often a cut is made through this figure at a particular value of ω (and δ), and the shifts from the unperturbed atomic eigenenergies are displayed. This is shown in Fig. 3.6(b).

The figure shows that oscillating fields exactly on resonance split the states symmetrically, whereas off-resonant fields do not. The splitting of a particular state is always ω'_R, and the difference between the upper (and lower) components of the shifted states is ω, the frequency of the applied field. The solid circles represent the steady-state populations in the two dressed states. The populations in the two states are equal on resonance, which implies that the **average** energy shift is zero. On the other hand, the occupancy of the states is different off resonance, with a larger value for the state which is closer to the unperturbed atomic state. In fact, the occupancy of this state approaches unity as the detuning is increased.

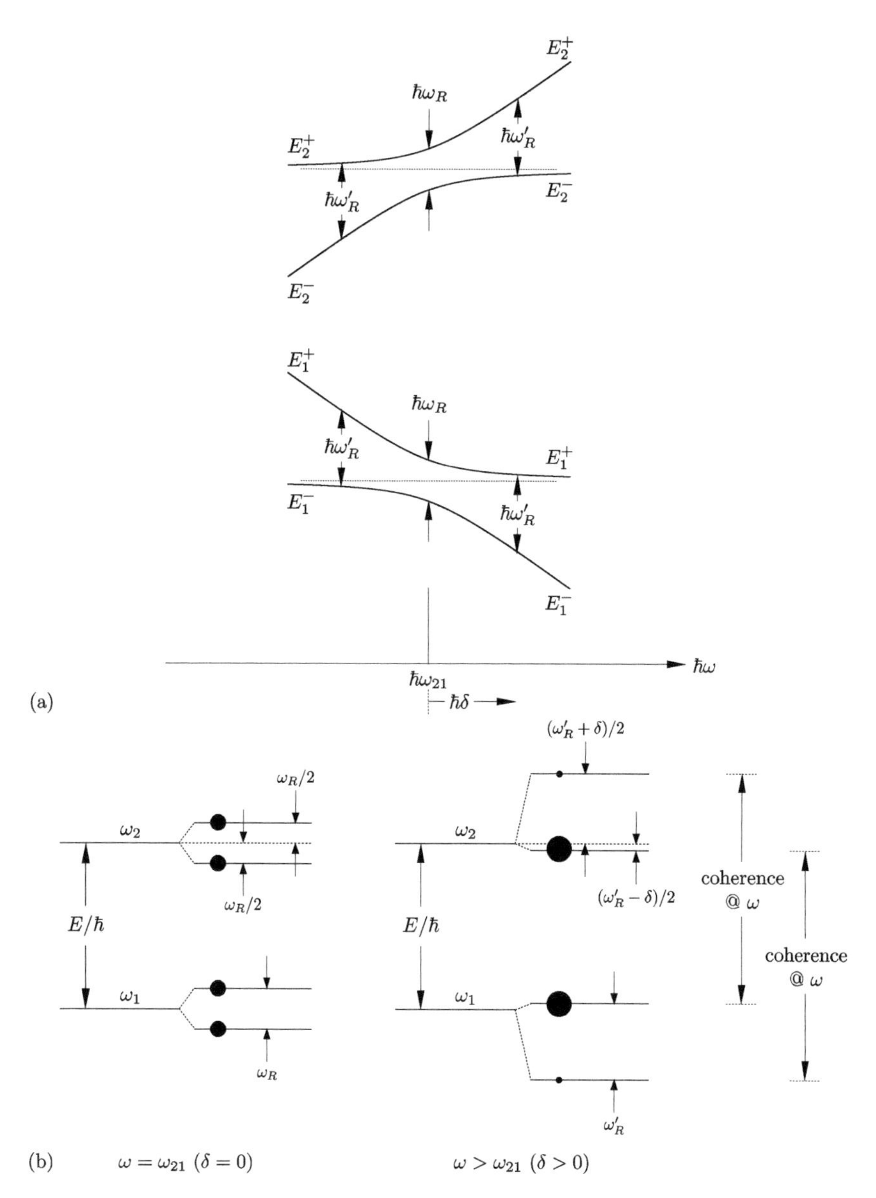

Figure 3.6: (a) Dressed-state energies as a function of detuning for fixed field strength. The separation between E_2^- and E_1^- (and between E_2^+ and E_1^+) is always $\hbar\omega$. The zero of $\hbar\delta$ is at $\hbar\omega_{21}$. (b) Cut through figure in (a) showing shifted states for the $\delta = 0$ and $\delta > 0$ cases. The solid circles represent the steady-state population in each state, equal on resonance and unequal off resonance.

From the expressions for the eigenenergies in (3.27), we see that the shifts of the dressed states from the unperturbed atomic states are (for both signs of detuning)

$$W^{\pm} = \hbar \frac{\omega'_R \pm |\delta|}{2}$$

Since the occupancy of the state with the smaller shift approaches unity at large detunings, the average shift is the shift of this dressed state. Thus the AC Stark shift of the atomic state using the dressed atom picture is

$$\Delta E = \hbar \frac{\omega'_R - |\delta|}{2} = \hbar \frac{\sqrt{\omega_R^2 + \delta^2} - |\delta|}{2} \approx \frac{\omega_R^2}{4|\delta|} \qquad \text{for } |\delta| \gg \omega_R \quad (3.28)$$

which matches the oscillator strength expression given in Eq. (3.23) in this limit.

6. Atom plus field basis states

In discussing the dressed atom we have concentrated on the atom, even though we should really be talking about the atom plus field system. This is the source of the "ambiguity" by $m\hbar$ in the eigenenergy, and it suggests using a basis comprised of the field with m photons plus the atom in $|1\rangle$ or $|2\rangle$. Then the basis states become $|m, 1\rangle$ or $|m, 2\rangle$ where m is the number of photons in the field. The associated eigenenergies are denoted by $E_m^{1,2}$.

Due to the quantum nature of the field, m is not knowable exactly (since the field strength is). However, it is possible to circumvent this problem by writing $m = n + 1$ or $n + 2$ or $n - 1$, where n is a large unknowable number, but the *differences* between various eigenenergies are accounted for explicitly. The $E_m^{2,1}$ are then written with the energy $n\hbar$ subtracted off. Thus

$$E_{n+k}^{2,1} = \hbar\omega_{2,1} + k\hbar\omega$$

For example, when the resonance condition of $\omega = \omega_{21}$ is met, E_{n+1}^{1} will have the same energy as E_n^2.

7. Spectrum of fluorescence from dressed atom

If the driving field is weak ($\omega_R \ll \Gamma$), then the system absorbs and emits only one photon at a time and must emit radiation at exactly the same frequency as the driving field in order to conserve energy (Doppler shift and atomic recoil are neglected for now). However, when the driving field is strong ($\omega_R \gg \Gamma$), then several photons may be absorbed and emitted simultaneously and energy conservation restricts only the sum of the energies of the emitted photons. This implies, for example, that the spectrum of two simultaneous fluorescent photons must be symmetrical about the driving frequency. The frequencies of peaks in the spectrum may be found from the positions of the induced energy levels in the dressed atom picture.

When the driving field is at resonance, we have seen in Fig. 3.6 that the splittings are symmetric about the unperturbed states. Thus all four components of the fluorescence have the same intensity, leading to a spectrum with twice the intensity in the central peak as in the side peaks. Off-resonance excitation produces a non-broadened δ-function spectral component at the driving frequency (Rayleigh component), in addition to the symmetric peaks about the driving frequency. The two kinds of spectra are shown in Fig. 3.7.

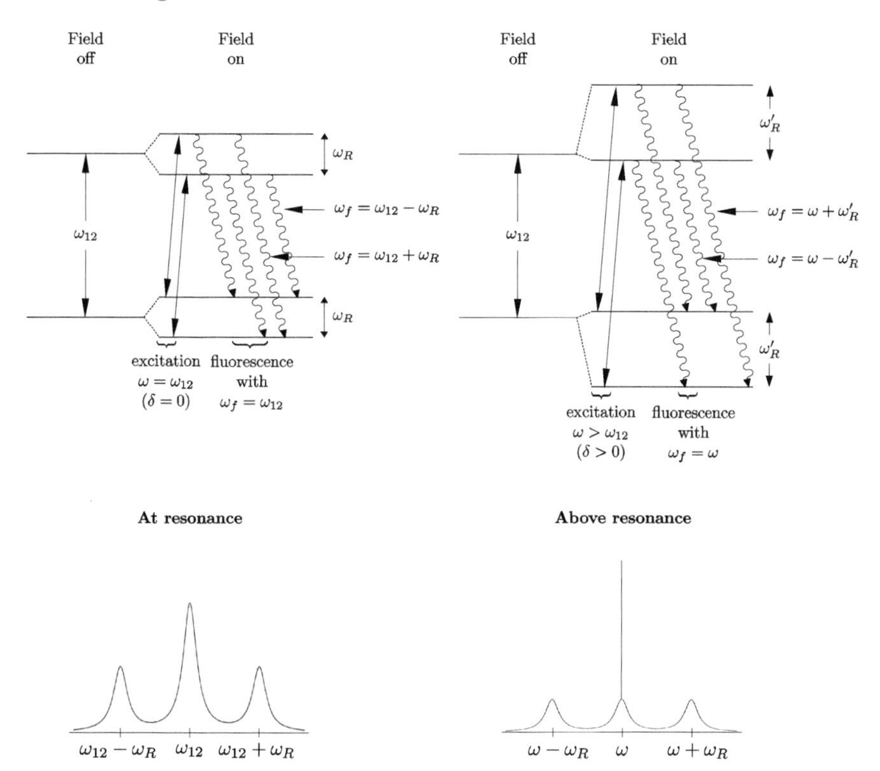

Figure 3.7: Spectra of spontaneous fluorescence from strongly driven atom, for the cases of driving on resonance and above resonance.

H. Problems

1. Quantum defect

Use the data given in the table below to calculate the quantum defect for the following levels of sodium: $n = 3, 4, 6, 10$ and $\ell = 0, 1, 2, 3$. For each ℓ plot δ_ℓ as a function of n.

SODIUM

11 electrons $Z = 11$

Ground state $1s^2\ 2s^2\ 2p^6\ 3s\ ^2S_{1/2}$

$3s\ ^2S_{1/2}$ 41449.65 cm^{-1} Ionization Potential: 5.138 volts

Config.	Term	J	Level	Config.	Term	J	Level
$3s$	$3s\ ^2S$	$1/2$	0.00	$6h$	$6h\ ^2H^o$	$\{4^1/_2, 5^1/_2\}$	38403.4
$3p$	$3p\ ^2P^o$	$1/2$ $1^1/_2$	16956.183 16973.379	$7p$	$7p\ ^2P^o$	$1/2$ $1^1/_2$	38540.40 38541.14
$4s$	$4s\ ^2S$	$1/2$	25739.86	$8s$	$8s\ ^2S$	$1/2$	38968.35
$3d$	$3d\ ^2D$	$2^1/_2$ $1^1/_2$	29172.855 29172.904	$7d$	$7d\ ^2D$	$2^1/_2$ $1^1/_2$	39200.962 39200.963
$4p$	$4p\ ^2P^o$	$1/2$ $1^1/_2$	30266.88 30272.51	$7f$	$7f\ ^2F^o$	$\{2^1/_2, 3^1/_2\}$	39209.2
$5s$	$5s\ ^2S$	$1/2$	33200.696	$8p$	$8p\ ^2P^o$	$1/2$ $1^1/_2$	39298.54 39299.01
$4d$	$4d\ ^2D$	$2^1/_2$ $1^1/_2$	34548.754 34548.789	$9s$	$9s\ ^2S$	$1/2$	39574.51
$4f$	$4f\ ^2F^o$	$\{2^1/_2, 3^1/_2\}$	34588.6	$8d$	$8d\ ^2D$	$\{2^1/_2, 1^1/_2\}$	39729.00
$5p$	$5p\ ^2P^o$	$1/2$ $1^1/_2$	35040.27 35042.79	$8f$	$8f\ ^2F^o$	$\{2^1/_2, 3^1/_2\}$	[39734.0]
$6s$	$6s\ ^2S$	$1/2$	36372.647	$9p$	$9p\ ^2P^o$	$1/2$ $1^1/_2$	39794.53 39795.00
$5d$	$5d\ ^2D$	$2^1/_2$ $1^1/_2$	37036.781 37036.805	$10s$	$10s\ ^2S$	$1/2$	39983.0
$5f$	$5f\ ^2F^o$	$\{2^1/_2, 3^1/_2\}$	37057.6	$9d$	$9d\ ^2D$	$\{2^1/_2, 1^1/_2\}$	40090.57
$5g$	$5g\ ^2G$	$\{3^1/_2, 4^1/_2\}$	37060.2	$9f$	$9f\ ^2F^o$	$\{2^1/_2, 3^1/_2\}$	40093.2
$6p$	$6p\ ^2P^o$	$1/2$ $1^1/_2$	37296.51 37297.76	$10p$	$10p\ ^2P^o$	$\{1/2, 1^1/_2\}$	40137.23
$7s$	$7s\ ^2S$	$1/2$	38012.074	$11s$	$11s\ ^2S$	$1/2$	40273.5
$6d$	$6d\ ^2D$	$2^1/_2$ $1^1/_2$	38387.287 38387.300	$10d$	$10d\ ^2D$	$\{2^1/_2, 1^1/_2\}$	40349.17
$6f$	$6f\ ^2F^o$	$\{2^1/_2, 3^1/_2\}$	38400.1	$10f$	$10f\ ^2F^o$	$\{2^1/_2, 3^1/_2\}$	40350.9

The data are reproduced from NIST Atomic energy level tables. The energy of each level corresponding to a term is given in cm^{-1} above the ground state (3S). For levels with fine structure use the average value of the energies when weighted by the m_j multiplicity. This is the "center-of-gravity" of the level. Use the sodium Rydberg constant

$$R_{\text{Na}} = 109\,734.69\ \text{cm}^{-1}$$

Report the quantum defects to a number of significant figures sufficient to see the trends in the data.

Solution

The quantum defect is defined using

$$E_{n,\ell} = -\frac{R_{\text{Na}}}{(n-\delta_\ell)^2} \qquad \Longrightarrow \qquad \delta_\ell = n - \sqrt{\frac{R_{\text{Na}}}{|E_{n,\ell}|}}$$

$E_{n,\ell}$ can be found from the energy level data. Since the levels are given in terms of the shift from the ground state $\Delta E_{n,\ell}$ it is related to the energy (negative because these are bound states) as

$$E_{n,\ell} = W_{n,\ell} - \text{IP}$$

where

$$W_{n,\ell} = \frac{\sum (2J+1)\,\Delta E_{n,\ell}}{\sum (2J+1)}$$

is the weighted average for levels with fine structure and IP is the ionization potential.

Using the values

$$R_{\text{Na}} = 109\,734.69\ \text{cm}^{-1} \qquad \text{and} \qquad \text{IP} = 41\,449.65\ \text{cm}^{-1}$$

we get the quantum defects as

n	δ_ℓ			
	δ_0	δ_1	δ_2	δ_3
3	1.372910	0.882865	0.0102848	does not exist
4	1.357063	0.8669295	0.01232579	0.0007673
6	1.350907	0.8592408	0.01389949	0.00134215
10	1.350147	0.856014	0.0142453	0.006387

The figure below shows a plot of the quantum defects versus n.

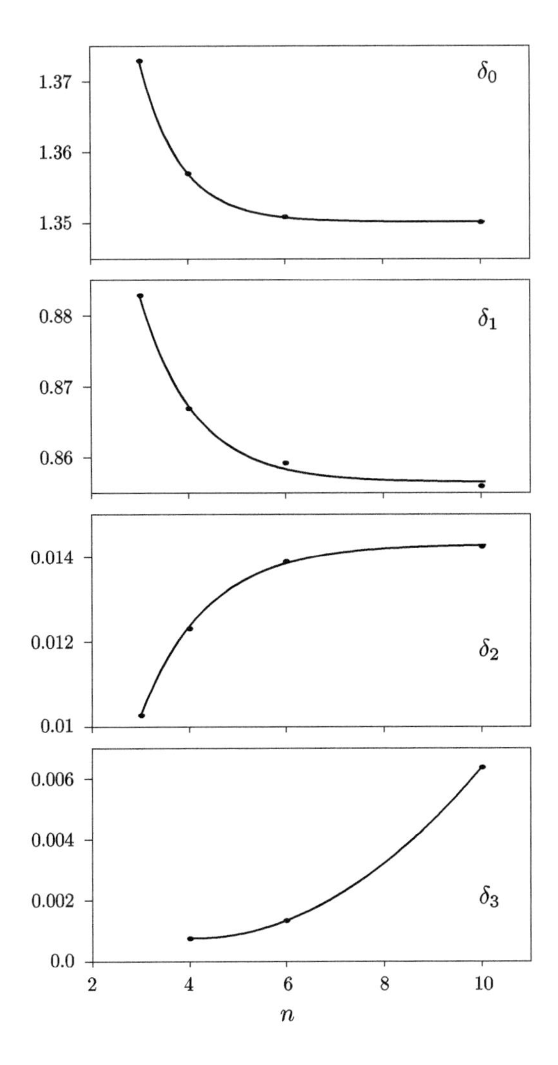

2. Classical electron bound by a harmonic potential

Consider a classical electron bound by the harmonic potential $U = \frac{1}{2}kx^2$.

(a) Find the DC polarizability in the presence of a static electric field $\mathcal{E}_\circ$.

(b) Find the oscillator strength of this system from the DC polarizability.

(c) Show that the energy $E = U_{\text{field}} + U_{\text{int}} = \frac{1}{2}U_{\text{field}}$.

(d) Find the AC polarizability $\alpha(\omega)$ in the presence of an AC field of the form $\mathcal{E}_\circ \cos\omega t$, for ω not close to $\omega_\circ$.

Solution

(a) The equilibrium position of the electron shifts by Δx in the presence of the electric field, and this induces a dipole moment $e\,\Delta x$.

Equating the harmonic force to the electric force we have

$$k\,\Delta x = -e\,\mathcal{E}_\circ \qquad \Longrightarrow \qquad \Delta x = -\frac{e}{k}\,\mathcal{E}_\circ$$

This gives the induced dipole moment as

$$\vec{P} = -e\,\Delta x = \frac{e^2}{k}\,\mathcal{E}_\circ$$

which yields the DC polarizability

$$\alpha_\circ = \frac{e^2}{k} = \frac{e^2}{m\omega_\circ^2}$$

(b) From oscillator strength theory

$$\alpha_\circ = \frac{e^2}{m}\frac{f}{\omega_\circ^2}$$

This gives

$$f = \frac{m\omega_\circ^2}{e^2}\,\alpha_\circ = \frac{m\omega_\circ^2}{e^2}\left(\frac{e^2}{m\omega_\circ^2}\right) = 1$$

which is expected from the definition of oscillator strength.

(c) The two energies are

$$U_{\text{field}} = -\vec{P}\cdot\vec{E} = -\left(\frac{e^2}{k}\mathcal{E}_\circ\right)\mathcal{E}_\circ = -\frac{e^2}{k}\mathcal{E}_\circ^2$$

$$U_{\text{int}} = \frac{1}{2}k\,(\Delta x)^2 = \frac{1}{2}k\left(\frac{e^2\mathcal{E}_\circ^2}{k^2}\right) = \frac{e^2}{2k}\mathcal{E}_\circ^2$$

Therefore the total energy is

$$E = -\frac{e^2}{k}\mathcal{E}_\circ^2 + \frac{e^2}{2k}\mathcal{E}_\circ^2 = -\frac{e^2}{2k}\mathcal{E}_\circ^2 = \frac{1}{2}U_{\text{field}}$$

(d) The equation of motion for the electron in the AC field is

$$\ddot{x} + \omega_\circ^2 x = -\frac{e\,\mathcal{E}_\circ}{m}\cos\omega t$$

For ω not close to $\omega_\circ$ (far from resonance), the response of the driven oscillator is such that the phase difference from the drive is either 0 (in phase) or π (out of phase). Therefore the motion is of the form

$$x = \pm x_\circ \cos\omega t$$

Substituting into the equation of motion, we get

$$-x_\circ\,\omega^2 + x_\circ\,\omega_\circ^2 = -\frac{e\mathcal{E}_\circ}{m} \qquad \Longrightarrow \qquad x_\circ = \frac{e\mathcal{E}_\circ}{m\,(\omega^2 - \omega_\circ^2)}$$

Therefore the induced dipole moment is

$$\vec{P} = -ex_\circ\cos\omega t = -\frac{e^2}{m(\omega^2 - \omega_\circ^2)}\mathcal{E}_\circ\cos\omega t$$

which gives the AC polarizability as

$$\alpha(\omega) = -\frac{e^2}{m(\omega^2 - \omega_\circ^2)}$$

3. Oscillator strength

(a) Consider an alkali atom with essentially all of the oscillator strength from the ground state saturated by the resonance $ns \to np$ transition (i.e. $f_{ns,np} = 1$ and $f_{ns,np' \neq p} \sim 0$). If this transition has frequency ω_{ps} find the critical electric field $\mathcal{E}_{\text{crit}}$ beyond which the Stark effect is no longer quadratic (take this to be where $|\langle np|H'|ns\rangle| = \hbar\omega_{ps}$). Express the result in V/cm where ω_{ps} is expressed in atomic units.

(b) What is $\mathcal{E}_{\text{crit}}$ for Na with transition wavelength of 589 nm?

Solution

(a) The critical electric field $\mathcal{E}_{\text{crit}}$ is defined such that

$$|\langle np|H'|ns\rangle| = \hbar\omega_{ps} \qquad \text{where } H' = ez\mathcal{E}_{\text{crit}}$$

This implies that

$$\mathcal{E}_{\text{crit}} = \frac{\hbar\omega_{ps}}{|\langle np|ez|ns\rangle|}$$

From the expression for oscillator strength

$$f_{ns,np} = \frac{2m\omega_{ps}}{e^2\hbar}\,|\langle np|ez|ns\rangle|^2 = 1 \quad \Longrightarrow \quad |\langle np|ez|ns\rangle| = \sqrt{\frac{e^2\hbar}{2m\omega_{ps}}}$$

Therefore

$$\mathcal{E}_{\text{crit}} = \sqrt{\frac{2m\hbar\omega_{ps}^3}{e^2}}$$

The atomic unit of frequency is $me^4/\hbar^3$, thus the critical field when ω_{ps} is expressed in atomic units is

$$\mathcal{E}_{\text{crit}} = \sqrt{\frac{2m\hbar}{e^2}}\left(\frac{me^4}{\hbar^3}\right)^{3/2}\left(\omega_{ps}^{\text{au}}\right)^{3/2} = 7.27 \times 10^9\,\left(\omega_{ps}^{\text{au}}\right)^{3/2}\ \text{V/cm}$$

(b) For the $3p \to 3s$ transition in Na, $\lambda = 589$ nm, therefore

$$\omega_{ps} = 3.2 \times 10^{15}\ \text{rad/s} \qquad \text{and} \qquad \omega_{ps}^{au} = \frac{3.2 \times 10^{15}}{4.1 \times 10^{16}} = 0.078$$

Hence the critical field is

$$\mathcal{E}_{\text{crit}} = 1.58 \times 10^8\ \text{V/cm}$$

4. Dressed atoms

Show that the dressed atom gives the same lowest eigenfunction as first-order perturbation theory at low field intensity (pick $\delta < 0$).

Solution

We start with the Hamiltonian

$$H = \frac{\hbar}{2}\begin{bmatrix} -\omega_{21} & \omega_R e^{i\omega t} \\ \omega_R e^{-i\omega t} & \omega_{21} \end{bmatrix}$$

(i) Dressed atom

Guessing solutions of the form

$$\phi(t) = \begin{bmatrix} c_1 e^{+i\omega t/2} \\ c_2 e^{-i\omega t/2} \end{bmatrix}$$

and substituting into Schrödinger's equation, we get

$$\begin{bmatrix} \dot{c}_1 e^{+i\omega t/2} + c_1 \frac{i\omega}{2} e^{+i\omega t/2} \\ \dot{c}_2 e^{-i\omega t/2} - c_2 \frac{i\omega}{2} e^{-i\omega t/2} \end{bmatrix} = -\frac{i}{2}\begin{bmatrix} -\omega_{21} c_1 e^{+i\omega t/2} + \omega_R c_2 e^{+i\omega t/2} \\ \omega_R c_1 e^{-i\omega t/2} + \omega_{21} c_2 e^{-i\omega t/2} \end{bmatrix}$$

This gives

$$\dot{c}_1 + \frac{i c_1 \omega}{2} = \frac{i\omega_{21} c_1}{2} - \frac{i}{2}\omega_R c_2$$
$$\dot{c}_2 - \frac{i c_2 \omega}{2} = \frac{-i\omega_R c_1}{2} - \frac{i}{2}\omega_{21} c_2$$

or

$$\dot{c}_1 = \frac{i}{2}\left(-\delta c_1 - \omega_R c_2\right)$$
$$\dot{c}_2 = \frac{i}{2}\left(-\omega_R c_1 + \delta c_2\right)$$

Written in matrix form, this is

$$i\hbar \begin{bmatrix} \dot{c}_1 \\ \dot{c}_2 \end{bmatrix} = \frac{\hbar}{2}\begin{bmatrix} \delta & \omega_R \\ \omega_R & -\delta \end{bmatrix}\begin{bmatrix} c_1 \\ c_2 \end{bmatrix}$$

Eigenvalues are obtained by diagonalizing the matrix

$$\begin{bmatrix} \delta & \omega_R \\ \omega_R & -\delta \end{bmatrix}$$

which gives

$$\begin{vmatrix} \delta-\lambda & \omega_R \\ \omega_R & -\delta-\lambda \end{vmatrix} = 0 \qquad \Longrightarrow \qquad \lambda^2 - \delta^2 - \omega_R^2 = 0$$

This yields the eigenvalues as

$$\lambda = \pm\sqrt{\omega_R^2 + \delta^2}$$

The lowest eigenfunction is given by $\lambda = -\sqrt{\omega_R^2 + \delta^2}$, which gives the coefficients as

$$\delta c_1 + \omega_R c_2 = -\sqrt{\omega_R^2 + \delta^2}\, c_1 \qquad \Longrightarrow \qquad c_2 = -\frac{\delta + \sqrt{\omega_R^2 + \delta^2}}{\omega_R}\, c_1$$

Thus the eigenfunction is

$$|\phi(t)\rangle = e^{+i\omega t/2}\,|1\rangle - \frac{\delta + \sqrt{\omega_R^2 + \delta^2}}{\omega_R} e^{-i\omega t/2}\,|2\rangle$$

For $\delta < 0$ and weak field ($\omega_R \ll |\delta|$), the eigenfunction becomes

$$\begin{aligned} |\phi\rangle_{\text{dressed}} &= e^{+i\omega t/2}\,|1\rangle - \frac{-|\delta| + |\delta| + \omega_R^2/2|\delta|}{\omega_R} e^{-i\omega t/2}\,|2\rangle \\ &= e^{+i\omega t/2}\,|1\rangle - \frac{\omega_R}{2|\delta|} e^{-i\omega t/2}\,|2\rangle \\ &= e^{+i\omega t/2}\,|1\rangle + \frac{\omega_R}{2\delta} e^{-i\omega t/2}\,|2\rangle \end{aligned}$$

(ii) First-order perturbation theory

From the Hamiltonian we have

$$\dot{a}_2^{(1)}(t) = \frac{1}{i\hbar}\frac{\hbar\omega_R}{2} e^{-i\omega t} a_1^{(0)}(t)\, e^{i\omega_{21} t} = \frac{\omega_R}{2i} e^{-i\delta t} a_1^{(0)}(t)$$

Using the initial conditions of $a_1^{(0)}(0) = 1$ and $a_2^{(0)}(0) = 0$, we get

$$a_2^{(1)}(t) = \frac{\omega_R}{2i}\int_0^t e^{-i\delta t'}\,dt' = \frac{\omega_R}{2i}\frac{e^{-i\delta t'}}{-i\delta}\bigg|_0^t = \frac{\omega_R}{2\delta} e^{-i\delta t}$$

Therefore the coefficients are

$$\begin{aligned} a_1(t) &= (1 + 0 + \ldots) \\ a_2(t) &= \left(0 + \frac{\omega_R}{2\delta} e^{-i\delta t} + \ldots\right) \end{aligned}$$

and the first-order eigenfunction is

$$\begin{aligned}
|\phi^{(1)}\rangle_{\text{pert theory}} &= e^{-i\omega_1 t}|1\rangle + \frac{\omega_R}{2\delta}e^{-i(\omega_2+\delta)t}|2\rangle \\
&= e^{-i(\omega-2\omega_1)t/2}\left[e^{+i\omega t/2}|1\rangle + \frac{\omega_R}{2\delta}e^{+i(\omega+2\omega_1-2\omega_2-2\delta)t/2}|2\rangle\right] \\
&= e^{-i(\omega-2\omega_1)t/2}\left[e^{+i\omega t/2}|1\rangle + \frac{\omega_R}{2\delta}e^{-i\omega t/2}|2\rangle\right] \\
&= e^{-i(\omega-2\omega_1)t/2}|\phi\rangle_{\text{dressed}}
\end{aligned}$$

This shows that the two wavefunctions are the same (up to an irrelevant phase factor).

Chapter 4

Nucleus

IN the previous chapter, we have discussed atoms as if the nuclei were point charges with no structure and infinite mass. Real nuclei have mass, possibly non-zero angular momentum $\vec{I}$, and a charge which is spread out over a finite volume. As a result, they possess a magnetic dipole moment and an electric quadrupole moment coupled to the electronic angular momentum, and possibly higher moments as well. All of these properties affect the atomic energy levels, at a level about 10^{-5} Ry—these effects will be discussed in this chapter.

The finite size of the nucleus produces only a small shift of the spectral line, and thus the only quantity accessible to measurement is the variation of the line position between different isotopes of the same element. Laser spectroscopy makes it possible to measure such "isotope shifts" to at least 10^{-9} Ry, or 10^{-3} to 10^{-4} of the shift. The moments of the nucleus couple to its spin which interacts with the angular momentum of the rest of the atom. This splits the energy levels of the atom according to the value of F, the quantum number corresponding to the total angular momentum ($\vec{F} = \vec{I} + \vec{J}$). The resulting hyperfine structure can be measured with almost limitless precision (certainly $< 10^{-18}$ Ry) using the techniques of RF spectroscopy. Hyperfine transitions in Cs and H are currently the best available time and frequency standards. Generally speaking magnetic dipole interactions predominate in atoms and electric quadrupole interactions in molecules.

With the exception of the mass shift, the manifestations of nuclear structure in atomic spectra provide important information on the static properties of nuclei which are among the most precise information about nuclei. Unfortunately the great precision of the atomic measurements is generally lost in deducing information about nuclear structure because the core electrons affect the magnetic and electric interactions of the valence electrons with the nucleus.

A. Isotope effects

When comparing the spectral lines originating from atoms whose nuclei differ only in the number of neutrons (i.e. different isotopes of the same element), effects due to the finite mass and volume of the nucleus become apparent; even neglecting hyperfine structure (by taking the center of gravity of the observed splitting) the spectral lines of the different isotopes vary slightly in position—generally at the many parts per million level. The difference between the lines of the various isotopes is referred to as the isotope shift: it is observed to have both positive (i.e. a heavier isotope has a higher energy spacing) and negative values.

General speaking, light elements ($A < 40$) have positive frequency shift, whereas heavy elements ($A > 60$) have negative shifts. This reflects the contribution of two distinct physical processes to the shift—the finite mass shift (almost always positive), and the nuclear volume shift (almost always negative). These will be discussed separately.

1. Mass effect

The origin of the mass effect is obvious from the Bohr energy level formula

$$E_N = E_n^\circ \left[\frac{M}{M+m}\right] \approx E_n^\circ \left[1 - \frac{m}{M}\right]$$

where the term involving m/M comes from solving the two body electron-nucleus (of mass M) system using the relative coordinate and the associated reduced mass. Obviously increasing M increases E_N.

In two (or more) electron atoms the situation becomes more complicated due to the relative motion of the electrons. It would, for example, be possible to arrange the electrons symmetrically on opposite sides of the nucleus in which case there would be zero isotope effect. The virial theorem assures us that the mean value of the kinetic energy equals minus the total energy, so if we treat the nuclear motion as a perturbation on a fixed nucleus solution, the mass effect will be

$$\begin{aligned} \Delta E_{N,M} &\approx -\frac{P^2}{2M} = -\frac{1}{2M}\left[\sum_i \vec{p}_i\right]^2 \\ &= -\frac{m}{M}\left[\frac{1}{2m}\sum_i p_i^2 + \frac{1}{2m}\sum_{i\neq j} \vec{p}_i \cdot \vec{p}_j\right] \end{aligned}$$

The first term is called the normal shift since (using the virial theorem again) it is

$$\Delta E_{N,M}^{\text{Normal}} = -\frac{m}{M} E_N^{\circ}$$

The second term is called the specific shift because it depends on the particular atomic state. $\Delta E^{\text{specific}} = 0$ unless there are two or more valence electrons. For configuration $s\ell$ one gets

$$\Delta E_{ns,n'\ell}^{\text{specific}} = (1-2s)\frac{m}{M}\frac{3f_{ns,n'\ell}}{2}\hbar\omega_{n's',n\ell}$$

where $f_{ns,n'\ell}$ is the oscillator strength.

Thus the specific shift has opposite signs for $s = 1$ and $s = 0$ states—a reflection of the fact that the specific isotope shift is closely related to the exchange interaction. The above equation also reflects the general result that $\Delta E^{\text{specific}}$ is zero unless the two electrons are connected by an electric dipole allowed transition (otherwise f will vanish). Furthermore the specific isotope shift is of the same order of magnitude as the normal isotope shift—for $f > 2/3$, in fact, it can be larger (reversing the sign of the mass dependence of the isotope effect.)

The preceding discussion shows that the fractional energy shift of a level due to the mass of the nucleus decreases rapidly with increasing mass of the nucleus. The normal part of this shift has a variation in the fractional magnitude due to a change ΔM in the mass of the isotope of

$$\frac{\Delta E_{N,M+\Delta M}^{\text{Normal}} - \Delta E_{N,M}^{\text{Normal}}}{E_N} = \left[\frac{m}{M+\Delta M} - \frac{m}{M}\right] \approx \frac{m}{M}\left[\frac{\Delta M}{M}\right]$$

which decreases as M^{-2}, reaching 10 parts per million for a nucleus with $A = 54$ (assuming $\Delta M = 1$).

2. Volume effect

Inside the nucleus, the electrostatic potential no longer behaves like Ze/r, but is reduced from this value. If the valence electron(s) penetrate significantly into this region (e.g. for s electrons) then its energy will rise (relative to the value for a point nucleus) because of this reduced potential. Adding neutrons to the nucleus generally spreads out the charge distribution, causing a further rise in its energy. This reduction in the binding energy results in a decrease of the transition energy and therefore to a negative isotope shift (assuming that the s state is the lower energy state involved in the transition).

For an s state the density of the electron probability distribution at the nucleus is given by the semi-empirical Fermi–Segŕe formula

$$|\psi_s(0)|^2 = \frac{Z_a^2 Z}{\pi a_\circ^3 n^{*3}}\left[1 + \left|\frac{\partial \delta_s}{\partial n}\right|\right] \tag{4.1}$$

where δ_s is the quantum defect and $Z_a e$ is the charge of the atomic core.

Combining this with a model of the nuclear charge cloud results in the following nuclear volume correction to the energy (of an s electron)

$$\Delta E_n^V = Z_a^2 \frac{R_\infty}{n^{*3}} \left[1 + \left| \frac{\partial \delta_s}{\partial n} \right| \right] C \tag{4.2}$$

with

$$C = \frac{4(\gamma + 1)}{[\Gamma(2\gamma + 1)]^2} B(\gamma) \left[\frac{2Zr_\circ}{a_\circ} \right]^{2\gamma} \frac{\delta r_\circ}{r_\circ}$$

where

$$\gamma = \sqrt{1 - \alpha^2 Z^2}$$

and Γ is a gamma function $\Gamma(N+1) = N!$ and $B(\gamma)$ is a factor that depends on the nuclear charge distribution

$$B(\gamma) = (2\gamma + 1)^{-1} \qquad \text{for a charge shell}$$

$$= (2\gamma + 1)^{-1} \left[\frac{3}{2\gamma + 3} \right] \qquad \text{for a uniform charge}$$

The nuclear radius is taken as (for atomic number A)

$$r_\circ = 1.15 \times 10^{-13} A^{1/2}$$

so that

$$\frac{\delta r_\circ}{r_\circ} = \frac{\delta A}{3A}$$

There are obviously a number of assumptions in Eq. (4.2), and it should not be expected to work as well as expression for the nuclear mass shift. The observed shift is generally 1/2 to 3/4 of the one given above except for non-spherical nuclei (e.g. rare-earth nuclei) which have anomalously large shifts.

B. Hyperfine structure

The fact that the nucleus is a charge cloud with angular momentum suggests the possibility that it might possess magnetic and electric moments. Time reversal and parity invariance restrict the possible magnetic moments to dipole, octupole, ..., i.e. odd values of ℓ when the multipole moments are expanded in the standard manner in terms of the spherical harmonics $Y_{\ell m}$'s; and the possible electric moments to monopole (total charge Ze), quadrupole, hexadecapole, ..., i.e. even values of ℓ.

The hyperfine energy shift for a particular state is then written in a series expansion (with progressively smaller corrections) as

$$W^{\text{hfs}} = A\,K_1 + B\,K_2 + C\,K_3 + D\,K_4 + \ldots$$

where A is the magnetic dipole hyperfine coupling constant, B is the electric quadrupole hyperfine coupling constant, C is the magnetic octupole hyperfine coupling constant, D is the electric hexadecapole hyperfine coupling constant, and so on. K's are factors that depend on the quantum numbers of the state.

The magnetic dipole and electric quadrupole interaction are dominant in the hyperfine interaction. The magnetic dipole moment can be measured only if the nucleus has $I \geq 1/2$ and it splits only those levels for which $J \geq 1/2$. Similarly, the electric quadrupole interaction is observable only when both I and J are ≥ 1.

1. Magnetic dipole

The magnetic moment of the nucleus is generally expressed in terms of the nuclear magneton

$$\mu_N = \frac{e\hbar}{2m_p c} = \frac{\mu_B}{1836} = 0.762 \text{ kHz/G}$$

and the nuclear g factor

$$\vec{\mu}_N = g_I \mu_N \vec{I}/\hbar$$

To emphasize the fact that the nuclei are complex particles we note that the g factors of proton and neutron are

$$g_p = +5.586 \qquad \text{and} \qquad g_n = -3.826$$

neither one of which is close to a simple integer.

The magnetic moment of the nucleus couples to the magnetic field produced at the nucleus by the electrons in the atom. As a result $\vec{J}$ and $\vec{I}$ are coupled together to form $\vec{F}$, the total angular momentum of the entire atom

$$\vec{F} = \vec{J} + \vec{I}$$

The magnetic coupling between $\vec{J}$ and $\vec{I}$ adds a term to the Hamiltonian for the magnetic dipole hyperfine structure

$$H_{\mathrm{M}}^{\mathrm{hfs}} = -\vec{\mu}_N \cdot \vec{B}_j$$

$\vec{B}_j$ is proportional to $\vec{J}$, and is written in terms of a constant a as

$$\vec{B}_j = \frac{a}{\mu_N} \vec{J}/\hbar$$

Thus the hyperfine structure Hamiltonian becomes

$$H_{\mathrm{M}}^{\mathrm{hfs}} = -g_I a \, \vec{I} \cdot \vec{J}/\hbar^2 = A \, \vec{I} \cdot \vec{J}/\hbar^2 \tag{4.3}$$

Using a procedure similar to what was done for the case of spin-orbit coupling in calculating the Landé g factor, we can show that

$$\vec{I} \cdot \vec{J} = \frac{1}{2}\left(|\vec{F}|^2 - |\vec{I}|^2 - |\vec{J}|^2\right)$$

Thus the energy shift for a state with quantum numbers F, I, and J is

$$W_{\mathrm{MD}}^{\mathrm{hfs}} = A \, \frac{F(F+1) - I(I+1) - J(J+1)}{2} \tag{4.4}$$

The most important case of magnetic hyperfine structure occurs for atoms with s electrons both because of their preponderance among materials that are easy to handle experimentally and because the magnetic hyperfine interaction is largest for them. In this case the electron cloud looks like an isotropic region of magnetization. Using the spin magnetic moment of the electron, the magnetization at a radius r is

$$\vec{M}(r) = \vec{\mu}_e \, |\psi_s(r)|^2 = -g_s \mu_B \, |\psi_s(r)|^2 \, \vec{S}/\hbar$$

The isotropic magnetization can be viewed as arising from the sum of a small uniform sphere at the origin plus a hollow sphere containing the remainder of the magnetization, as shown in Fig. 4.1.

A uniformly magnetized sphere produces a uniform B field inside

$$\vec{B} = \frac{8\pi}{3} \vec{M}$$

This fact may be used to show (by superposition) that a uniformly magnetized spherical shell has $B = 0$ inside, which means that the field due to the

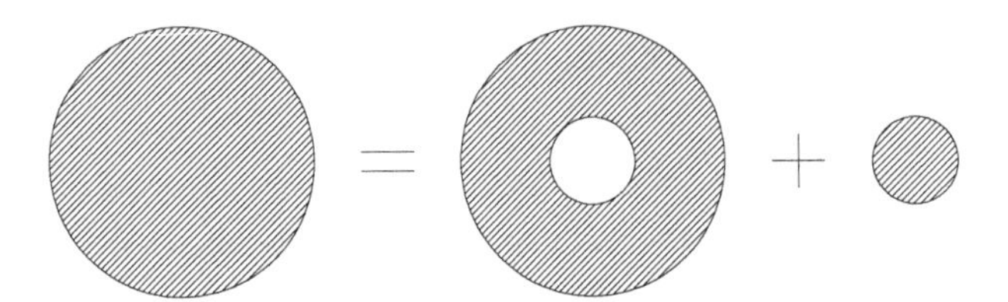

Figure 4.1: Decomposition of a spherically symmetric cloud of magnetization for finding the magnetic field at its center.

hollow sphere part in the decomposition shown in Fig. 4.1 vanishes. Hence the field is

$$\vec{B}_j = -\frac{8\pi}{3} g_s \mu_B \, |\psi_s(0)|^2 \, \vec{S}/\hbar$$

so that the energy shift (taking $g_s = 2$) is

$$\begin{aligned} W_{\text{MDs}}^{\text{hfs}} &= g_I \mu_N \, \frac{8\pi}{3} 2\mu_B \, |\psi_s(0)|^2 \, \vec{I} \cdot \vec{S}/\hbar^2 \\ &= \frac{16\pi}{3} g_I \, \frac{m}{m_p} \mu_B^2 \, |\psi_s(0)|^2 \, \vec{I} \cdot \vec{S}/\hbar^2 \\ &= \frac{8\pi}{3} g_I \, \frac{m}{m_p} \alpha^2 R_\infty a_\circ^3 \, |\psi_s(0)|^2 \, \vec{I} \cdot \vec{S}/\hbar^2 \end{aligned}$$

Using Eq. (4.1) for $|\psi_s(0)|^2$ one obtains

$$W_{\text{MDs}}^{\text{hfs}} = \frac{8\alpha^2 g_I Z_a^2 Z}{3n^{*3}} \, \frac{m}{m_p} \, R_\infty \left[1 + \left| \frac{\partial \delta_s}{\partial n} \right| \right]$$

This s-state interaction is sometimes called the "contact" (in the sense of touch) term.

Now we consider the hyperfine interaction for states with $\ell \neq 0$. These states have zeros of the wave function at the origin (recall $\psi \sim r^\ell$), so the magnetic field at the nucleus is generated by both the orbital motion and the intrinsic magnetic moment of the electron

$$\vec{B}_j = \vec{B}_\ell + \vec{B}_s$$

For simplicity, we do a calculation for a single electron. We evaluate the fields using classical techniques since the quantum mechanical result is identical.

$$\vec{B}_\ell(0) \equiv \frac{1}{c}\int \frac{I d\vec{r} \times (-\vec{r})}{r^3}$$

$$= \frac{-e}{c}\int \frac{\vec{v} \times (-\vec{r})}{r^3} \qquad \leftarrow \text{for an orbiting electron}$$

$$= -\frac{e}{mc}\int \frac{\vec{r} \times m\vec{v}}{r^3}$$

$$= -\frac{2\mu_B}{\hbar}\langle r^{-3}\rangle \vec{L}$$

$$\vec{B}_s(0) = -\frac{1}{r^3}\left[\vec{\mu}_e - 3\left(\vec{\mu}_e \cdot \hat{r}\right)\hat{r}\right] \qquad \leftarrow \text{magnetic field of a dipole}$$

$$= -\frac{g_s\mu_B}{\hbar}\langle r^{-3}\rangle \left[-\vec{S} + 3(\vec{S}\cdot\hat{r})\hat{r}\right]$$

so that the total field is

$$\vec{B}_j(0) = -\frac{2\mu_B}{\hbar}\langle r^{-3}\rangle \left[\vec{L} - \frac{g_s}{2}\vec{S} + \frac{3}{2}(\vec{S}\cdot\hat{r})\hat{r}\right]$$

We need to evaluate the projection of $\vec{B}_j$ along $\vec{J}$. (The following calculation is much like the calculation for the Landé g factor.)

$$\langle \vec{J}\cdot\vec{B}_j\rangle = \langle(\vec{L}+\vec{S})\cdot\vec{B}_j\rangle$$

$$= -\frac{2\mu_B}{\hbar}\langle r^{-3}\rangle \frac{1}{\hbar^2}\langle|\vec{L}|^2 - |\vec{S}|^2 + 3(\vec{S}\cdot\hat{r})\hat{r}\cdot(\vec{L}+\vec{S})\rangle$$

$$= -\frac{2\mu_B}{\hbar}\langle r^{-3}\rangle \left[\ell(\ell+1) - s(s+1) + \frac{3}{\hbar^2}\langle(\vec{S}\cdot\hat{r})^2 + (\hat{r}\cdot\vec{L})\rangle\right]$$

where we have taken $g_s = 2$ exactly. Using

$$s(s+1) = 3/4 \qquad \langle(\vec{S}\cdot\hat{r})^2\rangle = \hbar^2/4 \qquad \hat{r}\cdot\vec{L} = 0$$

we obtain

$$\langle \vec{J}\cdot\vec{B}_j\rangle = -\frac{2\mu_B}{\hbar}\langle r^{-3}\rangle \ell(\ell+1)$$

so after all this work, the field turns out to depend only on ℓ.

To find the constant A in the Hamiltonian in Eq. (4.3) we have

$$H_{\mathrm{MD}}^{\mathrm{hfs}} = A\vec{I}\cdot\vec{J}/\hbar^2$$

$$= -\frac{g_I\mu_N}{\hbar^2}\frac{\vec{I}\cdot\vec{J}}{|\vec{J}|}\langle\frac{\vec{J}\cdot\vec{B}_j}{|\vec{J}|}\rangle$$

Therefore

$$A = 2g_I\mu_N\mu_B \langle r^{-3}\rangle \frac{\ell(\ell+1)}{J(J+1)}$$

$$= 2g_I \frac{m}{m_p}\frac{\mu_B^2}{a_\circ^3}\langle\frac{a_\circ^3}{r^3}\rangle\frac{\ell(\ell+1)}{J(J+1)}$$

$$= g_I \frac{m}{m_p}\alpha^2 R_\infty \langle\frac{a_\circ^3}{r^3}\rangle\frac{\ell(\ell+1)}{J(J+1)}$$

Using the following relation which is appropriate for an atom with one valence electron

$$\langle\frac{a_\circ^3}{r^3}\rangle = \frac{Z_a^2 Z}{n^{*3}\,\ell(\ell+1)(\ell+1/2)} \tag{4.5}$$

one gets

$$A = \frac{g_I Z_a^2 Z}{n^{*3}(\ell+1/2)J(J+1)}\frac{m}{m_p}\alpha^2 R_\infty \tag{4.6}$$

The above expressions for the hyperfine shifts do not contain corrections for nuclear size ($\sim 10^{-4}$), and do not contain relativistic corrections which can be 10% for $Z = 30$ and a factor of 2 for $Z = 80$. In addition the magnetic moment of the nucleus causes a slight unpairing of the core electrons (i.e. mixes in other core configuration) which is called **core polarization** and which changes the magnitude dipole interaction.

2. Electric quadrupole

If the nucleus does not have a spherically symmetric charge distribution it probably has a non-zero electric quadrupole moment

$$Q = \frac{1}{e}\int dV\,\rho(\vec{r})[3z^2 - r^2]$$

which is < 0 for an oblate charge distribution. In contrast to the nuclear magnetic dipole, which is predominantly determined by the nucleons, Q is sensitive to collective deformations of the nucleus. Some nuclei are observed with 30% differences between polar and equatorial axes, then Q can be comparable to $\langle r^2\rangle$, i.e. $\approx 10^{-24}$ cm^2.

The interaction energy of the quadrupole moment Q with the electron can be found by expanding the term $|\vec{r}_e - \vec{r}_N|^{-1}$ in spherical harmonics and evaluating the resulting expressions in terms of Clebsch–Gordan coefficients. The resulting energy shifts are then

$$W_{\mathrm{EQ}}^{\mathrm{hfs}} = B\,K(K+1)$$

where

$$B = \frac{3(Q/a_\circ^2)}{8I(2I-1)J(J+1)} \langle r^{-3} \rangle \, R_\infty \tag{4.7}$$

and

$$K = [F(F+1) - I(I+1) - J(J+1)] \tag{4.8}$$

[Note that the constant appearing in Eq. (4.4) for the hyperfine shift due the magnetic dipole interaction is $K/2$.]

The term $\langle r^{-3} \rangle$ can be estimated from Eq. (4.5).

The preceding expressions, like the corresponding ones for the magnetic interactions, have several significant omissions. The most important are relativistic corrections and core shielding corrections.

3. Order-of-magnitude of hyperfine structure

The expression for the magnetic hyperfine structure in Eq. (4.6) is quite similar to the fine structure expression in Eq. (3.11) in Chapter 3. Their ratio is

$$\frac{W_{\mathrm{MD}}^{\mathrm{hfs}}}{W^{\mathrm{fs}}} \approx g_I \, \frac{m}{m_p} \, \frac{1}{Z}$$

which is typically 10^{-3} to 10^{-4}.

For a neutral atom ($Z_a = 1$), Eq. (4.6) gives

$$\begin{aligned} \frac{W_{\mathrm{MD}}^{\mathrm{hfs}}}{\hbar} &= g_I \, \frac{m}{m_p} \, \frac{Z}{(\ell + 1/2)J(J+1)} \, \alpha^2 \frac{E_n}{n^*} \\ &\approx (\ell + 1/2)^{-3} \text{ GHz} \end{aligned}$$

with a factor of 10 spread in either direction.

The quadrupole interaction is generally considerably smaller. Using $\langle r^{-3} \rangle$ from Eq. (4.5) in Eq. (4.7), one gets

$$\begin{aligned} B &= \frac{3Z}{8I(2I-1)J(J+1)} \, \frac{Q}{a_\circ^2} \, \frac{E_n}{n^* \ell(\ell+1)(\ell+1/2)} \\ &\approx Z \frac{10^{-24}}{10^{-16}} E_n (\ell + 1/2)^{-3} \end{aligned}$$

Hence

$$B/\hbar \approx 0.01 \, Z(\ell + 1/2)^{-3} \text{ GHz}$$

Thus one generally expects that magnetic hyperfine structure dominates hyperfine structure in atoms. The opposite is generally true in molecules for two reasons—(i) unpaired electrons are relatively rare, and (ii) the molecular binding mechanism can create large electric field gradients at the sites of the nuclei.

In concluding this discussion of hyperfine structure in atoms, it is important to realize that the preceding formulae are (except in hydrogenic atoms) only approximations and never permit one to extract the nuclear dipole or quadrupole moments with the full accuracy of laser spectroscopy experiments—let alone RF spectroscopy experiments. Thus the constants $A, B, \ldots$ in the combined hyperfine energy formula

$$\begin{aligned} W^{\text{hfs}} &= W^{\text{hfs}}_{\text{MD}} + W^{\text{hfs}}_{\text{EQ}} + \ldots \\ &= A\,K/2 + B\,K(K+1) + \ldots \end{aligned}$$

[with K defined in Eq. (4.8)] should be regarded primarily as empirical constants from the standpoint of atomic physics. Even if the problems of connecting A and B with the nuclear moments could be solved, the principal result would be better measurements of nuclear properties.

4. Zeeman shift in weak magnetic field

The treatment of the energy shift of an atom with hyperfine structure in a magnetic field—called the Zeeman shift—closely resembles what was done in the LS-coupling scheme.

In the presence of a weak magnetic field, the hyperfine interaction couples $\vec{J}$ and $\vec{I}$ together to form $\vec{F}$, the total angular momentum of the atom. Thus the good quantum numbers are F and m_F, together with J and I. The total magnetic moment of the atom is the sum of the electronic and the nuclear moments

$$\vec{\mu}_a = g_j \mu_B \vec{J} - g_I \mu_N \vec{I}$$

where the minus sign is because the electron and nuclear charges have opposite signs.

Thus the Zeeman shift of an $|F, m_F\rangle$ state in the presence of a weak magnetic field of strength $B_\circ$ is

$$W_B = g_F \mu_B m_F B_\circ \tag{4.9}$$

where g_F is the Landé g factor of the atom. Using a procedure similar to what was done for calculating g_j in Chapter 3, "Atoms," this factor can be derived as

$$g_F = \left[g_j - g_I \frac{\mu_N}{\mu_B} \right] \frac{F(F+1) + J(J+1) - I(I+1)}{2F(F+1)}$$

If g_j is not zero, the second term in the round brackets can be neglected because of the factor $\mu_N/\mu_B = 1/1836$.

5. Decoupling of hyperfine interaction by magnetic field

If a sufficiently strong magnetic field is applied to the atom, it attempts to decouple $\vec{J}$ and $\vec{I}$ from each other and to force them to precess about its direction. F is no longer a good quantum number, and m_j and m_I become almost good. For the hyperfine interaction, this decoupling is called the Back–Goudsmit effect. Obviously, this has the same physics as the Paschen–Back effect—the decoupling, also by a magnetic field, of $\vec{L}$ and $\vec{S}$ from each other and the concomitant destruction of J as a good quantum number. Owing to the fact that $\vec{L}$ and $\vec{S}$ are coupled by an order of $\mu_B^2/a_\circ^3$, whereas $\vec{J}$ and $\vec{I}$ are coupled by an order of $\mu_N\mu_B/a_\circ^3$, the hyperfine interaction is $\sim$ $\mu_B/\mu_N = 1836$ times weaker than the spin orbit coupling and consequently can be decoupled by much weaker magnetic fields. Setting the interaction energy $\mu_B B_{\text{decoup}} = a^2 H(\mu_N/\mu_B)$ gives $B_{\text{decoup}} = 136$ G as the order of magnitude of the field needed to decouple hyperfine interactions. Owing to the fact that I is often greater than 1 and $g_I \sim 4$, fields several times larger are generally required.

We now discuss the decoupling of the hyperfine structure in detail. To begin with, the interaction Hamiltonian in the presence of a magnetic field $B_\circ\hat{z}$ is

$$H' = ah\vec{I}\cdot\vec{J}/\hbar^2 + (g_j\mu_B m_j - g_I'\mu_B m_I)\,B_\circ$$

where the constant a is in frequency units (Hz), and g_I' is defined as

$$g_I' = g_I\frac{m}{m_p} \qquad \Longrightarrow \qquad \vec{\mu}_N = g_I'\mu_B\vec{I}/\hbar$$

Thus g_I' is 1836 times smaller than the usual nuclear g factor due to the replacement of μ_N by μ_B—a replacement which facilitates seeing the decoupling of $\vec{I}$ and $\vec{J}$.

If we now define

$$\text{Field parameter} \qquad x' \equiv g_j\mu_B B_\circ/ah$$

$$\text{Relative nuclear moment} \qquad \alpha \equiv g_I/g_j$$

and note that the projection of the total angular momentum $m = m_I + m_j$ is always a good quantum number, then we get for each value of m

$$\begin{aligned} H_m' &= ah\left[\vec{I}\cdot\vec{J}/\hbar^2 + x'\,(m_j - \alpha m_I)\right] \\ &= ah\left[-\alpha m x' + \vec{I}\cdot\vec{J}/\hbar^2 + m_j\,(\alpha+1)\,x'\right] \end{aligned}$$

This may be broken into diagonal and off-diagonal parts using

$$\vec{I} \cdot \vec{J} = I_z J_z + \frac{1}{2}\left(I^+ J^- + I^- J^+\right)$$

$$\frac{H'_m}{ah} = -x'\alpha m + m m_j - m_j^2 + x'(1+\alpha)\, m_j + \frac{1}{2\hbar^2}\left[I^+ J^- + I^- J^+\right]$$

where we have used the standard raising and lowering operators for angular momentum quantization, which obey the relationship

$$J^{\pm}\left|J, m\right\rangle = \sqrt{J(J+1) - m(m \pm 1)}\, \hbar \left|J, m \pm 1\right\rangle$$

thus giving the off-diagonal matrix elements.

Simplification to $J = 1/2$

For $J = 1/2$, there are only two states for each m except for $|m| = I + J$ which is the non-degenerate stretch state. The problem therefore reduces to that of the standard 2×2 Hamiltonian

$$\frac{H'_m}{ah} = -x^2 \alpha m - \frac{1}{4} + \frac{1}{2}\begin{bmatrix} -d & v \\ v & d \end{bmatrix}$$

where

$$d = m + x'(1+\alpha) \qquad \text{and} \qquad v = \sqrt{I(I+1) - m^2 + 1/4}$$

It is convenient to define the hyperfine level separation

$$\Delta W = (I + 1/2)\, ah$$

so that

$$x = \frac{1+\alpha}{I + 1/2} x' = (g_J + g'_I) \frac{\mu_B B_\circ}{\Delta W}$$

Physically x is the ratio of the magnetic interaction (Zeeman energy) to the hyperfine separation.

Using these we get the Breit–Rabi formula

$$E_m^{\pm} = \frac{\Delta W}{2}\left[-\frac{1}{2I+1} \pm \sqrt{1 + \frac{4m}{2I+1} x + x^2}\right] - g'_I m \mu_B B_\circ$$

where the $+$ sign is for $F = I + 1/2$ and $-$ sign is for $F = I - 1/2$.

There are several noteworthy points about this formula.

(i) The center of gravity of all the lines is zero.

(ii) For $m = 0$, the field dependence is quadratic as $x \to 0$.

(iii) For the stretch states the square root factors exactly, giving a linear Zeeman structure at all field strengths.

The Breit–Rabi formula will be seen in detail in the problem at the end of this chapter. We will consider the case of $I = 3/2$ and $J = 1/2$, which are typical values for many alkali atoms on the D_1 line.

6. Hyperfine anomaly

The hyperfine anomaly—commonly denoted by ε—arises due to the effects of an extended nucleus, which is manifested by the differences between the hypothetical point-like and actual hyperfine interaction. The modification is due to two effects:

(i) The modification of the electron wavefunction by the extended nuclear charge distribution, called the Breit–Rosenthal–Crawford–Schawlow correction $\varepsilon_{\mathrm{BR}}$.

(ii) The extended nuclear magnetization. This effect was first anticipated by Kopfermann and thought to be too small to be observed. However, it was experimentally observed by Bitter; following this it was calculated by A. Bohr and V. Weisskopf, and is hence known as the Bohr–Weisskopf effect $\varepsilon_{\mathrm{BW}}$.

If we denote the point-like hyperfine interaction constant by a_{pt}, then

$$a = a_{\mathrm{pt}}\,(1+\varepsilon_{\mathrm{BR}})\,(1+\varepsilon_{\mathrm{BW}})$$

Thus the ratio of the interaction constant for two isotopes is

$$\frac{a_1}{a_2} = \frac{g_I(1)\,[1+\varepsilon_{\mathrm{BR}}(1)]\,[1+\varepsilon_{\mathrm{BW}}(1)]}{g_I(2)\,[1+\varepsilon_{\mathrm{BR}}(2)]\,[1+\varepsilon_{\mathrm{BW}}(2)]}$$

Using the fact that the correction is small, the above expression is written as

$$\frac{a_1}{a_2} = \frac{g_I(1)}{g_I(2)}\,\left[1+{}^1\Delta^2\right]$$

where we have defined the differential anomaly between the two isotopes as

$$\left[1+{}^1\Delta^2\right] = \frac{[1+\varepsilon_{\mathrm{BR}}(1)]\,[1+\varepsilon_{\mathrm{BW}}(1)]}{[1+\varepsilon_{\mathrm{BR}}(2)]\,[1+\varepsilon_{\mathrm{BW}}(2)]}$$

Precise values of the hyperfine interaction constants and independently measured g factors in the two isotopes are thus needed to obtain the differential hyperfine anomaly.

C. Problems

1. Hyperfine coupling investigated

The Hamiltonian for an atom in a magnetic field $B_\circ \hat{z}$ may be written as

$$H = ah\, \vec{I} \cdot \vec{J}/\hbar^2 + (g_j \mu_B m_j - g_I' \mu_B m_I)\, B_\circ$$

First restrict attention to the case $J = 1/2$ but arbitrary I.

(a) Low field ($x \approx 0$)

(i) Show that the spacing between the levels

$$F^+ = I + 1/2 \qquad \text{and} \qquad F^- = I - 1/2$$

is $\Delta W = a\hbar F^+$.

(ii) Find the center-of-mass of all states defined as

$$E_{\text{cm}} = \frac{\sum_i g_i E_i}{\sum_i g_i}$$

where g_i is the multiplicity of the level.

Defining

$$x = (g_I' + g_j)\, \mu_B\, B_\circ / \Delta W$$

we get the Breit–Rabi formula for the eigenenergies ($m = m_I + m_j$)

$$E_m^{\pm} = \frac{a\hbar F^+}{2} \left[-\frac{1}{2F^+} \pm \sqrt{1 + \frac{2mx}{F^+} + x^2} \right] - g_I' m \mu_B\, B_\circ$$

Now specialize to $I = 3/2$ (for rest of problem).

(b) The two F values are 1 and 2. Find the corresponding g factors g_1 and g_2 (assume $g_I' \ll g_j$).

(c) High field ($x \gg 1$)

(i) What are the good quantum numbers here?

(ii) Find the energies of the corresponding states.

(iii) Does the center-of-mass rule [in (ii) above] still hold?

(d) Show on a figure the energies vs x. Use the non-crossing rule to advantage (levels of the same m do not cross). Label the lines with quantum numbers at low and high fields and indicate m.

(e) Field independent transitions (useful in experiments)

(i) Show on your figure the three (magnetic) dipole transitions (selection rules $\Delta m = 0, \pm 1$) which are field independent at intermediate field ($0 < x < 1$).

(ii) Show two more at very high field ($x \gg 1$).

(iii) For Na ($I = 3/2$, $g_I' = 2.22\,\mu_N/\mu_B$), find the magnetic field for one low field transition.

(iv) Can you find a dipole transition whose frequency is independent of the hyperfine separation $2ah$? Indicate it in your figure.

Solution

(a) (i) In the low field limit, the hyperfine interaction couples $\vec{I}$ and $\vec{J}$ to form $\vec{F}$

$$\vec{F} = \vec{I} + \vec{J}$$

Using

$$\vec{F} \cdot \vec{F} = \left(\vec{I} + \vec{J}\right) \cdot \left(\vec{I} + \vec{J}\right) = \vec{I} \cdot \vec{I} + 2\vec{I} \cdot \vec{J} + \vec{J} \cdot \vec{J}$$

we get

$$\vec{I} \cdot \vec{J} = \frac{1}{2}\left(\vec{F}^2 - \vec{I}^2 - \vec{J}^2\right)$$

and the hyperfine interaction part of the Hamiltonian as

$$H'_{\text{hyp}} = \frac{ah}{2\hbar^2}\left(\vec{F}^2 - \vec{I}^2 - \vec{J}^2\right)$$

Thus the energy shift of a level with quantum numbers I, J, F is

$$E_{\text{hyp}} = \frac{ah}{2}\left[F(F+1) - I(I+1) - J(J+1)\right]$$

For $J = 1/2$, the two hyperfine levels have F values of

$$F^+ = I + \frac{1}{2} \qquad \text{and} \qquad F^- = I - \frac{1}{2}$$

Hence their separation is

$$\begin{aligned}
\Delta W &= \frac{ah}{2}\left[\left(I + \frac{1}{2}\right)\left(I + \frac{3}{2}\right) - \left(I - \frac{1}{2}\right)\left(I + \frac{1}{2}\right)\right] \\
&= \frac{ah}{2} 2\left(I + \frac{1}{2}\right) \\
&= ahF^+
\end{aligned}$$

(ii) For $J = 1/2$, the energies can be written as

$$E^{+} = \frac{ah}{2}\left[\left(I+\frac{1}{2}\right)\left(I+\frac{3}{2}\right) - I\left(I+1\right) - \frac{1}{2}\left(\frac{3}{2}\right)\right]$$
$$E^{-} = \frac{ah}{2}\left[\left(I-\frac{1}{2}\right)\left(I+\frac{1}{2}\right) - I\left(I+1\right) - \frac{1}{2}\left(\frac{3}{2}\right)\right]$$

Noting that the degeneracy of level $|F\rangle$ is $g_i = 2F+1$, we get the center-of-mass energy

$$E_{\text{cm}} = \frac{\left(2F^{+}+1\right)E^{+} + \left(2F^{-}+1\right)E^{-}}{2\left(F^{+}+F^{-}+1\right)}$$

The numerator of the above expression is

$$\begin{aligned}\text{Num} &= \left[2\left(I+\frac{1}{2}\right)+1\right] + \left[2\left(I-\frac{1}{2}\right)+1\right]E^{-}\\ &= \left(2I+2\right)E^{+} + 2I\left(E^{-}\right)\\ &= 2I\left(E^{+}-E^{-}\right) + 2E^{+}\\ &= 2I\cdot\frac{ah}{2}\left[I+\left(-I-1\right)\right] + 2\cdot\frac{ah}{2}\left[I\right]\\ &= -Iah + Iah\\ &= 0\end{aligned}$$

Thus the center-of-mass energy is 0.

(b) The field parameter is defined as

$$x = \left(g_I' + g_j\right)\mu_B\, B_\circ/\Delta W$$

For low field ($x \ll 1$), the Breit–Rabi formula can be approximated as

$$\begin{aligned}E_m^{\pm} &\approx -\frac{ah}{4} \pm \frac{ahF^{+}}{2}\left(1+\frac{mx}{F^{+}}\right) - g_I' m\mu_B B_\circ\\ &= -\frac{ah}{4} \pm \frac{ahF^{+}}{2} \pm \frac{ahm}{2}\left(g_I'+g_j\right)\frac{\mu_B B_\circ}{ahF^{+}} - g_I' m\mu_B B_\circ\end{aligned}$$

Neglecting g_I' this shows that

$$g_{F^{\pm}} = \pm\frac{g_j}{2F^{+}}$$

Using $s = 1/2$ and $\ell = 0$, we get

$$g_j = 1 + \frac{J(J+1) + s(s+1) - \ell(\ell+1)}{2J(J+1)} = 2$$

Therefore

$$g_1 = -1/2 \qquad \text{and} \qquad g_2 = +1/2$$

(c) (i) In the high field regime ($x \gg 1$), the magnetic field interaction dominates over the $\vec{I} \cdot \vec{J}$ hyperfine interaction. Hence F is also no longer a good quantum number, and the good quantum numbers are I, J, m_I, m_j, and $m = m_I + m_j$.

(ii) In this regime of $x \gg 1$, the Breit–Rabi formula can be approximated as

$$E_m^{\pm} \approx \frac{ahF^+}{2}\left[\frac{1}{2F^+} \pm x\left(1 + \frac{2m}{2xF^+}\right)\right] - g_I' m\mu_B B_\circ$$

Specializing to the case of $I = 3/2$ or $F^+ = 2$, we get

$$E_m^{\pm} = ah\left[-\frac{1}{4} \pm x\left(1 + \frac{m}{2x}\right)\right] - g_I' m\mu_B B_\circ$$

For $m = \pm 2$, the square root term in the Breit–Rabi formula factorizes completely

$$\begin{aligned} E(2, \pm 2) &= ah\left[-\frac{1}{4} + \sqrt{1 \pm 2x + x^2}\right] \pm 2g_I' \mu_B B_\circ \\ &= ah\left[-\frac{1}{4} + (1 \pm x)\right] \pm 2g_I' \mu_B B_\circ \end{aligned}$$

to give the energies of these states as

$$E(2, \pm 2) = \frac{3ah}{4} \pm xah \pm 2g_I' \mu_B B_\circ$$

The energies of the other states are

$$F = 2: \qquad E(2, m) = \frac{ah}{2}\left(m - \frac{1}{2}\right) + xah - g_I' m\mu_B B_\circ$$

$$F = 1: \qquad E(1, m) = -\frac{ah}{2}\left(m + \frac{1}{2}\right) - xah - g_I' m\mu_B B_\circ$$

Using these relations we get the energies of all the states as

$$E(2,+2) = +\frac{3}{4}ah + ahx - 2g_I'\mu_B B_\circ$$
$$E(2,+1) = +\frac{1}{4}ah + ahx - g_I'\mu_B B_\circ$$
$$E(2,0) = -\frac{1}{4}ah + ahx$$
$$E(2,-1) = -\frac{3}{4}ah + ahx + g_I'\mu_B B_\circ$$
$$E(2,-2) = +\frac{3}{4}ah - ahx + 2g_I'\mu_B B_\circ$$

$$E(1,+1) = -\frac{3}{4}ah - ahx - g_I'\mu_B B_\circ$$
$$E(1,0) = -\frac{1}{4}ah - ahx$$
$$E(1,-1) = +\frac{1}{4}ah - ahx + g_I'\mu_B B_\circ$$

(iii) Since the magnetic sublevels are non-degenerate, $g_i = 1$ for all the above states. The center-of-mass energy is thus just the sum of the above energies, which its easy to see is 0. This implies that the center-of-mass rule still holds.

(d) The figure below shows a plot of the Breit–Rabi formula as a function of x, for the case of $I = 3/2$ and $J = 1/2$ considered here.

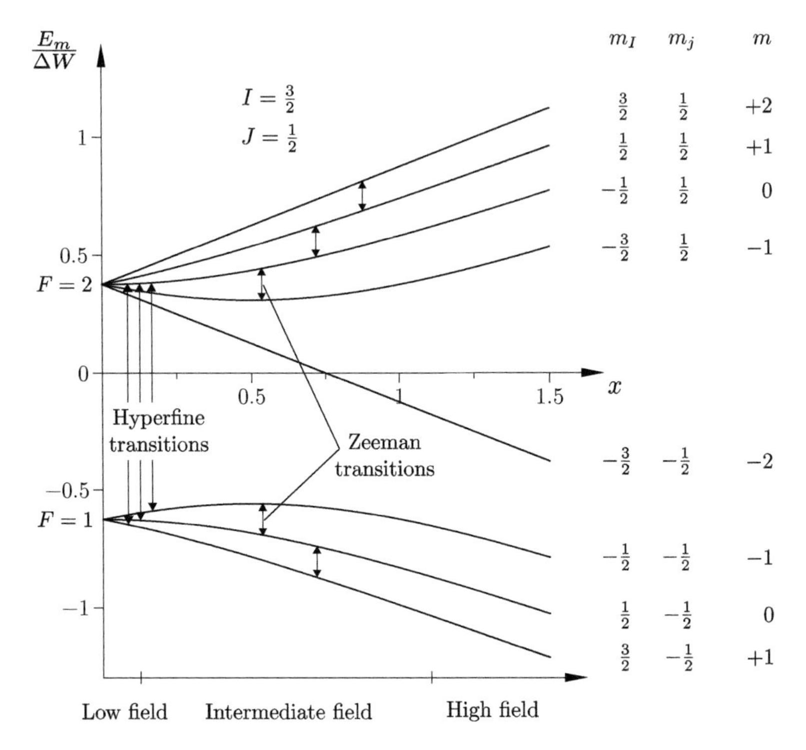

Each energy level is labeled with the quantum numbers m_I, m_j, and $m = m_I + m_j$. The quantum number m remains good throughout, but m_I and m_j are good only at large x (high field).

As expected from the factoring of the $m = \pm 2$ (stretch) states of the upper manifold, these lines are linear in x throughout.

(e) **Field independent transitions**

(i) Field independent transitions occur when the lines are parallel.

Let us find transitions of the form $|F = 1, m_1\rangle \rightarrow |F = 2, m_2\rangle$ that are field independent.

For small x

$$\Delta E = E(2, m_2) - E(1, m_1) = \frac{ah}{2}(m_2 - m_1)x - g_I' \mu_B (m_2 - m_1) B_\circ$$

Field independence requires

$$\frac{d\Delta E}{dx} = 0$$

which happens if $m_1 = m_2$. Therefore an example of a field independent transition is

$$|1,0\rangle \to |2,0\rangle \qquad \text{with selection rule } \Delta m = 0$$

At intermediate x, $\dfrac{d\Delta E}{dx} = 0$ happens for

$$|1,-1\rangle \to |2,0\rangle \qquad \text{with selection rule } \Delta m = 1$$
$$|1,0\rangle \to |2,-1\rangle \qquad \text{with selection rule } \Delta m = -1$$

Both these transitions occur when $x \approx 0.27$.

These three transitions are shown in the figure as hyperfine transitions.

(ii) At high field, the figure shows that transitions within the same Zeeman manifold will be field independent, such as transitions $|2,1\rangle \to |2,2\rangle$ or $|2,0\rangle \to |2,1\rangle$.

For the first case

$$E(2,2) - E(2,1) = ah\left[(1+x) - \sqrt{1+x+x^2}\right] - g_I'\mu_B\frac{xah}{(g_I' + g_j)\mu_B}$$

Therefore

$$\frac{d\Delta E}{dx} = 0 \qquad \Longrightarrow \qquad ah\left[1 - \frac{1+2x}{2\sqrt{1+x+x^2}} - \frac{2g_I'}{(g_I' - g_j)}\right] = 0$$

which gives

$$x \approx \frac{1}{4}\sqrt{\frac{g_j}{g_I'}} = 10.8$$

For the second case

$$E(2,1) - E(2,0) = ah\left[\sqrt{1+x+x^2} - \sqrt{1+x}\right] - g_I'\mu_B\frac{xah}{(g_I' + g_j)\mu_B}$$

Therefore

$$\frac{d\Delta E}{dx} = 0 \qquad \Longrightarrow \qquad \left[\frac{1+2x}{2\sqrt{1+x+x^2}} - \frac{1}{2\sqrt{1+x}} - \frac{2g_I'}{g_j + g_I'}\right] = 0$$

which gives

$$x \approx \frac{1}{4}\sqrt{\frac{3g_j}{g_I'}} = 18.5$$

Examples of such transitions are shown in the figure as Zeeman transitions.

(iii) For Na in the ground state

$$g_j = 2 \qquad \text{and} \qquad g_I' = 2.22\,\frac{\mu_N}{\mu_B} = \frac{2.22}{1836} = 1.2 \times 10^{-3}$$

and the hyperfine separation is

$$\Delta W = 2ah = 1.77 \text{ GHz}$$

The magnetic field for the field independent transition $|1,0\rangle \rightarrow |2,-1\rangle$ is

$$x \approx 0.27 = \frac{(g_j + g_I')\mu_B B_\circ}{2ah} \qquad \Longrightarrow \qquad B_\circ \approx 170 \text{ G}$$

(iv) In order to find that a transition that is independent of ah, we try the transition $|2,-1\rangle \rightarrow |1,-1\rangle$. The energy separation for this is

$$\Delta E = 2ah\sqrt{1 - x + x^2}$$

Independence of ah requires

$$\frac{d\Delta E}{d(ah)} = 0$$

which implies

$$2\sqrt{1 - x + x^2} + 2ah\frac{d}{dx}\sqrt{1 - x + x^2}\,\frac{dx}{d(ah)} = 0$$

and yields

$$x - 2 = 0 \qquad \Longrightarrow \qquad x = 2$$

Thus the transition $|2,-1\rangle \rightarrow |1,-1\rangle$ is independent of ah if we work in a field that produces $x = 2$ or $B \approx 1260$ G.

Chapter 5

Resonance

THIS chapter is about the interaction of a two-state system with a sinusoidally oscillating field (either electric or magnetic), whose oscillation frequency is nearly **resonant** with the natural frequency of the system, i.e. the difference in energies between the two states. The oscillating field will be treated classically, and the linewidth of both states will be taken as zero until near the end of the chapter, where relaxation will be treated phenomenologically. The organization of the material is historical because this happens to be a logical order of presentation—magnetic resonance of a classical moment is discussed before taking up a quantized spin; the density matrix is introduced last and used to treat systems with damping, which is a useful prelude to the application of resonance ideas at optical frequencies, and to the many real systems which have damping.

A. Introduction

The cornerstone of contemporary atomic, molecular, and optical physics (AMO physics) is the study of atomic and molecular systems and their interactions through their resonant interaction with applied oscillating fields. The thrust of these studies has evolved continuously since Rabi performed the first resonance experiments in 1938. In the decade following World War II, the edifice of quantum electrodynamics was constructed largely in response to resonance measurements of unprecedented accuracy on the properties of the electron, and the fine and hyperfine structure of simple atoms like hydrogen. At the same time, nuclear magnetic resonance (NMR) and electron paramagnetic resonance (EPR) techniques were developed, and quickly became essential research tools for chemists and solid state physicists. Molecular beam magnetic and electric resonance studies yielded a wealth of information on the properties of nuclei and molecules, and provided invaluable data for the nuclear physicist and physical chemist. This work continues: the elucidation of basic theories such as quantum mechanics, tests of quantum electrodynamics, the development of new techniques, the application of old techniques to more systems, and the universal move to ever higher precision. Molecular beam studies, periodically invigorated by new sources of higher intensity or new species (e.g. clusters) are carried out in numerous laboratories—chemical as well as physical—and new methods for applying the techniques of NMR are being developed.

Properly practiced, resonance techniques controllably alter the quantum mechanical state of a system without adding any uncertainty. Thus resonance techniques may be used not only to learn about the structure of a system, but also to prepare it in a particular way for further use or study. Because of these two facets, resonance studies have led physicists through a fundamental change in attitude—from the passive study of atoms to the active control of their internal quantum state and their interactions with the radiation field. This active approach is embodied generally in the study and creation of coherence phenomena (**coherent control**), with one particular example being the field of **laser cooling** where the external (translational) degrees of freedom of the atom are controlled.

However, the chief legacy of the early work on resonance spectroscopy is the family of lasers which have sprung up like the proverbial brooms of the sorcerer's apprentice. The scientific applications of these devices have been prodigious. They have caused the resurrection of physical optics—now freshly christened **quantum optics**—and turned it into one of the liveliest fields in physics. They have had a similar impact on atomic and molecular spectroscopy. In addition, they have led to new families of physical studies such as single particle spectroscopy, multi-photon excitation, cavity quantum electrodynamics, laser cooling and trapping, Bose–Einstein condensation, to name but a few of the many developments.

1. Resonance measurements and QED

One characteristic of atomic resonance is that the results, if you can obtain them at all, are generally of very high accuracy, so high that the information is qualitatively different from other types.

The hydrogen fine structure is a good example. In the late 1930s, there was extensive investigation of the Balmer series of hydrogen

$$|\, n > 2\,\rangle \rightarrow |\, n = 2\,\rangle$$

The Dirac theory was thought to be in good shape, but some doubts were arising. Careful study of the Balmer-alpha line

$$|\, n = 3\,\rangle \rightarrow |\, n = 2\,\rangle$$

showed that the line shape might not be given accurately by the Dirac theory. Pasternack (in 1939) studied the spectrum, and suggested that the $2\,^2\mathrm{S}_{1/2}$ and the $2\,^2\mathrm{P}_{1/2}$ states were not exactly degenerate, but that the energy of the $^2\mathrm{S}_{1/2}$ state was greater than the Dirac value by $\sim$1200 MHz. However, there was no rush to throw out the Dirac theory on such flimsy evidence.

In 1947, Lamb found a splitting between the $^2\mathrm{S}_{1/2}$ and the $^2\mathrm{P}_{1/2}$ states using a resonance method. The experiment is one of the classics of physics. Although his very first observation was relatively crude, it was nevertheless accurate to one percent. He found

$$\mathrm{LS_H} = \frac{1}{h}\left[E(^2\mathrm{S}_{1/2}) - E(^2\mathrm{P}_{1/2})\right] = 1050(10)\,\mathrm{MHz}$$

The inadequacy of the Dirac theory was inescapably demonstrated. Today the Lamb shift is known to an error of less than 10 kHz, a precision of approximately 10 parts per million.

The magnetic moment of the electron offers another example. As we saw in Chapter 3, "Atoms," Uhlenbeck and Goudsmit in 1925 suggested that the electron has intrinsic spin angular momentum $s = 1/2$, and a g factor of 2 exactly as predicted by the Dirac theory. However, experiments by Kusch showed that the g factor was slightly larger, a fact which could only be explained by the theory of quantum electrodynamics (QED).

The Lamb shift and the departure of g from 2 resulted in the award of the 1955 Nobel prize to Lamb and Kusch. The two results provided the experimental basis for the theory of QED, for which Feynman, Schwinger, and Tomonaga received the Nobel Prize in 1965.

2. Experimental precision

In Chapter 2, "Preliminaries," we discussed the case of a driven damped harmonic oscillator in the classical regime. Recall that the average power dissipated near resonance ($\omega \approx \omega_\circ$) was a universal resonance curve called a **Lorentzian curve**, with a linewidth of $\Delta\omega = \Gamma$.

Noting that the P vs ω curve is the Fourier transform of the P vs t curve, the decay time and the linewidth obey

$$\tau\,\Delta\omega = 1$$

This can be regarded as an uncertainty relation.

Assuming that energy and frequency are related by $E = \hbar\omega$, then the uncertainty in energy is $\Delta E = \hbar\Delta\omega$ and

$$\tau\,\Delta E = \hbar$$

It is important to realize that the Uncertainty Principle merely characterizes the spread of individual measurements. Ultimate precision depends on the experimenter's skill; the Uncertainty Principle essentially sets the scale of difficulty for his or her efforts.

The precision of a resonance measurement is determined by how well one can **split** the resonance line. This depends on the signal-to-noise ratio (SNR). As a rule of thumb, the uncertainty $\delta\omega$ in the location of the center of the line is

$$\delta\omega = \frac{\Delta\omega}{\text{SNR}}$$

In principle, one can make $\delta\omega$ arbitrary small by acquiring enough data to achieve the required **statistical** accuracy. In practice, **systematic** errors eventually limit the precision. Splitting a line by a factor of 10^4 is a formidable task which has only been achieved a few times, most notably in the measurement of the Lamb shift. A factor of 10^3, however, is not uncommon, and 10^2 is child's play.

B. Magnetic resonance

The two-state system is basic to atomic physics because it approximates accurately many physical systems, particularly systems involving resonance phenomena. All two-state systems obey the same dynamical equations; thus to know one is to know all. The archetype two-state system is a spin 1/2 particle (such as an electron, proton, or neutron) in a magnetic field, where the spin motion displays the total range of phenomena in a two-state system.

1. Classical motion of magnetic moment in a static B field

In a static magnetic field, a magnetic moment that is tipped at an angle with respect to the field will precess around it. This is just like the motion of a spinning top which precesses around the direction of gravity (say, along the z axis), when its axis of rotation is tipped from z, as shown in Fig. 5.1.

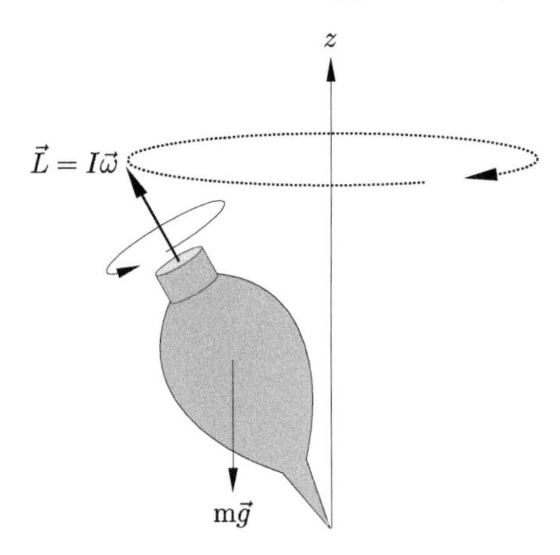

Figure 5.1: Precession of a spinning top around the direction of gravity when its axis of rotation is tipped.

In order to see this for the moment, consider its interaction energy with the field

$$U = -\vec{\mu} \cdot \vec{B}$$

This results in the following force

$$\vec{F} \equiv -\nabla U = \nabla(\vec{\mu} \cdot \vec{B})$$

and torque

$$\vec{N} \equiv \vec{r} \times \vec{F} = \vec{\mu} \times \vec{B}$$

In a uniform field, $\vec{F} = 0$, but the torque causes the angular momentum to evolve according to the following equation of motion

$$\frac{d\vec{L}}{dt} = \vec{N} = \vec{\mu} \times \vec{B}$$

From Chapter 3, "Atoms," we know that $\vec{\mu} = \gamma\,\vec{L}$. Therefore, the above equation becomes

$$\frac{d\vec{L}}{dt} = \gamma\,\vec{L} \times \vec{B} = -\gamma\,\vec{B} \times \vec{L} \tag{5.1}$$

This results in a pure precession of $\vec{L}$ about $\vec{B}$ when the field is static. To see this, choose $\vec{B}$ to be constant and along the z axis

$$\vec{B} = B_\circ\,\hat{z}$$

and $\vec{L}$ in a direction given by the usual spherical polar coordinates, i.e. it is tipped at an angle of θ from the z axis, and rotated at an angle of ϕ from the x axis.

With this choice, the torque $-\gamma\vec{B} \times \vec{L}$ has no $\hat{r}$ component (because $\vec{L}$ is along r), nor $\hat{\theta}$ component (because the $\vec{L}\vec{B}$ plane contains θ), and hence

$$\vec{N} = -\gamma\,B_\circ|\vec{L}|\sin\theta\,\hat{\phi}$$

This implies that $\vec{L}$ maintains a constant magnitude, and a constant tipping angle θ. The only thing that evolves with time is the angle ϕ. Noting that the $\hat{\phi}$ component of $d\vec{L}/dt$ is $|\vec{L}|\sin\theta\,d\phi/dt$, Eq. (5.1) implies

$$\phi(t) = -\gamma\,B_\circ t$$

This shows that the moment precesses about $\vec{B}$ with a constant angular velocity

$$\Omega_L = \gamma B_\circ$$

Ω_L is called the **Larmor frequency**. By convention, Ω_L is always taken to be positive, because its sign only determines the direction of rotation. Thus the above analysis shows that the moment precesses in the $-\phi$ direction.

Note that Planck's constant does not appear in the equation of motion: the motion is classical.

2. Rotating coordinate transformation

A second way to find the motion is to look at the problem in a rotating coordinate system.

If some vector $\vec{A}$ rotates with angular velocity $\vec{\Omega}$, then

$$\frac{d\vec{A}}{dt} = \vec{\Omega} \times \vec{A}$$

The rate of change of the vector in the two systems is related as

$$\left[\frac{d\vec{A}}{dt}\right]_{\text{in}} = \left[\frac{d\vec{A}}{dt}\right]_{\text{rot}} + \vec{\Omega} \times \vec{A}$$

Thus the operator prescription for transforming a vector from an inertial to a rotating system is

$$\left[\frac{d}{dt}\right]_{\text{in}} = \left[\frac{d}{dt}\right]_{\text{rot}} + \vec{\Omega} \times \tag{5.2}$$

Applying this to Eq. (5.1) gives

$$\left[\frac{d\vec{L}}{dt}\right]_{\text{rot}} = \gamma\vec{L} \times \vec{B} - \vec{\Omega} \times \vec{L} = \gamma\vec{L} \times (\vec{B} + \vec{\Omega}/\gamma) \tag{5.3}$$

If we define

$$\vec{B}_{\text{eff}} = \vec{B} + \vec{\Omega}/\gamma$$

Eq. (5.3) becomes

$$\left[\frac{d\vec{L}}{dt}\right]_{\text{rot}} = \gamma\vec{L} \times \vec{B}_{\text{eff}}$$

If $\vec{B}_{\text{eff}} = 0$, then $\vec{L}$ is constant is the rotating system. The condition for this is

$$\vec{\Omega} = -\gamma\vec{B} \implies \Omega = -\gamma B_{\circ}$$

as we have found previously.

3. Larmor's theorem

Treating the effects of a magnetic field on a magnetic moment by transforming to a rotating coordinate system is closely related to Larmor's theorem, which asserts that the effect of a magnetic field on a free charge can be eliminated by a suitable rotating coordinate transformation.

Consider a particle of mass m and charge q, under the influence of a force comprising of an applied force $\vec{F}_\circ$ and the Lorentz force due to a static field $\vec{B}$

$$\vec{F} = \vec{F}_\circ + \frac{q}{c}\vec{v} \times \vec{B} \tag{5.4}$$

Now we consider the motion in a rotating coordinate system. By applying Eq. (5.2) twice to $\vec{r}$, we have

$$(\ddot{\vec{r}})_{\rm rot} = (\ddot{\vec{r}})_{\rm in} - 2\vec{\Omega} \times \vec{v}_{\rm rot} - \vec{\Omega} \times (\vec{\Omega} \times \vec{r})$$

so that the force transforms as

$$\vec{F}_{\rm rot} = \vec{F}_{\rm in} - 2m\,(\vec{\Omega} \times \vec{v}_{\rm rot}) - m\,\vec{\Omega} \times (\vec{\Omega} \times \vec{r})$$

where $\vec{F}_{\rm rot}$ is the apparent force in the rotating system, and $\vec{F}_{\rm in}$ is the true or inertial force. Substituting the inertial force from Eq. (5.4) gives

$$\vec{F}_{\rm rot} = \vec{F}_\circ + \frac{q}{c}\vec{v} \times \vec{B} + 2m\,\vec{v} \times \vec{\Omega} - m\,\vec{\Omega} \times (\vec{\Omega} \times \vec{r})$$

If we choose $\vec{B} = B_\circ\hat{z}$ and $\vec{\Omega} = -(qB_\circ/2mc)\hat{z} = -\gamma B_\circ\hat{z}$, we have

$$\vec{F}_{\rm rot} = \vec{F}_\circ - m\,\Omega^2\,\hat{z} \times (\hat{z} \times \vec{r})$$

The last term is usually small. If we drop it, we have

$$\vec{F}_{\rm rot} = \vec{F}_\circ$$

Thus we see that the effect of the magnetic field is removed by going into a system rotating at the Larmor frequency of $\gamma B_\circ$.

Although Larmor's theorem is suggestive of the rotating coordinate transformation, it is important to realize that the two transformations apply to fundamentally different systems. A magnetic moment is not necessarily charged—e.g. a neutral atom can have a net magnetic moment, and the neutron possesses a magnetic moment in spite of being neutral—and it experiences no net force in a uniform magnetic field. Furthermore, the rotating coordinate transformation is exact for a magnetic moment, whereas Larmor's theorem for the motion of a charged particle is only valid when the Ω^2 term is neglected.

4. Motion in a rotating B field

We have already seen that the moment precesses because of the torque $\vec{\mu} \times \vec{B}$, which means that the component parallel to $\vec{B}$ is a constant of the motion. Thus if $\vec{B}$ is chosen to be along the z axis, then the $\hat{z}$ component of the moment does not change. If we want to change it, then we have to introduce a field in the xy plane. This is the role of B_1 in the following.

(i) On-resonance behavior

Consider a moment $\vec{\mu}$ precessing about a static field $\vec{B}_\circ$. We take $\vec{B}_\circ = B_\circ\, \hat{z}$, so that the moment precesses in the $-\phi$ direction with a constant θ. Thus its components are

$$\begin{aligned}
\mu_x &= \mu \sin\theta \cos\omega_\circ t \\
\mu_y &= -\mu \sin\theta \sin\omega_\circ t \\
\mu_z &= \mu \cos\theta
\end{aligned}$$

where $\omega_\circ = \Omega_L = \gamma B_\circ$ is the Larmor frequency.

Suppose we now introduce a rotating magnetic field $\vec{B}_1$, which rotates in the xy plane in the same direction as the moment, and at the Larmor frequency with no detuning. Then the total magnetic field is

$$\vec{B}(t) = B_1(\hat{x} \cos\omega_\circ t - \hat{y} \sin\omega_\circ t) + B_\circ \hat{z}$$

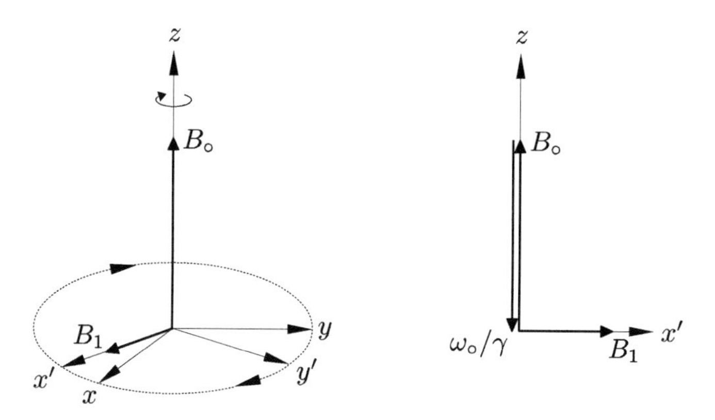

Figure 5.2: Rotating coordinate transformation to the primed system that is co-rotating with B_1 at ω, with x' chosen to lie along B_1. For the exact resonance case of $\omega = \omega_\circ$ considered here, the effective field around which the moment precesses is equal to B_1.

Fig. 5.2 above shows that the motion of $\vec{\mu}$ is simple in a rotating coordinate system. Define a coordinate system (x', y', z') which co-rotates with B_1

around the z axis of the (x, y, z) system at a rate $\omega_\circ$, so that $z' = z$. In the rotating system, the field $\vec{B}_1$ is stationary, and if x' is chosen to lie along $\vec{B}_1$, we have

$$\vec{B}_{\text{eff}}(t) = \vec{B}(t) - (\omega_\circ/\gamma)\,\hat{z} = B_1\hat{x}' + (B_\circ - \omega_\circ/\gamma)\hat{z} = B_1\hat{x}'$$

Thus, the effective field is static and has the value of B_1. The moment precesses about this field at a rate

$$\omega_R = \gamma B_1$$

and in the same direction as for the static field. This equation contains a lot of history; the RF magnetic resonance community conventionally calls this frequency ω_1, but the laser resonance community calls it the **Rabi frequency** ω_R in honor of Rabi's invention of the resonance technique.

If the moment initially lies along the $+z$ axis, then its tip traces a circle in the yz plane. At time t it has precessed through an angle $\phi = \omega_R t$. Thus the $\hat{z}$ component of the moment is given by

$$\mu_z(t) = \mu \cos \omega_R t$$

At time $T = \pi/\omega_R$, the moment points along the $-z$ axis; it has "turned over."

(ii) Off-resonance behavior

Now suppose that the field B_1 rotates at a frequency that is detuned from resonance by $\delta \equiv \omega - \omega_\circ$. In a coordinate frame rotating with B_1, the effective field is

$$\vec{B}_{\text{eff}} = B_1\hat{x}' + (B_\circ - \omega/\gamma)\hat{z}$$

The effective field lies at angle θ with the z axis, as shown in Fig. 5.3. The field is static, and the moment precesses about it at rate (called the **effective Rabi frequency**)

$$\omega'_R = \gamma B_{\text{eff}} = \gamma\sqrt{(B_\circ - \omega/\gamma)^2 + B_1^2} = \sqrt{(\omega_\circ - \omega)^2 + \omega_R^2} = \sqrt{\delta^2 + \omega_R^2}$$

where $\omega_\circ = \gamma B_\circ$ and $\omega_R = \gamma B_1$, as before.

Assume that $\vec{\mu}$ points initially along the $+z$ axis. Finding $\mu_z(t)$ is a straightforward problem in geometry. The moment precesses about B_{eff} at rate ω'_R, sweeping a circle as shown in the figure. The radius of the circle is $\mu \sin\theta$, where

$$\sin\theta = \frac{B_1}{\sqrt{(B_\circ - \omega/\gamma)^2 + B_1^2}} = \frac{\omega_R}{\sqrt{(\omega - \omega_\circ)^2 + \omega_R^2}} = \frac{\omega_R}{\omega'_R}$$

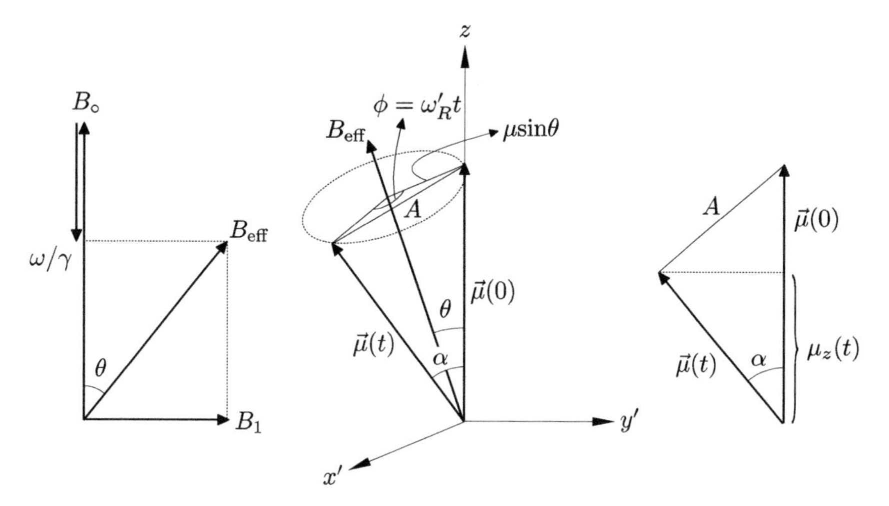

Figure 5.3: Rotating coordinate transformation to the primed system that is rotating with B_1 at ω, for the off resonant case of $\omega < \omega_\circ$. Hence the effective field is not along B_1 but tipped at an angle θ with respect to the z axis. The moment makes an angle α with the z axis after it has precessed through an angle $\phi = \omega'_R t$.

In time t the tip sweeps through angle $\phi = \omega'_R t$. The $\hat{z}$ component of the moment is $\mu \cos\alpha$, where α is the angle between the moment and the z axis after it has precessed through angle ϕ. As the figure shows, $\cos\alpha$ is found by solving

$$A^2 = 2\mu^2(1 - \cos\alpha)$$

Since $A = 2\mu \sin\theta \sin(\omega'_R t/2)$, we have

$$\begin{aligned} \mu_z(t) &= \mu \cos\alpha \\ &= \mu \left[1 - 2\sin^2\theta \, \sin^2\left(\frac{\omega'_R t}{2}\right)\right] \\ &= \mu \left[1 - 2\,\frac{\omega_R^2}{\omega_R'^2}\, \sin^2\left(\frac{\omega'_R t}{2}\right)\right] \end{aligned} \tag{5.5}$$

The $\hat{z}$ component of $\vec{\mu}$ oscillates in time, but unless $\omega = \omega_\circ$, the moment never completely inverts. The rate of oscillation depends on the magnitude of the rotating field and the detuning; the amplitude of oscillation depends on the detuning relative to ω_R. The quantum mechanical result is identical.

5. Adiabatic rapid passage and the Landau–Zener crossing

Adiabatic rapid passage is a technique for inverting a spin population by sweeping the system through resonance. Usually, the frequency of the oscillating field is varied with time. The principle is qualitatively simple in the rotating coordinate system. The problem can also be solved analytically. In this section we give the qualitative argument and then present the analytic quantum result. The solution is of quite general interest because this physical situation arises frequently, for example in inelastic scattering where it is called a **curve crossing**.

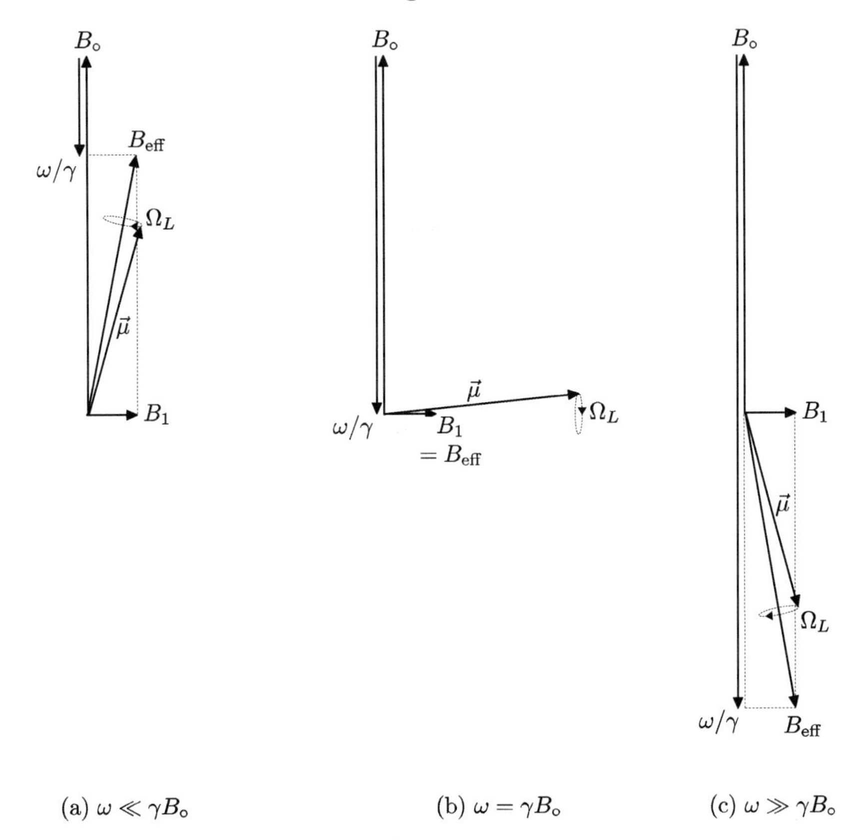

Figure 5.4: Change in direction of effective field around which the moment precesses as the frequency is swept through resonance. (a) $B_{\rm eff}$ is nearly parallel to $B_\circ$ when $\omega \ll \gamma B_\circ$. (b) $B_{\rm eff} = B_1$ on resonance when $\omega = \gamma B_\circ$. (c) $B_{\rm eff}$ is almost in the $-z$ direction when $\omega \gg \gamma B_\circ$.

Consider a moment $\vec{\mu}$ in the presence of a static magnetic field $\vec{B}_\circ$ and a perpendicular rotating field $\vec{B}_1$ at some frequency ω, as shown in Fig. 5.4. Initially, let the frequency of the rotating field be far below resonance

$\omega \ll \gamma B_\circ$. In the frame rotating with $\vec{B}_1$ the magnetic moment "sees" an effective field $\vec{B}_{\rm eff}$ whose direction is nearly parallel to $\vec{B}_\circ$. A magnetic moment $\vec{\mu}$ initially parallel to $\vec{B}_\circ$ will precess around $\vec{B}_{\rm eff}$, making only a small angle with $\vec{B}_{\rm eff}$, as shown in Fig. 5.4(a).

If ω is swept **slowly** through resonance, $\vec{\mu}$ will continue to precess tightly around $\vec{B}_{\rm eff}$, as shown in Fig. 5.4(b) and (c), and follow its direction adiabatically. In Fig. 5.4(c), the effective field now points in the $-z$ direction, because $\omega \gg \gamma B_\circ$. Since the spin still precesses tightly around $\vec{B}_{\rm eff}$, its direction in the laboratory system has "flipped" from $+z$ to $-z$. The laboratory field $\vec{B}_\circ$ remains unchanged, so this represents a transition from spin up to spin down.

The requirement for $\vec{\mu}$ to follow the effective field $\vec{B}_{\rm eff}(t)$ is that the Larmor frequency $\Omega_L = \gamma B_{\rm eff}$ be large compared to $\dot{\theta}$, the rate at which $B_{\rm eff}(t)$ is changing direction. This requirement is most severe near exact resonance where $\theta = \pi/2$. Using $B_{\rm eff}(t) = B_\circ - \omega/\gamma$, we have in this case (from geometry)

$$|\dot{\theta}_{\rm max}| = \frac{1}{B_1}\frac{dB_{\rm eff}(t)}{dt} = \frac{1}{\gamma B_1}\frac{d\omega}{dt} \gg \gamma B_1$$

Or, using $\omega_R = \gamma B_1$,

$$\frac{d\omega}{dt} \ll \omega_R^2 \tag{5.6}$$

In this example we have shown that a slow change from $\omega \ll \gamma B_\circ$ to $\omega \gg \gamma B_\circ$ will flip the spin; the same argument shows that the reverse direction of slow change will also flip the spin.

For a two-state system the problem can be solved rigorously. Consider a spin 1/2 system in a magnetic field $\vec{B}_{\rm eff}$ with energies*

$$E_\pm = \pm\frac{1}{2}\hbar\gamma B_{\rm eff} = \pm\frac{1}{2}\hbar\omega_R'$$

where we have chosen the zero of the energy to be the mean energy of the two states. The two energy levels are sketched in Fig. 5.5.

In the absence of the rotating field B_1, the effective field in the rotating frame is $B_\circ - \omega/\gamma$, and

$$E_\pm^\circ = \pm\frac{1}{2}\hbar(\omega - \omega_\circ)$$

where $\omega_\circ = \gamma B_\circ$. As ω is swept through resonance, the two states move along the energies shown as dashed lines in the figure. The energies change,

*This is the same as the eigenvalues of H_2 in the dressed atom picture dealt with in Chapter 3, "Atoms."

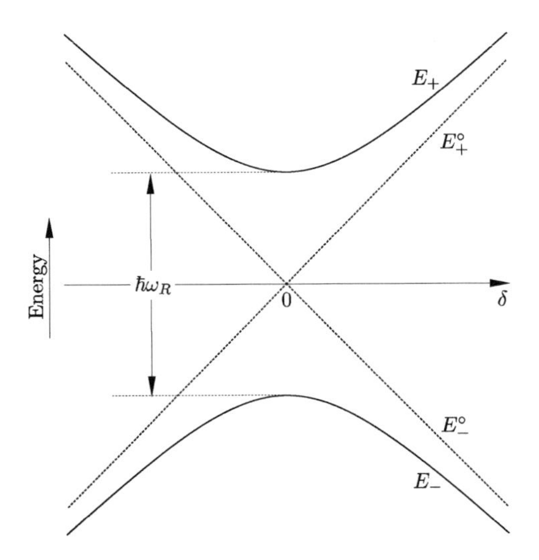

Figure 5.5: Avoided crossing of energy levels in a two-state system due to coupling. Intersecting dashed lines in the absence of coupling become non-intersecting hyperbolas in the presence of coupling.

but the states do not. There is no coupling between the states, so a spin initially in one or the other will remain so indefinitely.

In the presence of a rotating field B_1, however, the (dressed) energy levels look quite different. Instead of intersecting lines they form non-intersecting hyperbolas separated by energy $\hbar\omega_R'$ (or a minimum of $\hbar\omega_R$ on resonance), shown as solid lines in the figure. If the system moves along these hyperbolas, then $|\uparrow\rangle \to |\downarrow\rangle$ and $|\downarrow\rangle \to |\uparrow\rangle$.

(i) Quantum treatment

Whether or not the system follows an energy level adiabatically depends on how rapidly the energy is changed, compared to the minimum energy separation. To cast the problem in quantum mechanical terms, imagine two non-interacting states whose energy separation $\hbar\omega$ depends on some parameter x which varies linearly in time, and vanishes for some value $x_\circ$. Now add a perturbation having an off-diagonal matrix element V which is independent of x, so that the energies at $x_\circ$ are $\pm V$ (corresponding to the point $\delta = 0$ in the figure). The probability that the system will "jump" from one state to the other after passing through the "avoided crossing" (i.e. the probability of non-adiabatic behavior) is

$$P_{\text{na}} = \exp(-2\pi\Gamma) \qquad \text{with} \qquad \Gamma = \frac{|V|^2}{\hbar^2}\left[\frac{d\omega}{dt}\right]^{-1}$$

This result was originally obtained by Landau and Zener. The jumping of a system as it travels across an avoided crossing is called the Landau–Zener effect.

Inserting the parameters for our magnetic field problem, we have

$$P_{\mathrm{na}} = \exp\left[-\frac{\pi}{2}\frac{\omega_R^2}{d\omega/dT}\right]$$

Note that the negative factor in the exponential is related to the inequality in Eq. (5.6). When that equation is satisfied, the exponent is large and the probability of non-adiabatic behavior is exponentially small.

Incidentally, the term "rapid" in adiabatic rapid passage is something of a misnomer. The technique was originally developed in NMR in which thermal relaxation effects destroy the spin polarization if one does not invert the population sufficiently rapidly. In the absence of such relaxation processes, one can take as long as one pleases to traverse the anti-crossing; in fact, the slower the rate the less the probability of jumping.

C. Magnetic resonance of quantized spin 1/2

1. Pauli spin matrices

A spin 1/2 system in quantum mechanics is represented by the angular momentum operator $\vec{S}$ with components S_x, S_y, and S_z. They are commonly written in terms of the Pauli spin matrices, which are the same except that they are multiplied by $2/\hbar$ to make them dimensionless

$$\vec{\sigma} = \frac{2}{\hbar}\vec{S}$$

To find the matrices, we choose the standard basis with eigenstates $|s,m\rangle$ that are simultaneous eigenstates of S^2 and S_z

$$S^2\,|s,m\rangle = s(s+1)\hbar^2\,|s,m\rangle$$
$$S_z\,|s,m\rangle = m\hbar\,|s,m\rangle$$

with eigenvalues $s = 1/2$ and $m = \pm 1/2$.

The eigenstates are usually written as column vectors, and identified as follows

$$|1\rangle \equiv |1/2,+1/2\rangle = \begin{bmatrix} 1 \\ 0 \end{bmatrix} \qquad |2\rangle \equiv |1/2,-1/2\rangle = \begin{bmatrix} 0 \\ 1 \end{bmatrix} \tag{5.7}$$

In this basis, the mm' element of the matrix representing S_i is

$$S_{i,mm'} = \langle 1/2, m|S_i|1/2, m'\rangle$$

It is clear that S_z is diagonal because the basis vectors are eigenstates of S_z; and its matrix elements are

$$S_{z,mm'} = m\hbar\,\delta_{m,m'}$$

where δ is the Kronecker delta function.

Finding the representations of S_x and S_y is a bit more involved, and is best done using standard raising and lowering operators defined for quantizing angular momentum

$$S_+ = S_x + iS_y \qquad \text{and} \qquad S_- = S_x - iS_y$$

The action of these operators on the eigenstates is

$$S_\pm\,|s,m\rangle = \hbar\sqrt{s(s+1) - m(m\pm 1)}\,|s,m\pm 1\rangle$$

Thus their matrix elements are

$$S_{\pm,mm'} = \hbar\,\delta_{m,m'\pm 1}$$

Using

$$S_x = \frac{S_+ + S_-}{2} \qquad \text{and} \qquad S_y = \frac{S_+ - S_-}{2i}$$

one gets the components of the spin operator as

$$S_x = \frac{\hbar}{2}\begin{bmatrix} 0 & 1 \\ 1 & 0 \end{bmatrix} \qquad S_y = \frac{\hbar}{2}\begin{bmatrix} 0 & -i \\ i & 0 \end{bmatrix} \qquad S_z = \frac{\hbar}{2}\begin{bmatrix} 1 & 0 \\ 0 & -1 \end{bmatrix}$$

Thus the three Pauli spin matrices are

$$\sigma_x = \begin{bmatrix} 0 & 1 \\ 1 & 0 \end{bmatrix} \qquad \sigma_y = \begin{bmatrix} 0 & -i \\ i & 0 \end{bmatrix} \qquad \sigma_z = \begin{bmatrix} 1 & 0 \\ 0 & -1 \end{bmatrix} \tag{5.8}$$

A bit of algebra shows that the components obey the usual **commutation relations** for any angular momentum operator

$$[S_x, S_y] = i\hbar\, S_z \qquad [S_y, S_z] = i\hbar\, S_x \qquad [S_z, S_x] = i\hbar\, S_y \tag{5.9}$$

which is a shorthand way of writing $[S_i, S_j] = \epsilon_{ijk}\, i\hbar\, S_k$, where ϵ_{ijk} is the totally antisymmetric tensor.

Similarly, the Pauli matrices obey the commutation relations

$$[\sigma_x, \sigma_y] = 2i\, \sigma_z \qquad [\sigma_y, \sigma_z] = 2i\, \sigma_x \qquad [\sigma_z, \sigma_x] = 2i\, \sigma_y \tag{5.10}$$

Furthermore, they satisfy ${\sigma_i}^2 = \mathbb{1}$ and are Hermitian (remember they represent observables). This means that all three have real eigenvalues equal to ± 1.

2. Expectation value of quantized moment

Before solving the quantum mechanical problem of a magnetic moment in a time varying field, it is worthwhile demonstrating that its motion is classical. By "its motion is classical" we mean that the time evolution of the expectation value of the magnetic moment operator obeys the classical equation of motion. Specifically, we shall show that

$$\frac{d}{dt}\langle \vec{\mu}_{\text{op}} \rangle = \gamma\, \langle \vec{\mu}_{\text{op}} \rangle \times \vec{B} \tag{5.11}$$

Proof: Recall that the time evolution of the expectation value of any quantum operator O obeys the equation of motion

$$\frac{d}{dt}\langle O \rangle = \frac{i}{\hbar}\langle [H, O] \rangle + \langle \frac{\partial O}{\partial t} \rangle \tag{5.12}$$

If the operator is not explicitly time dependent the last term vanishes.

The interaction Hamiltonian of the magnetic moment with a magnetic field is

$$H = -\vec{\mu}_{\text{op}} \cdot \vec{B}$$

From Chapter 3, "Atoms," we know that $\vec{\mu}_{\text{op}} = \gamma \vec{J}$, where γ is the gyromagnetic ratio and $\vec{J}$ is the quantum angular momentum operator. If we choose the quantizing field to be static and along z, i.e. $\vec{B} = B_\circ \, \hat{z}$, then the interaction Hamiltonian becomes

$$H = -\gamma \vec{J} \cdot \vec{B} = -\gamma J_z B_\circ$$

Substituting into the equation of motion in Eq. (5.12), we get

$$\frac{d}{dt} \langle \vec{J} \rangle = -\frac{i}{\hbar} \gamma B_\circ \langle [J_z, \vec{J}] \rangle$$

The commutation relations for $\vec{J}$ [as we have seen in (5.9)] are

$$[J_x, J_y] = i\hbar \, J_z \qquad [J_y, J_z] = i\hbar \, J_x \qquad [J_z, J_x] = i\hbar \, J_y$$

This implies that

$$\langle \dot{J}_x \rangle = \gamma B_\circ \langle J_y \rangle \qquad \langle \dot{J}_y \rangle = -\gamma B_\circ \langle J_x \rangle \qquad \langle \dot{J}_z \rangle = 0$$

which describes a uniform precession of $\langle \vec{J} \rangle$ about the z axis at a rate $-\gamma B_\circ$. Thus

$$\frac{d}{dt} \langle \vec{J} \rangle = \gamma \langle \vec{J} \rangle \times \vec{B}$$

and since $\vec{\mu}_{\text{op}} = \gamma \vec{J}$, this directly yields Eq. (5.11)

$$\frac{d}{dt} \langle \vec{\mu}_{\text{op}} \rangle = \gamma \langle \vec{\mu}_{\text{op}} \rangle \times \vec{B}$$

Thus the quantum mechanical and classical equation of motion are identical.

3. The Rabi transition probability

For a spin 1/2 particle in a magnetic field, we can push the classical solution further and obtain the amplitude and probabilities for each state. Consider

$$\frac{\langle \mu_z \rangle}{\hbar} = \gamma \langle J_z \rangle = \gamma \langle m \rangle$$

where m is the usual "magnetic" quantum number.

For a spin 1/2 particle m has the value $+1/2$ or $-1/2$. Let the probabilities for having these values be P_+ and P_-, respectively, with $P_+ + P_- = 1$. Then

$$\langle m \rangle = \frac{1}{2}P_+ - \frac{1}{2}P_- = \frac{1}{2}(1 - 2P_-)$$

which implies that

$$\langle \mu_z \rangle = \frac{1}{2}\gamma\hbar\,(1 - 2P_-)$$

If $\vec{\mu}$ lies along the z axis at $t = 0$, then $\mu_z(0) = \gamma\hbar/2$, and we have

$$\mu_z(t) = \mu_z(0)(1 - 2P_-)$$

In this case, P_- is the probability that a spin in state $m = +1/2$ at $t = 0$ has made a transition to $m = -1/2$ at time t, denoted by $P_{\uparrow\rightarrow\downarrow}(t)$. Comparing with Eq. (5.5) we see

$$\begin{aligned} P_{\uparrow\rightarrow\downarrow}(t) &= \frac{\omega_R^2}{\omega_R^2 + (\omega - \omega_\circ)^2}\sin^2\left(\frac{1}{2}\sqrt{\omega_R^2 + (\omega - \omega_\circ)^2}\,t\right) \\ &= \frac{\omega_R^2}{\omega_R'^2}\sin^2\left(\frac{\omega_R' t}{2}\right) \end{aligned}$$

This result is known as the **Rabi transition probability**. It is important enough to memorize. We have derived it from a classical correspondence argument, but it can also be derived quantum mechanically. In fact, such a treatment (as done in the next section) is essential for a complete understanding of the system.

4. Wavefunctions for quantized spin 1/2

Now we investigate the time dependence of the wavefunctions for a quantized spin 1/2 system with moment $\vec{\mu} = \gamma \vec{S}$ placed in the time varying field discussed earlier

$$\vec{B}(t) = B_1(\hat{x}\cos\omega t - \hat{y}\sin\omega t) + B_\circ \hat{z}$$

i.e. a static field $\vec{B}_\circ$ along z, and a time dependent field $\vec{B}_1(t)$ which rotates in the xy plane with frequency ω starting at $t = 0$. Because we are now dealing with a quantum system, we must use Schrödinger's equation rather than the laws of classical Electricity and Magnetism to discuss the dynamics of the system.

If we choose the $|1\rangle$ and $|2\rangle$ basis defined in (5.7), then

$$|\psi\rangle = a_1 |1\rangle + a_2 |2\rangle \tag{5.13}$$

Using $\omega_\circ = \gamma B_\circ$, the unperturbed Hamiltonian is

$$\begin{aligned} H_\circ &= -\vec{\mu} \cdot \vec{B}_\circ \\ &= -\gamma \vec{S} \cdot \vec{B}_\circ \\ &= -\gamma(\omega_\circ/\gamma)\, S_z \\ &= -\frac{\hbar\omega_\circ}{2} \begin{bmatrix} 1 & 0 \\ 0 & -1 \end{bmatrix} \end{aligned}$$

where in the last step we have used the Pauli spin matrix σ_z from (5.8).

The eigenenergies are

$$\begin{aligned} E_1 &= -\hbar\omega_\circ/2 \\ E_2 &= +\hbar\omega_\circ/2 \end{aligned}$$

which shows that state $|1\rangle$ has the lower energy, and $\omega_{21} = \omega_\circ$.

The perturbation Hamiltonian is

$$H'(t) = -\vec{\mu} \cdot \vec{B}_1(t) = -\gamma \vec{S} \cdot \vec{B}_1(t)$$

We choose $\omega_R = -\gamma B_1$, so that ω_R contains both the magnitude and sense of precession. (The sense of precession around $B_\circ$ is already accounted for in the direction of rotation of $\vec{B}_1$.) Then, the perturbation Hamiltonian becomes

$$\begin{aligned} H'(t) &= -\gamma(-\omega_R/\gamma)\,(S_x \cos\omega t - S_y \sin\omega t) \\ &= \frac{\hbar\omega_R}{2}\left(\begin{bmatrix} 0 & 1 \\ 1 & 0 \end{bmatrix} \cos\omega t - \begin{bmatrix} 0 & -i \\ i & 0 \end{bmatrix} \sin\omega t \right) \\ &= \frac{\hbar\omega_R}{2} \begin{bmatrix} 0 & e^{+i\omega t} \\ e^{-i\omega t} & 0 \end{bmatrix} \end{aligned}$$

using the Pauli spin matrices σ_x and σ_y.

Thus the total Hamiltonian is

$$\begin{aligned} H &= H_\circ + H' \\ &= \frac{\hbar}{2}\begin{bmatrix} -\omega_{21} & \omega_R e^{+i\omega t} \\ \omega_R e^{-i\omega t} & \omega_{21} \end{bmatrix} \end{aligned} \tag{5.14}$$

This Hamiltonian is identical to the one which was used in the dressed-atom picture in Chapter 3, "Atoms." Therefore, the general time dependent solution is what was derived earlier; all we have to do is find a_w and a_s from the initial conditions, which are $a_1 = 1$ and $a_2 = 0$ at $t = 0$. Thus we get

$$a_w = \cos\theta \qquad \text{and} \qquad a_s = \sin\theta$$

and the wavefunction becomes

$$\begin{aligned} |\psi(t)\rangle &= \left(\cos^2\theta\, e^{-iE_1^+ t/\hbar} + \sin^2\theta\, e^{-iE_1^- t/\hbar}\right)|1\rangle \\ &\quad + \cos\theta\sin\theta\left(e^{-iE_2^+ t/\hbar} - e^{-iE_2^- t/\hbar}\right)|2\rangle \\ &= e^{-i(\omega_1-\delta/2)t}\left(\cos\frac{\omega'_R t}{2} - i\frac{\delta}{\omega'_R}\sin\frac{\omega'_R t}{2}\right)|1\rangle \\ &\quad - ie^{-i(\omega_2+\delta/2)t}\left(\frac{\omega_R}{\omega'_R}\sin\frac{\omega'_R t}{2}\right)|2\rangle \end{aligned}$$

Thus the probability of transition from state $|1\rangle$ to state $|2\rangle$ is $|a_2|^2$ or

$$P_{1\rightarrow 2}(t) = \frac{\omega_R^2}{\omega_R'^2}\sin^2\left(\frac{\omega'_R t}{2}\right)$$

which is exactly the same result as we obtained classically for $P_{\uparrow\rightarrow\downarrow}(t)$.

5. Separated oscillatory fields — SOF

The separated oscillatory fields (SOF) technique is one of the most powerful methods of precision spectroscopy. As the name suggests, it involves the sequential application of the transition-producing fields to the system under study with an interval in between. This technique was originally conceived by Norman Ramsey in 1949 for application in RF studies of molecular beams using two separated resonance coils through which the molecular beams passed sequentially.* It represents the first deliberate exploitation of a quantum superposition state. Subsequently it has been extended to high frequencies where the RF regions were in the optical regime, to two-photon transitions, to rapidly decaying systems, and to experiments where the two regions were temporally (rather than spatially) separated. It is routinely used to push measurements to the highest possible precision (e.g. in the Cs beam time standard apparatus). Ramsey won the 1990 Nobel Prize for inventing this method.

The SOF technique is based on an interference between the excitations produced at two separated fields—thus it is sensitive to the phase difference (coherence) of the oscillating fields. The method is most easily understood by consideration of the classical spin undergoing magnetic resonance in the two RF regions. The parameters of the system which influence the transition probability are

- ω, $\omega_\circ$, δ, ω_R, ω'_R as defined previously
- $\theta = \sin^{-1} \omega_R/\omega'_R = \cos^{-1} \delta/\omega'_R$
- τ = length of time in each resonance region
- T = length of time with no oscillating field

To maximize the interference between the two oscillating fields, we want there to be a probability of 1/2 for a transition in each resonance region. This is achieved by adjusting both field intensities (ω_R above) so that $\omega_R \tau = \pi/2$—an interaction with this property is termed a "$\pi/2$ pulse" if the system is at resonance, in which case the spin's orientation is now along the y axis in the rotating coordinate system [quantum mechanically

*Ramsey came up with this idea when explaining to students that the resolution of a lens in an optical telescope can be improved by blackening the central portion of the lens. This is because the lens forms an image by taking a **spatial** Fourier transform of the object—blackening the central portion is equivalent to removing the low-frequency components from the center while keeping the high-frequency components near the edge, which are the ones contributing to the resolution. In the SOF method, the blackening of the lens corresponds to having a dark region between the two OF regions. Of course, this blackening means a loss of signal—this kind of trade-off between signal and resolution is a common feature of all precision measurements.

this equalizes the magnitudes of the coefficients a_1 and a_2 in Eq. (5.13) in the two-state system].

During the field-free time T, the spin precesses merrily about $B_\circ$. When it encounters the second OF, it receives a second interaction equal to the first. If the system is exactly on resonance, this second OF interaction will just complete the inversion of the spin. If, on the other hand, the system is off resonance just enough so that $\delta T = \pi$ then the spin will have precessed about the z axis an angle π less far than the oscillating field. It will consequently lie in the $-y'$ direction rather than in the $+y'$ direction in the coordinate system rotating with the second OF, and as a result the second OF will precess the spin back to $+z$, its original direction, and the probability of transition will be 0! A little more thought shows that the transition probability will oscillate sinusoidally with period $\Delta\omega = 2\pi/T$. The central maximum of this interference pattern (in ω space) is centered at $\omega_\circ$ and its full width at half maximum is π/T. The central maximum can be made arbitrarily sharp simply by increasing T. In fact, SOF can be used in this fashion to produce linewidths for decaying particles which are narrower than the reciprocal of the natural linewidth! (This does not violate the uncertainty principle because SOF is a way of selecting only those few particles which have lived for time T longer than the average decay time; of course there will be fewer atoms of this kind, so one pays in signal strength for what one gains in sensitivity, which as explained in the previous footnote is a common trade-off in precision measurements.)

Results for transition amplitudes and probabilities for SOF can be derived as follows. For initial conditions of $a_1(0) = 1$ and $a_2(0) = 0$, at the final time of $T_f = 2\tau + T$ at the end of the second OF region we get

$$
\begin{aligned}
a_2(T_f) = 2i\sin\theta &\left[\cos\theta\sin^2\left(\frac{\omega'_R\tau}{2}\right)\sin\left(\frac{\delta T}{2}\right) - \frac{1}{2}\sin(\omega'_R\tau)\cos\left(\frac{\delta T}{2}\right)\right] \\
&\times \exp\left[-i\omega T_f/2\right]
\end{aligned}
$$

The probability of transition to state $|2\rangle$ is $|a_2(T_f)|^2$, and is therefore

$$
\begin{aligned}
P_{1\to2} = {} & 4\sin^2\theta\sin^2\left(\frac{\omega'_R\tau}{2}\right) \\
&\times\left[\cos\left(\frac{\delta T}{2}\right)\cos\left(\frac{\omega'_R\tau}{2}\right) - \cos\theta\sin\left(\frac{\delta T}{2}\right)\sin\left(\frac{\omega'_R\tau}{2}\right)\right]^2
\end{aligned}
$$

The first term in this expression is just four times the probability of transition for a spin passing through one of the OFs. All interference terms (which must involve T) are contained in the second term, which also prevents $P_{1\to2}$ from exceeding 1. The quantity δT is the phase difference accumulated by the spin in the OF-free region relative to the phase in the first OF region. If the phase in the second OF region differs from the phase in the first OF, the above results must be modified by adding the difference to δT.

As mentioned previously, the best SOF interference pattern is obtained when $\omega_R\tau = \pi/2$, in which case the resonance values are $\theta = \pi/2$, $\sin\omega'_R\tau/2 = \cos\omega'_R\tau/2 = 1/\sqrt{2}$, and $P_{1\to 2} = 1$.

In Fig. 5.6 we compare the resonance patterns obtained with the SOF technique (Ramsey method) and the single field technique (Rabi method). Note that to achieve $P_{1\to 2} = 1$ at resonance requires $\omega_R T_s = \pi$ in a single OF measurement (where T_s is the total time for the measurement), whereas $\omega_R\tau = \pi/2$ in a SOF measurement. If we choose $T = n\tau$ for the SOF measurement, then the total time for the measurement is $T_f = (n+2)\tau$. Thus a fair comparison would be to set the total measurement times to be equal, i.e. $T_f = (n+2)\tau = T_s$. The curves are shown as a function of the dimensionless detuning parameter $x = \delta/\omega_R$. In terms of x, the two curves (with appropriately chosen ω_R) are

$$
\begin{aligned}
P_{1\mathrm{OF}} &= \frac{1}{1+x^2}\sin^2\left(\frac{\pi}{2}\sqrt{1+x^2}\right)\\
P_{\mathrm{SOF}} &= 4\frac{1}{1+x^2}\sin^2\left(\frac{\pi}{4}\sqrt{1+x^2}\right)\\
&\quad\times\left[\cos\left(\frac{\pi}{4}nx\right)\cos\left(\frac{\pi}{4}\sqrt{1+x^2}\right)\right.\\
&\quad\left.-\frac{x}{\sqrt{1+x^2}}\sin\left(\frac{\pi}{4}nx\right)\sin\left(\frac{\pi}{4}\sqrt{1+x^2}\right)\right]^2
\end{aligned}
$$

As expected, the two expressions become the same when $n = 0$ (which means $T = 0$).

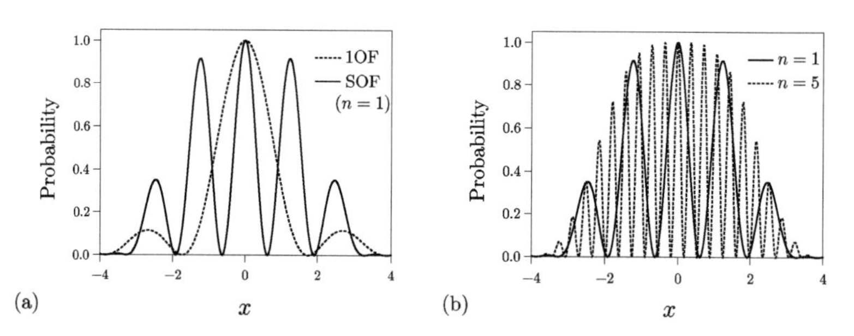

Figure 5.6: (a) Transition probability for single and separated oscillatory fields as a function of the detuning parameter $x = \delta/\omega_R$. To get probability of 1 on resonance, $\omega_R T_s = \pi$ for 1OF, while $\omega_R\tau = \pi/2$ for SOF. SOF curve shown for $n = 1$. (b) SOF curves as in (a) for $n = 1$ and 5.

In Fig. 5.6(a) we compare the 1OF pattern with SOF pattern for $n = 1$.

Even for this simple case

$$\Gamma_{1\text{OF}} = 1.59\,\pi/T_s$$
$$\Gamma_{\text{SOF}} = 0.65\,\pi/T_s$$

showing that the SOF method gives significantly narrower linewidth. In Fig. 5.6(b), we show the SOF resonance patterns for $n = 1$ and $n = 5$, which shows that the number of fringes increases with n but there is an overall envelope that remains the same.

The SOF method offers the following other advantages over the single OF method besides narrow linewidth.

(i) Perturbations in the resonance frequency $\omega_\circ$ which occur in the OF-free region (as long as they average to zero) do not decrease $P_{1\to 2}$ as they would if they occurred while the OF were on.

(ii) Power broadening and shifts (especially in multi-photon processes) can be reduced by making $\tau \ll T$.

(iii) Long-lived particles can be selectively studied, allowing one to potentially go below the natural linewidth.

(iv) Precision measurements can be made with short pulsed sources.

(v) There is no restriction to do the measurement in a beam—the two OF pulses need to be separated only in time, so the measurement can be done with a localized cloud of atoms.

An experiment in 1975 showing the power of SOF techniques resulted in the first real improvement in the Lamb shift in H beyond Lamb's original measurement.

D. Resonance in a two-state system

In the previous section, we considered resonance of a spin 1/2 particle in a magnetic field—a strong static field that sets the quantization axis, and a perpendicular rotating field that induces transitions. We will now see that this situation is identical to the resonance of a general two-state system $|1\rangle$ and $|2\rangle$ coupled with an oscillating field—either electric or magnetic—but only when the rotating wave approximation is made. States coupled by an electric field have transitions that are called **electric dipole** allowed, while those coupled by a magnetic field have transitions that are called **magnetic dipole** allowed. We will discuss them in more detail, particularly the selection rules for these transitions, in Chapter 6, "Interaction."

1. Rotating wave approximation

We first consider states that are coupled by an oscillating electric field. Let us take the field to be linearly polarized along the z axis, so that it has the form

$$\vec{E} = \mathcal{E}\cos\omega t\,\hat{z} = \frac{1}{2}\mathcal{E}\left[e^{i\omega t} + e^{-i\omega t}\right]\hat{z}$$

If we define the Rabi frequency as

$$\omega_R \equiv \frac{1}{\hbar}\langle 2|e\mathcal{E}z|1\rangle \tag{5.15}$$

then the off-diagonal perturbation matrix element (due to the $H' = -\vec{d}\cdot\vec{E}$ interaction) is

$$\langle 2|H'|1\rangle = \frac{\hbar\omega_R}{2}\left[e^{i\omega t} + e^{-i\omega t}\right]$$

Thus if $|\psi(t)\rangle$ is written (in the interaction representation) as

$$|\psi(t)\rangle = a_1(t)e^{-i\omega_{21}t}|1\rangle + a_2(t)e^{i\omega_{21}t}|2\rangle$$

then substitution in Schrödinger's equation leads to the following coupled equations for a_1 and a_2

$$\dot{a}_1 = -\frac{i\omega_R}{2}\left[e^{+i(\omega-\omega_{21})t} + e^{-i(\omega+\omega_{21})t}\right]a_2$$
$$\dot{a}_2 = -\frac{i\omega_R}{2}\left[e^{-i(\omega-\omega_{21})t} + e^{+i(\omega+\omega_{21})t}\right]a_1$$

We now make what is called the **rotating wave approximation**, which amounts to ignoring the counter-rotating $(\omega + \omega_{21})$ term. Then we get

$$\dot{a}_1 = -\frac{i\omega_R}{2}e^{+i\delta t}a_2$$

$$\dot{a}_2 = -\frac{i\omega_R}{2}e^{-i\delta t}a_1$$

These are the same equations as Eqs. (3.24) in Chapter 3 used in the dressed-atom picture. The comparison becomes easier if we write the equations in Hamiltonian form

$$H = \frac{\hbar}{2}\begin{bmatrix} -\omega_{21} & \omega_R e^{+i\omega t} \\ \omega_R e^{-i\omega t} & \omega_{21} \end{bmatrix}$$

which is also the same as the Hamiltonian for magnetic resonance of spin 1/2 that we saw in Eq. (5.14).

The analysis for an oscillating magnetic field is similar except that the interaction is due to $H' = -\vec{\mu}\cdot\vec{B}$. The only condition is that the oscillating field be orthogonal to the quantization axis. If the quantization axis is set by the direction of propagation of light and the transition is coupled by the same radiation, then the condition is automatically satisfied because all fields are transverse in the radiation zone. In fact, the SOF technique, mentioned previously, is used to make precise measurements of the clock transition in Cs, which is a magnetic dipole transition between the two hyperfine levels of the ground state where an oscillating magnetic field is used to drive the transition.

Thus the resonance conditions are the same for a quantized spin 1/2 in a rotating magnetic field, or of a two-state system coupled by an oscillating field after making the rotating wave approximation.

2. Isomorphism with spin 1/2 in a magnetic field

The preceding discussion of the magnetic resonance of a quantized spin 1/2 system also made direct contact with classical resonance of a magnetic moment because the transition probability for the two was the same. This is a reflection of a deeper isomorphism which was first proved by Feynman, Vernon, and Hellwarth—that it is possible to map the dynamical behavior of any two-state quantum mechanical system onto the dynamical behavior of a spin 1/2 particle in the properly associated time-varying magnetic field.* The above similarities are but a special case of this relationship, which we will see more of in the next section on the density matrix. The importance of this isomorphism is that our intuition for the behavior of a spin 1/2 in a time-varying magnetic field can now be directly applied to any quantized two-state system.

*R. P. Feynman, F. L. Vernon, Jr, and R. W. Hellwarth, "Geometrical representation of the Schrödinger equation for solving maser problems," *J. Appl. Phys.* **28**, 49–52 (1957).

E. Density matrix

Probabilities enter quantum systems in two ways.

(i) The familiar $|c_n|^2 \equiv |\langle\psi_n|\psi\rangle|^2$, arising because there is a finite probability amplitude c_n for the wavefunction $|\psi\rangle$ to be in the eigenstate $|\psi_n\rangle$.

(ii) A statistical probability arising because the systems have a probability distribution, e.g. a thermal distribution of spin components for atoms coming out of an oven.

If only (i) is present, the system is said to be in a **pure quantum state**, while a system with (ii) also is said to be a **statistical mixture**. The usual time dependent expansion

$$|\psi(t)\rangle = \sum_n c_n(t)\,|\psi_n\rangle \tag{5.16}$$

plus Schrödinger equation is not well suited for dealing with systems where both are present. On the other hand, the **density matrix** approach is advantageous in such a situation because it treats both on an equal footing.

1. General results

The density matrix operator is defined as

$$\rho(t) \equiv \overline{|\psi(t)\rangle\,\langle\psi(t)|} \tag{5.17}$$

and its matrix elements are

$$\rho_{nm}(t) \equiv \langle\psi_n|\,\rho(t)\,|\psi_m\rangle = \overline{c_m^*(t)c_n(t)}$$

The bar here indicates an ensemble average over identically (but not necessarily completely) prepared systems. An ensemble average is essential to treat statistical probabilities (e.g. only the ensemble average of spin projections of atoms from an oven is zero), and an ensemble average is always implicit in using a density matrix (but for notational simplicity it will be eliminated from here on).

The diagonal elements of the matrix are of the form $c_n^* c_n$ and represent the **population** (or probability) in the state $|\psi_n\rangle$; whereas the off-diagonal elements are of the form $c_m^* c_n$ and represent the **coherence** (or relative phase) between the amplitudes for being in states $|\psi_m\rangle$ and $|\psi_n\rangle$. Thus the density matrix provides a convenient test for a properly normalized system, i.e. sum of all probabilities should be equal to unity, which implies

$$\mathrm{Tr}\{\rho\} = 1$$

where Tr is the trace—the sum of the diagonal elements. It also provides a test for a pure quantum state because

$$\mathrm{Tr}\{\rho^2\} \le \mathrm{Tr}\{\rho\}$$

and only the equality implies a pure state.

The density matrix provides a way to find the expectation value of operators which do not commute with the Hamiltonian. For an operator A, the expectation value at time t is

$$\langle A \rangle_t \equiv \langle \psi(t)| A |\psi(t)\rangle = \sum_{m,n} c_m^*(t) c_n(t) A_{nm}$$

which can be evaluated easily as

$$\langle A \rangle_t = \sum_{m,n} \rho_{mn}(t) A_{nm} = \mathrm{Tr}\{\rho(t) A\}$$

As mentioned before, the above expression really involves two sums—the ensemble average in the preparation of the systems, and the usual quantum mechanical sum over the basis to find the expectation value.

The time evolution of the density matrix is determined by a first order differential equation which is obtained by applying Schrödinger's equation to the time derivative of Eq. (5.17)

$$i\hbar\,\dot{\rho} = H\rho - \rho H \equiv [H, \rho]$$

This reflects changes in ρ due solely to interactions included in the Hamiltonian—additional terms may be added to account for incoherent processes such as damping, addition or subtraction of atoms from the system, or interactions with other quantized systems not accessible to measurement (e.g. collisions).

We should always keep in mind that ρ is to be used on a statistical ensemble of systems similarly prepared. If this preparation is sufficient to force the system into a pure state [so that Eq. (5.16) holds for each member of the ensemble], then the ensemble average is superfluous; if the preparation is insufficient, then there will be random phases between some of the c_n's in Eq. (5.16).

2. Density matrix for a two-state system

The density matrix for a two-state system is

$$\rho = \begin{bmatrix} \rho_{11} & \rho_{12} \\ \rho_{21} & \rho_{22} \end{bmatrix} \qquad \text{with} \qquad \rho_{12} = \rho_{21}^*$$

We shall consider a two-state system in which $E_1 = \hbar\omega_\circ/2$ and $E_2 = -\hbar\omega_\circ/2$. Thus the unperturbed Hamiltonian is

$$H_\circ = \frac{\hbar}{2}\begin{bmatrix} \omega_\circ & 0 \\ 0 & -\omega_\circ \end{bmatrix} = \frac{\hbar\omega_\circ}{2}\sigma_z$$

We now subject the system to an off-diagonal perturbation of arbitrary strength and time dependence: $\langle 1|H'|2\rangle = (V_1 - iV_2)/2$, so that the perturbation Hamiltonian is

$$H' = \frac{1}{2}\begin{bmatrix} 0 & V_1 - iV_2 \\ V_1 + iV_2 & 0 \end{bmatrix} = \frac{V_1}{2}\sigma_x + \frac{V_2}{2}\sigma_y$$

Thus the total Hamiltonian is

$$H = \frac{1}{2}\begin{bmatrix} \hbar\omega_\circ & V_1 - iV_2 \\ V_1 + iV_2 & -\hbar\omega_\circ \end{bmatrix} = \frac{1}{2}\left[V_1\,\sigma_x + V_2\,\sigma_y + \hbar\omega_\circ\,\sigma_z\right] \tag{5.18}$$

This is a general enough system to encompass most two-state systems which are encountered in resonance physics.

Before solving for ρ we shall change variables in the density matrix

$$\rho = \frac{1}{2}\begin{bmatrix} r_\circ + r_3 & r_1 - ir_2 \\ r_1 + ir_2 & r_\circ - r_3 \end{bmatrix} = \frac{1}{2}\left[r_\circ \mathbb{1} + r_1\,\sigma_x + r_2\,\sigma_y + r_3\,\sigma_z\right] \tag{5.19}$$

There is no loss of generality in this substitution—it has four independent quantities just as ρ does—but it makes the physical constraints on ρ manifest

$$\text{Tr}\{\rho\} = r_\circ = 1 \qquad \text{and} \qquad \rho_{12} = \rho_{21}^*$$

We can now solve the equation of motion for $\rho(t)$

$$\dot{\rho} = \frac{1}{i\hbar}[H,\rho]$$

Since we have expressed both H and ρ in terms of the Pauli spin matrices, we can use the commutation relations from (5.10), and then equate the coefficients of each σ to get the following

$$\begin{aligned} \sigma_x: &\quad \dot{r_1} = \frac{1}{\hbar}V_2 r_3 - \omega_\circ r_2 \\ \sigma_y: &\quad \dot{r_2} = \omega_\circ r_1 - \frac{1}{\hbar}V_1 r_3 \\ \sigma_z: &\quad \dot{r_3} = \frac{1}{\hbar}V_1 r_2 - \frac{1}{\hbar}V_2 r_1 \end{aligned} \tag{5.20}$$

These results can be summarized by using the following vector representation*

$$\vec{\omega} = \frac{1}{\hbar}V_1\,\hat{x} + \frac{1}{\hbar}V_2\,\hat{y} + \omega_\circ\,\hat{z} \qquad \text{and} \qquad \vec{r} = r_1\,\hat{x} + r_2\,\hat{y} + r_3\,\hat{z}$$

so that Eq. (5.20) becomes

$$\dot{\vec{r}} = \vec{\omega} \times \vec{r} \tag{5.21}$$

The above equation proves that the time evolution of the density matrix for our very general two-state system is isomorphic to the behavior of a classical magnetic moment in a magnetic field which points along $\vec{\omega}$. Our previous discussion showing that the quantum mechanical spin obeyed this equation also is therefore superfluous for a spin 1/2 system.

One consequence of Eq. (5.21) is that $\dot{\vec{r}}$ is always perpendicular to $\vec{r}$, so that $|\vec{r}|$ does not change with time. This implies that if ρ is initially a pure state, ρ remains forever in a pure state no matter how violently $\vec{\omega}$ is gyrated. This is because from Eq. (5.19) and using $\sigma_i^2 = \mathbb{1}$, we have

$$\text{Tr}\{\rho^2\} = r_\circ^2 + r_1^2 + r_2^2 + r_3^2 = r_\circ^2 + |\vec{r}|^2$$

which does not change with time if $|\vec{r}|^2$ does not change.

In general it is not possible to decrease the purity (coherence) of a system with a Hamiltonian like the one in Eq. (5.18). Since real coherences do in fact die out, we shall have to add relaxation processes to our description in order to approach reality. The density matrix formulation makes this easy to do, and this development will be done in the next section.

3. Phenomenological treatment of relaxation — Bloch equations

Statistical mechanics tells us the form that the density matrix will finally take, but it does not tell us how the system will get there or how long it will take. All we know is that ultimately we will reach a thermal equilibrium ρ_T at temperature T, given by

$$\rho_T = \frac{1}{Z}e^{-H_\circ/kT}$$

where Z is the partition function.

Since the interactions that ultimately bring thermal equilibrium are incoherent processes, the density matrix formulation seems like a natural way to treat them. Unfortunately in most cases these interactions are sufficiently

*Due to Feynman, Vernon, and Hellwarth. See footnote on p. 167 for the full reference.

complex that this is done phenomenologically. For example, the equation of motion for the density matrix might be modified by the addition of a damping term

$$\dot{\rho} = \frac{1}{i\hbar}[H, \rho] - (\rho - \rho_T)/T_e$$

which would (in the absence of a source of non-equilibrium interactions) drive the system to equilibrium with time constant T_e.

This equation is not sufficiently general to describe the behavior of most systems studied in resonance physics. The reason is that most systems exhibit different decay times for the populations and the phase coherences, called T_1 and T_2 respectively.

T_1 — decay time for population differences between non-degenerate states, e.g. for r_3 used in Eq. (5.19). This is also called the energy decay time.

T_2 — decay time for coherences between either degenerate or non-degenerate states, e.g. for r_1 or r_2.

In general it requires a weaker interaction to destroy coherence than to destroy the population difference, so $T_2 < T_1$.

The effects of thermal relaxation with the two decay times described above are easily incorporated into the vector model for the two-state system, since r_z represents the population differences, while r_x and r_y represent coherences. The results are

$$\begin{aligned}
\dot{\vec{r}}_x &= \frac{1}{\hbar}(\vec{\omega} \times \vec{r})_x - (r_x - r_{xT})/T_2 \\
\dot{\vec{r}}_y &= \frac{1}{\hbar}(\vec{\omega} \times \vec{r})_y - (r_y - r_{yT})/T_2 \\
\dot{\vec{r}}_z &= \frac{1}{\hbar}(\vec{\omega} \times \vec{r})_z - (r_z - r_{zT})/T_1
\end{aligned}$$

For a magnetic spin system $\vec{r}$ corresponds directly to the magnetic moment $\vec{\mu}$. The above equations were first introduced by Bloch in this context and are therefore known as the Bloch equations.

The addition of phenomenological decay times does not generalize the density matrix enough to cover situations where atoms (possibly state-selected) are added or lost to a system. This situation can be covered by the addition of further terms to $\dot{\rho}$. Thus a calculation on a resonance experiment in which state-selected atoms are added to a two-state system through a tube which also permits atoms to leave (e.g. a hydrogen maser) might look like (while noting that the off-diagonal coherence terms decay to 0 at equilibrium)

$$\dot{\rho} = \frac{1}{i\hbar}[\rho, H] - \left[\begin{array}{cc} (\rho_{11} - \rho_{11T})/T_1 & \rho_{12}/T_2 \\ \rho_{21}/T_2 & (\rho_{22} - \rho_{22T})/T_1 \end{array} \right]$$
$$+ R \left[\begin{array}{cc} 0 & 0 \\ 0 & 1 \end{array} \right] - \rho/T_{\text{esc}} - \rho/T_{\text{col}}$$

where

- The first term represents interaction with static and oscillating fields.
- The second term is Bloch relaxation.
- R is the rate of addition of state-selected atoms.
- The last two terms express losses due to atoms escaping from the system and due to collisions (e.g. spin exchange) that cannot be incorporated in T_1 and T_2.

The terms representing addition or loss of atoms will not have zero trace, and consequently will not maintain $\text{Tr}\{\rho\} = 1$. Physically this is reasonable for systems that gain or lose atoms; the application of the density matrix to this case shows its power to deal with complicated situations. In most applications of the above equation, one looks for a steady-state solution with $\dot{\rho} = 0$, so this does not cause problems.

F. Resonance of a realistic two-state system

In this section we apply density matrix techniques to a physically realistic two-state "atom," with states labeled $|a\rangle$ and $|b\rangle$ respectively. The populations in the two states decay at a rate $\Gamma_1 = 1/T_1$, while the coherence between them decays at a rate $\Gamma_2 = 1/T_2$. They are coupled by an oscillating field. With all the decay terms, we must also have a source term; we presume that atoms are added to the lower state $|a\rangle$ at a rate $\Gamma_1 n_\circ$, so there will be $n_\circ$ atoms in the system on average. We shall study the resonance behavior of this system by finding the steady-state solution for the population of the upper state. In a dilute gas of these two-state systems, the spontaneous radiation is proportional to this population; often this radiation is the experimental signal.

We use the same expressions for H and ρ as in the previous sections with the sinusoidal applied field. This gives

$$H = \frac{\hbar}{2}\begin{bmatrix} -\omega_\circ & \omega_R e^{+i\omega t} \\ \omega_R e^{-i\omega t} & +\omega_\circ \end{bmatrix}$$

Thus the resonance conditions are the same as those of a quantized spin 1/2 in a rotating magnetic field, or of a coupled two-state system after making the rotating wave approximation.

We now write the basic equation for time evolution

$$\dot{\rho} = \frac{1}{i\hbar}[H,\rho] - \begin{bmatrix} \Gamma_1\rho_{aa} & \Gamma_2\rho_{ab} \\ \Gamma_2\rho_{ba} & \Gamma_1\rho_{bb} \end{bmatrix} + \Gamma_1\begin{bmatrix} n_\circ & 0 \\ 0 & 0 \end{bmatrix}$$

↑ **relaxation** ↑ **source**

Consider first the steady-state solution with field off. The equation for time evolution simplifies to

$$\dot{\rho} = -\begin{bmatrix} \Gamma_1\rho_{aa} - \Gamma_1 n_\circ & \Gamma_2\rho_{ab} \\ \Gamma_2\rho_{ba} & \Gamma_1\rho_{bb} \end{bmatrix}$$

Steady state means that $\dot{\rho} = 0$, so the solution with the oscillating field off is

$$\rho_{\text{off}} = \begin{bmatrix} n_\circ & 0 \\ 0 & 0 \end{bmatrix}$$

which shows that the only non-zero element of the density matrix is the diagonal element corresponding to the population in $|a\rangle$, and that there are $n_\circ$ "atoms" in that state.

We now consider the situation with the field on. We can achieve a simplification by defining

$$\rho' = \rho - \frac{1}{2}\begin{bmatrix} n_\circ & 0 \\ 0 & n_\circ \end{bmatrix} \tag{5.22}$$

so that it satisfies an equation with a source term of σ_z symmetry

$$\dot{\rho}' = \frac{1}{i\hbar}[H, \rho'] - \begin{bmatrix} \Gamma_1 \rho'_{aa} & \Gamma_2 \rho'_{ab} \\ \Gamma_2 \rho'_{ba} & \Gamma_1 \rho'_{bb} \end{bmatrix} + \frac{\Gamma_1}{2} \begin{bmatrix} n_\circ & 0 \\ 0 & -n_\circ \end{bmatrix}$$

Note that this redefinition changes only the diagonal elements of the density matrix but leaves the off-diagonal elements the same. The redefinition is useful in solving the above equation since it will turn out to be most convenient to use the Pauli matrices only for the diagonal terms, while using ρ_{ab} and $\rho_{ba} = \rho^*_{ab}$ for the off-diagonal terms. This is because we are basically working the damped oscillator problem in which a complex ρ_{ab} provides the simplest way to deal with the phase angle between the driving field and the oscillatory response of the system (which appears in the off-diagonal matrix elements). Thus we use

$$\rho'_{aa} = -\rho'_{bb} = r_3/2$$

and anticipate that ρ_{ab} will oscillate at the driving frequency by substituting

$$\rho_{ab} = Ae^{+i\omega t} \qquad \Longrightarrow \qquad \dot{\rho}_{ab} = (\dot{A} + i\omega A)e^{+i\omega t}$$

Grinding out the commutator leads to only the following two independent equations because the diagonal entries are the negative of each other, while the off-diagonal entries are complex conjugates of each other

$$\dot{A} = \frac{i}{2}\omega_R r_3 - i(\omega - \omega_\circ)A - \Gamma_2 A \tag{5.23}$$

$$\dot{r}_3 = i\omega_R(A - A^*) - \Gamma_1 r_3 + \Gamma_1 n_\circ \tag{5.24}$$

1. Steady-state solution

The steady-state solution to these equations is found by setting $\dot{A} = \dot{r}_3 = 0$. (Note that this is not equivalent to setting the off-diagonal matrix elements of the density matrix equal to 0 since it does not imply $A = 0$.) So, we get

$$A = \frac{1}{2}\frac{i}{\Gamma_2 + i\delta}\,\omega_R\, r_3 \qquad \text{with} \qquad \delta = \omega - \omega_\circ$$

and

$$A - A^* = \frac{i\Gamma_2}{\delta^2 + \Gamma_2^2}\,\omega_R\, r_3$$

which implies

$$r_3 = n_\circ \left(1 + \frac{\Gamma_2}{\Gamma_1}\frac{\omega_R^2}{\delta^2 + \Gamma_2^2}\right)^{-1} = n_\circ \frac{\delta^2 + \Gamma_2^2}{\delta^2 + \Gamma_2^2 + (\Gamma_2/\Gamma_1)\omega_R^2}$$

Since we are interested in the population of the upper state, we use Eq. (5.22) to get

$$\rho_{bb} = \rho'_{bb} + \frac{n_\circ}{2} = -\frac{r_3}{2} + \frac{n_\circ}{2} \tag{5.25}$$

Thus the steady-state population in the upper state is

$$\rho_{bb} = \frac{n_\circ}{2} \frac{(\Gamma_2/\Gamma_1)\omega_R^2}{\delta^2 + \Gamma_2^2 + (\Gamma_2/\Gamma_1)\omega_R^2} \tag{5.26}$$

The most noteworthy feature of the solution is that it has a Lorentzian shape at all values of ω_R. The width of the Lorentzian (the full-width-at-half-maximum, or FWHM) is

$$\Gamma = 2\left(\Gamma_2^2 + \frac{\Gamma_2}{\Gamma_1}\omega_R^2\right)^{1/2}$$

which is $2\Gamma_2$ for small oscillating fields and $2\sqrt{\Gamma_2/\Gamma_1}\,\omega_R$ at very high oscillating fields. Thus it is T_2 and and not T_1 which determines the width of the resonance curve. (The only effect of T_1 is on the overall magnitude of the upper state population.) The increase of Γ with increased oscillating field strength is known as **power broadening** even though the width grows linearly with field, not the power.

2. Free induction decay

Now let us investigate what happens when an oscillating field that has been on for a long time is suddenly turned off at $t = 0$. The situation is called free induction decay, because the system is allowed to decay free from the perturbing field.

If the field has been on for a long time such that $t \gg T_1$ or T_2, then the system will be in a steady state, and at $t = 0$ one has

$$\rho_{bb}^\circ = \frac{n_\circ}{2} \frac{(\Gamma_2/\Gamma_1)\omega_R^2}{\delta^2 + \Gamma_2^2 + (\Gamma_2/\Gamma_1)\omega_R^2}$$

and

$$A^\circ = \frac{n_\circ}{2} \frac{(\delta + i\Gamma_2)\omega_R}{\delta^2 + \Gamma_2^2 + (\Gamma_2/\Gamma_1)\omega_R^2}$$

Turning the field off simplifies the equations of motion for the density matrix elements because $\omega_R = 0$.

Combining Eqs. (5.24) and (5.25), we get

$$\dot{\rho}_{bb} = -\Gamma_1 \rho_{bb}$$

which implies

$$\rho_{bb} = \rho_{bb}^{\circ} e^{-\Gamma_1 t}$$

showing that the population in the upper state decays "freely" from its steady-state value at the rate Γ_1 once the field is turned off.

The situation for A is complicated by the fact that ω is not defined in the absence of an external field. One knows, however, that a driven oscillator oscillates only at its resonance frequency after a non-resonant driving force is removed, so we can take $\omega = \omega_\circ$, which simplifies Eq. (5.23) to

$$\dot{A} = -\Gamma_2 A \qquad \Longrightarrow \qquad A(t) = A^\circ e^{-\Gamma_2 t}$$

so that

$$\rho_{ab}(t) = A^\circ e^{-\Gamma_2 t} e^{+i\omega_\circ t}$$

showing that the off-diagonal element not only decays freely from its steady-state value at the rate Γ_2 but also oscillates at $\omega_\circ$.

The decoupling of the motions of ρ_{bb} and ρ_{ab} in the absence of an oscillating field makes plausible the assignment of the different decay rates Γ_1 and Γ_2. Since the off-diagonal terms oscillate, they are the ones responsible for any coherent oscillatory behavior of the system during its decay (spontaneous emission requires only that $\rho_{bb} \neq 0$). Thus an experiment that is sensitive to the amplitude of this behavior will show a decay at Γ_2. An example is an NMR experiment where one measures the voltage induced by the time-varying magnetization. If the experiment measures the **intensity** of radiation from some oscillating moment, then the decay rate will be $2\Gamma_2$.

3. Damping of Rabi probability

We now consider the effects of damping on the Rabi probability in a two-state system. We consider a resonance region into which ground state atoms are fed at a rate Γ, and from which (partially excited) atoms leave at the rate Γ per atom present in the region. This means that there is, on average, one atom present in the resonance region, and that it has been there an average time Γ^{-1}. This average is crucial because, as we now show it will remove the temporal oscillations in the Rabi transition probability. Finally, this result will be checked against the density matrix result for this simple form of damping.

Imagine that the resonance region is bathed in an electromagnetic field oscillating away from resonance at $\omega = \omega_\circ + \delta$, and which couples the states at a rate ω_R. Then the average probability of finding the atom in the excited state $|b\rangle$ is

$$\bar{P}_b(\delta) = \int_{-\infty}^{0} \Gamma e^{+\Gamma t}\, dt\, P_2(\delta, t)$$

where $\Gamma e^{+\Gamma t}\, dt$ is the probability that the atom entered between $-t$ and $-t + dt$ and still remains in the system; and $P_2(\delta, t)$ is the Rabi transition probability for the atom to be in the upper state at t.

Changing the integration limits, the probability can be evaluated as

$$\begin{aligned}
\bar{P}_b(\delta) &= \frac{\Gamma \omega_R^2}{\omega_R'^2} \int_0^{\infty} dt\, e^{-\Gamma t} \sin^2(\omega_R' t/2) \\
&= \frac{\Gamma \omega_R^2}{\omega_R'^2} \int_0^{\infty} dt\, e^{-\Gamma t} [1 - \cos(\omega_R' t)]/2 \\
&= \frac{\Gamma \omega_R^2}{2\omega_R'^2} \left[\frac{1}{\Gamma} - \frac{\Gamma}{\Gamma^2 + \omega_R'^2} \right]
\end{aligned}$$

which gets simplified to

$$\bar{P}_b(\delta) = \frac{1}{2} \frac{\omega_R^2}{\delta^2 + \omega_R^2 + \Gamma^2}$$

One interesting result of the damping is immediately apparent: the oscillating structure present in $P_2(\delta, t)$ as a function of δ (at long fixed time) has been replaced by a smooth Lorentzian.

To check this against the density matrix result, we note that the damping rate Γ applies to all atoms, irrespective of their state of excitation. Therefore $\Gamma_1 = \Gamma_2 = \Gamma$. Also $n_\circ = 1$ since we want an average of one atom in the field at a time. So from Eq. (5.26)

$$P_b(\delta) = \rho_{bb} = \frac{n_\circ}{2} \frac{(\Gamma_2/\Gamma_1)\, \omega_R^2}{\delta^2 + \Gamma_2^2 + (\Gamma_2/\Gamma_1)\omega_R^2} = \frac{1}{2} \frac{\omega_R^2}{\delta^2 + \omega_R^2 + \Gamma^2}$$

Thus the density matrix treatment and the "by hand" treatment of this simple relaxation mechanism give the same result.

G. Problems

1. RF-induced magnetic transitions

Imagine that in an RF experiment you wish to induce Zeeman transitions in Na atoms ($I = 3/2$) trapped in the ground $2S_{1/2}$ state, from the $|F = 2, m_F = 2\rangle$ to the $|F = 2, m_F = 1\rangle$ sublevel. The atoms are in a uniform field of 10 G along the z axis. They are surrounded by two one-turn coils of 2 cm diameter carrying RF current of 1 ma (amplitude, not peak-to-peak or rms) arranged to produce a rotating field in the xy plane.

(a) What is the RF magnetic field (in G)?

(b) What is the resonance frequency $\omega_\circ$, and the Rabi frequency ω_R?

(Hint: The magnetic field is very weak, or equivalently the field parameter $x \ll 1$. To work out ω_R, assume that the interaction with the rotating field has the same g_F as the interaction with the static field.)

(c) For how long would you pulse the field to produce a π pulse (i.e. to cause all the atoms to change state)?

(d) What changes in (a) to (c) above if the DC field is doubled to 20 G?

Solution

(a) The RF field will be of the form

$$B_{\rm RF} = B_1 \cos \omega t$$

where B_1 is field at the center of a loop of radius a, and is given by

$$B_1 = \frac{2I\pi a^2}{c} \frac{1}{(a^2)^{3/2}} = \frac{2I\pi}{ac}$$

For the coil configuration given $B_1 = 0.63$ mG.

(b) In a weak magnetic field of strength $B_\circ$, the energy shift of an $|F, m_F\rangle$ state is

$$W_B = g_F \mu_B m_F B_\circ$$

Thus the energy difference between the $|F = 2, m_F = 2\rangle$ level and the $|F = 2, m_F = 1\rangle$ level is

$$\Delta E = g_F \mu_B B_\circ \equiv \hbar\omega_\circ$$

For the $F = 2$ state of Na, $\ell = 0$, $s = 1/2$, $J = 1/2$, and $I = 1/2$, hence the g factors are

$$g_j = 1 + \frac{J(J+1) + s(s+1) - \ell(\ell+1)}{2J(J+1)} = 2$$

$$g_F = \left[g_j - g_I \frac{\mu_N}{\mu_B}\right] \frac{F(F+1) + J(J+1) - I(I+1)}{2F(F+1)} \approx \frac{1}{2}$$

neglecting g_I. Therefore

$$\omega_\circ = \frac{\mu_B B_\circ}{2\hbar} = 2\pi \times 7 \text{ MHz}$$

We get the same answer from the Breit-Rabi formula

$$E_m^+ = ah\left[-\frac{1}{4} + \sqrt{1 + mx + x^2}\right] \approx ah\left[-\frac{1}{4} + 1 + \frac{mx}{2}\right] \qquad (x \ll 1)$$

This gives for a change in m from 1 to 2 of

$$\Delta E = \frac{1}{2}ahx = \frac{1}{2}ah \frac{(g_J + g_I')\,\mu_B B_\circ}{(I + 1/2)\,ah} \approx \frac{1}{2}\mu_B B_\circ$$

which implies

$$\omega_\circ = \frac{\mu_B B_\circ}{2\hbar}$$

Using the fact that the interactions with $B_\circ$ and B_1 are the same, we get

$$\omega_R = \frac{B_1}{B_\circ}\omega_\circ = 2\pi \times 441 \text{ Hz}$$

(c) If the length of the pulse is T_π for a π pulse, then

$$\omega_R T_\pi = \pi \qquad \Longrightarrow \qquad T_\pi = \frac{\pi}{\omega_R} = 1.13 \times 10^{-3} \text{ s}$$

(d) If the DC field is doubled (i.e. $B_\circ \to 2B_\circ$), then $\omega_\circ \to 2\omega_\circ$ because the new field is still in the weak field regime. Both the parameters ω_R and T_π are independent of $B_\circ$, and hence remain unchanged.

2. Rabi transition probability

(a) Plot the Rabi transition probability vs $\omega_R t$ for the following values of detuning—$\delta = 0, \omega_R/2, \omega_R, \sqrt{3}\omega_R$, and $3\omega_R$.

(b) You should find that there is universal behavior for small $\omega_R t$. Find an analytic form for this region using perturbation theory.

Solution

(a) The Rabi transition probability is given by

$$P_{21}(t) = \frac{\omega_R^2}{\omega_R^2 + \delta^2} \sin^2\left(\frac{\sqrt{\omega_R^2 + \delta^2}}{2}\, t\right)$$

The figure below shows a plot of this function for different values of detuning.

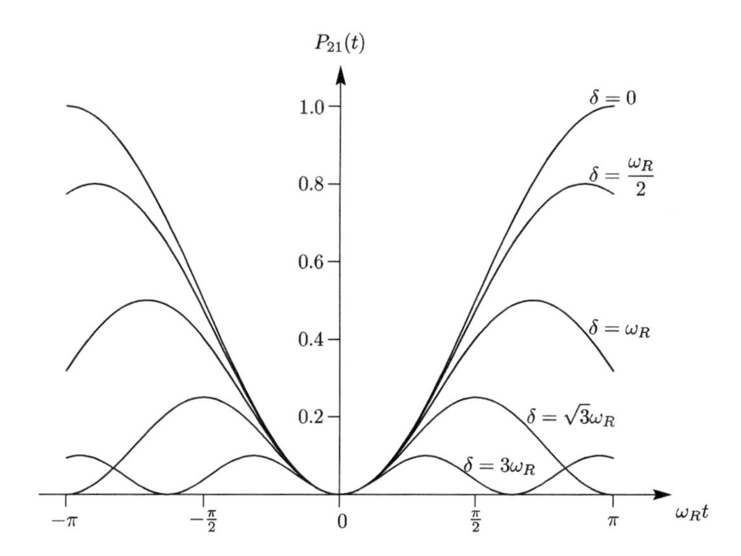

(b) As seen from the figure, all curves show the same behavior near the origin. To find an expression for this universal behavior using time dependent perturbation theory, we start with the magnetic resonance Hamiltonian

$$H = \frac{\hbar}{2} \begin{bmatrix} -\omega_\circ & \omega_R e^{+i\omega t} \\ \omega_R e^{-i\omega t} & \omega_\circ \end{bmatrix}$$

The $a_2(t)$ coefficient obeys the equation

$$\dot{a}_2^{(1)}(t) = \frac{1}{i\hbar}\frac{\hbar\omega_R}{2}e^{-i\omega t}e^{i\omega_\circ t}a_1^{(0)}(t) = \frac{\omega_R}{2i}e^{-i\delta t}a_1^{(0)}(t)$$

Using the initial conditions of $a_1^{(0)}(0) = 1$ and $a_2^{(0)}(0) = 0$, we get

$$a_2^{(1)}(t) = \frac{\omega_R}{2i}\int_0^t e^{-i\delta t'}dt' = \frac{\omega_R}{2i}\frac{e^{-i\delta t'}}{-i\delta}\bigg|_0^t = \frac{\omega_R}{2\delta}\left(e^{-i\delta t} - 1\right)$$

Expanding $e^{-i\delta t}$, we have

$$a_2^{(1)}(t) = \frac{\omega_R}{2\delta}\left(-i\delta t + \frac{\delta^2 t^2}{2} + \ldots\right)$$

Therefore to first order in $\omega_R t$

$$a_2^{(1)}(t) = -i\frac{\omega_R t}{2}$$

and the probability of transition is

$$P_{21}(t) = \left|a_2^{(1)}(t)\right|^2 = \frac{\omega_R^2 t^2}{4} \qquad \text{(independent of } \delta\text{)}$$

The answer can be verified from the Rabi formula. Defining $\delta = a\omega_R$ and approximating for small a, we have

$$\begin{aligned} P_{21}(t) &= \frac{1}{1+a^2}\sin^2\left(\frac{\sqrt{1+a^2}}{2}\omega_R t\right) \\ &\approx \frac{1}{1+a^2}\left(\frac{\sqrt{1+a^2}}{2}\omega_R t\right)^2 \\ &= \frac{\omega_R^2 t^2}{4} \end{aligned}$$

3. Steady-state solution for two-level system

Find ρ_{bb} [Eq. (5.26)] for a generalization of the situation considered in section F of the chapter, which includes the following:

(i) Decay rate out of system: ρ_{aa} decays at Γ_a and ρ_{bb} at Γ_b (the source term becomes $\Gamma_a n_\circ$).

(ii) Level $|b\rangle$ spontaneously decays to level $|a\rangle$ at a rate Γ_s.

Show that the new result for ρ_{bb} is

$$\rho_{bb} = \frac{n_\circ}{2} \frac{\left(\frac{\Gamma_2}{\Gamma_s + \Gamma_b}\right)\omega_R^2}{(\delta^2 + \Gamma_2^2) + \left(\frac{\Gamma_a + \Gamma_b}{2\Gamma_a}\right)\left(\frac{\Gamma_2}{\Gamma_s + \Gamma_b}\right)\omega_R^2}$$

Solution

The general decay rates can be incorporated into the time evolution of the density matrix elements by writing it as

$$\frac{d\rho}{dt} = \frac{1}{i\hbar}[H, \rho] - \begin{bmatrix} \Gamma_a \rho_{aa} - \Gamma_s \rho_{bb} & \Gamma_2 \rho_{ab} \\ \Gamma_2 \rho_{ba} & (\Gamma_b + \Gamma_s)\rho_{bb} \end{bmatrix} + \Gamma_a \begin{bmatrix} n_\circ & 0 \\ 0 & 0 \end{bmatrix}$$

where H is the resonance Hamiltonian used in the chapter

$$H = \frac{\hbar}{2} \begin{bmatrix} -\omega_\circ & \omega_R e^{+i\omega t} \\ \omega_R e^{-i\omega t} & +\omega_\circ \end{bmatrix}$$

Thus the off-diagonal terms continue to relax at the rate Γ_2, but Γ_2 includes a $\Gamma_s/2$ term for spontaneous decay. From the discussion in the chapter, the off-diagonal element is

$$\rho_{ab} = Ae^{+i\omega t}$$
$$\Longrightarrow \dot{\rho}_{ab} = (\dot{A} + i\omega A)e^{+i\omega t} \qquad \text{and} \qquad \dot{\rho}_{ba} = (\dot{A}^* - i\omega A)e^{-i\omega t}$$

Therefore the terms in the time evolution equation are

$$\frac{1}{i\hbar}[H, \rho] = -\frac{i}{2} \begin{bmatrix} -\omega_\circ \rho_{aa} + \omega_R e^{+i\omega t}\rho_{ba} & -\omega_\circ \rho_{ab} + \omega_R e^{+i\omega t}\rho_{bb} \\ +\omega_R e^{-i\omega t}\rho_{aa} + \omega_\circ \rho_{ba} & \omega_R e^{-i\omega t}\rho_{ab} + \omega_\circ \rho_{bb} \end{bmatrix}$$

$$+\frac{i}{2} \begin{bmatrix} -\rho_{aa}\omega_\circ + \rho_{ab}\omega_R e^{-i\omega t} & \rho_{aa}\omega_R e^{+i\omega t} + \rho_{ab}\omega_\circ \\ -\rho_{ba}\omega_\circ + \rho_{bb}\omega_R e^{-i\omega t} & \rho_{ba}\omega_R e^{+i\omega t} + \rho_{bb}\omega_\circ \end{bmatrix}$$

$$= -\frac{i}{2} \begin{bmatrix} A^*\omega_R - A\omega_R & \omega_R e^{+i\omega t}(\rho_{bb} - \rho_{aa}) - 2\omega_\circ \rho_{ab} \\ \omega_R e^{-i\omega t}(\rho_{aa} - \rho_{bb}) + 2\omega_\circ \rho_{ba} & A\omega_R - A^*\omega_R \end{bmatrix}$$

and

$$\begin{bmatrix} \dot{\rho}_{aa} & \left(\dot{A} + i\omega A\right) e^{+i\omega t} \\ \left(\dot{A}^* - i\omega A^*\right) e^{-i\omega t} & \dot{\rho}_{bb} \end{bmatrix}$$

$$= -\frac{i}{2} \begin{bmatrix} \omega_R(A^* - A) & [\omega_R(\rho_{bb} - \rho_{aa}) - 2\omega_\circ A]\, e^{+i\omega t} \\ [\omega_R(\rho_{aa} - \rho_{bb}) + 2\omega_\circ A^*]\, e^{-i\omega t} & \omega_R(A - A^*) \end{bmatrix}$$

$$- \begin{bmatrix} \Gamma_a \rho_{aa} - \Gamma_s \rho_{bb} - \Gamma_a n_\circ & \Gamma_2 A e^{-i\omega t} \\ \Gamma_2 A^* e^{i\omega t} & (\Gamma_b + \Gamma_s)\rho_{bb} \end{bmatrix}$$

which gives

$$\begin{aligned} \dot{A} + i\omega A &= \frac{i\omega_R}{2}\left(\rho_{aa} - \rho_{bb}\right) + i\omega_\circ A - \Gamma_2 A \\ \dot{A} &= \frac{i\omega_R}{2}\left(\rho_{aa} - \rho_{bb}\right) - \left(i\delta + \Gamma_2\right) A \\ \dot{\rho}_{aa} &= -\frac{i\omega_R}{2}\left(A^* - A\right) - \Gamma_a \rho_{aa} + \Gamma_s \rho_{bb} + \Gamma_a n_\circ \\ \dot{\rho}_{bb} &= -\frac{i\omega_R}{2}\left(A - A^*\right) - \left(\Gamma_b + \Gamma_s\right) \rho_{bb} \end{aligned}$$

Under steady state,

$$\dot{A} = \dot{\rho}_{aa} = \dot{\rho}_{bb} = 0$$

$$\Longrightarrow A = \frac{i\omega_R}{2}\frac{\left(\rho_{aa} - \rho_{bb}\right)}{\Gamma_2 + i\delta} \quad \text{and} \quad A^* = -\frac{i\omega_R}{2}\frac{\left(\rho_{aa} - \rho_{bb}\right)}{\Gamma_2 - i\delta}$$

$$\text{and} \qquad \dot{\rho}_{aa} + \dot{\rho}_{bb} = -\Gamma_b \rho_{bb} - \Gamma_a \rho_{aa} + \Gamma_a n_\circ = 0$$

Therefore

$$\rho_{aa} = \frac{\Gamma_a n_\circ - \Gamma_b \rho_{bb}}{\Gamma_a} = n_\circ - \frac{\Gamma_b}{\Gamma_a}\rho_{bb}$$

$$\rho_{bb} = \frac{i\omega_R}{2}\frac{\left(A^* - A\right)}{\Gamma_b + \Gamma_s}$$

Using

$$A^* - A = -\frac{i\omega_R}{2}\left[\frac{\rho_{aa}-\rho_{bb}}{\Gamma_2 - i\delta} + \frac{\rho_{aa}-\rho_{bb}}{\Gamma_2 + i\delta}\right]$$

$$= -\frac{i\omega_R}{2}\frac{(\rho_{aa}\Gamma_2 + i\delta\rho_{aa} - \rho_{bb}\Gamma_2 - i\delta\rho_{bb})}{\Gamma_2^2+\delta^2}$$

$$- \frac{i\omega_R}{2}\frac{(\rho_{aa}\Gamma_2 - i\delta\rho_{aa} - \rho_{bb}\Gamma_2 + i\delta\rho_{bb})}{\Gamma_2^2+\delta^2}$$

$$= -\frac{i\omega_R\Gamma_2}{\Gamma_2^2+\delta^2}(\rho_{aa}-\rho_{bb})$$

$$= -\frac{i\omega_R\Gamma_2}{\Gamma_2^2+\delta^2}\left(n_\circ - \frac{\Gamma_b}{\Gamma_a}\rho_{bb} - \rho_{bb}\right)$$

and substituting, we get

$$\rho_{bb} = \frac{i\omega_R}{2(\Gamma_b+\Gamma_s)}\frac{-i\omega_R\Gamma_2}{\Gamma_2^2+\delta^2}\left(n_\circ - \frac{\Gamma_b}{\Gamma_a}\rho_{bb} - \rho_{bb}\right)$$

$$= \frac{\omega_R^2\Gamma_2}{2(\Gamma_b+\Gamma_s)(\Gamma_2^2+\delta^2)}\left[n_\circ - \left(\frac{\Gamma_b}{\Gamma_a}+1\right)\rho_{bb}\right]$$

which implies

$$\rho_{bb}\left[\frac{2(\Gamma_b+\Gamma_s)(\Gamma_2^2+\delta^2)}{\omega_R^2\Gamma_2} + \frac{\Gamma_b}{\Gamma_a} + 1\right] = n_\circ$$

so that

$$\rho_{bb} = \frac{n_\circ}{\dfrac{2(\Gamma_b+\Gamma_s)(\Gamma_2^2+\delta^2)}{\omega_R^2\Gamma_2} + \dfrac{\Gamma_b}{\Gamma_a} + 1}$$

$$= \frac{n_\circ}{2}\frac{\left(\dfrac{\Gamma_2}{\Gamma_s+\Gamma_b}\right)\omega_R^2}{(\Gamma_2^2+\delta^2) + \left(\dfrac{\Gamma_b+\Gamma_a}{2\Gamma_a}\right)\left(\dfrac{\Gamma_2}{\Gamma_s+\Gamma_b}\right)\omega_R^2}$$

Quod erat demonstrandum.

Chapter 6

Interaction

IN this chapter, we treat the interaction of electromagnetic (EM) radiation with atomic systems using **semiclassical** techniques. The treatment is semiclassical because, as mentioned in Chapter 2, "Preliminaries," the field is treated classically and only the atomic system is quantized. We have already reviewed the relevant features of classical EM theory (in cgs units) in that chapter which we will need for the atom-field interaction Hamiltonian. When treating absorptive or scattering process, we shall not consider the atom as a source of radiation here—instead we shall use the equations for classical EM wave in a **vacuum** to describe the wave even though it interacts with an atom. Thus we shall not find any change in the intensity of the field as it passes the atom (even if the atom is excited by the field). This is consistent with our classical field approach—a classical field has many photons in it whereas the atomic system can add or subtract only a few. The focus in this chapter will be on transitions that are **single photon**, we reserve multiphoton processes for the next chapter.

A. Interaction of EM radiation with atoms

1. Hamiltonian

Recall from Chapter 2 that the vector potential that solves the classical EM wave equation, for a given value of the propagation vector $\vec{k}$ and polarization $\hat{\varepsilon}$, is

$$\vec{A}(\vec{r},t) = \frac{c\mathcal{E}}{2i\omega}\hat{\varepsilon}\left[e^{i(\vec{k}\cdot\vec{r}-\omega t)} + e^{-i(\vec{k}\cdot\vec{r}-\omega t)}\right] \tag{6.1}$$

where $\mathcal{E}$ is chosen so that it is the amplitude of the E field vector

$$\vec{E}(\vec{r},t) = -\frac{1}{c}\frac{\partial\vec{A}}{\partial t} = \mathcal{E}\hat{\varepsilon}\cos\left(\vec{k}\cdot\vec{r}-\omega t\right) \tag{6.2}$$

The corresponding B field is

$$\vec{B}(\vec{r},t) = \nabla\times\vec{A} = \mathcal{E}\left(\hat{k}\times\hat{\varepsilon}\right)\cos\left(\vec{k}\cdot\vec{r}-\omega t\right) \tag{6.3}$$

showing that it has the same magnitude as the E field.

The Hamiltonian for the atom-field interaction is obtained by replacing the mechanical momentum with the canonical momentum

$$\vec{p}_{\text{mech}} \qquad \rightarrow \qquad \vec{p}_{\text{canon}} = \vec{p}_{\text{mech}} + \frac{q}{c}\vec{A}$$

for each charge q in the atomic system. Assuming that the interaction with the nucleus is negligible in comparison with that for the electrons this is equivalent to the substitution in the Hamiltonian (for each electron, but we'll just treat one electron atoms for now)

$$\frac{p^2}{2m} \qquad \rightarrow \qquad \frac{p^2}{2m} - \frac{e}{mc}\vec{p}\cdot\vec{A} + \frac{e^2}{2mc^2}|\vec{A}|^2$$

Thus we may view the interaction Hamiltonian as

$$\begin{aligned} H^{\text{int}} &= -\frac{e}{mc}\vec{p}\cdot\vec{A} + \frac{e^2}{2mc^2}|\vec{A}|^2 + g_s\mu_B\vec{S}\cdot(\nabla\times\vec{A}) \\ &\equiv H' \qquad + H'^{(2)} \qquad + H'^{(s)} \end{aligned} \tag{6.4}$$

The $\vec{S}\cdot(\nabla\times\vec{A})$ term accounts for the interaction of the intrinsic electron spin with an external magnetic field. In general the second term $H'^{(2)}$ contributes significantly only at high intensities and for photon scattering processes.

All wavelengths we shall deal with are much longer than the size of an atom. Consequently, we shall never have to consider more than the first derivative in the spatial behavior of $\vec{A}(\vec{r},t)$. We expand from Eq. (6.1)

$$\vec{A}(\vec{r},t) = \frac{c\mathcal{E}}{2i\omega}\,\hat{\varepsilon}\left\{e^{-i\omega t}\left[1 + i\vec{k}\cdot\vec{r} - \frac{1}{2}\left(\vec{k}\cdot\vec{r}\right)^2 + \ldots\right] \quad + \text{ c.c.}\right\}$$

and keep only the first two terms back in H'. Then the matrix elements of H' are

$$\begin{aligned}\langle b|H'|a\rangle &= -\frac{e\mathcal{E}}{2i\omega}\left[e^{-i\omega t}\hat{\varepsilon}^* + e^{+i\omega t}\hat{\varepsilon}\right]\cdot\langle b|\dot{\vec{r}}|a\rangle \\ &- \frac{e\mathcal{E}}{2\omega}\left[e^{-i\omega t}\hat{\varepsilon}^*\cdot\langle b|\dot{\vec{r}}\vec{r}|a\rangle\cdot\vec{k} - e^{+i\omega t}\hat{\varepsilon}\cdot\langle b|\dot{\vec{r}}\vec{r}|a\rangle\cdot\vec{k}\right]\end{aligned} \tag{6.5}$$

where $\dot{\vec{r}}\vec{r}$ is a dyadic, and the * on $\hat{\varepsilon}$ is necessary to get the correct result for the scalar product.

2. Electric dipole approximation — E1

When the first term in Eq. (6.5) is non-zero, it dominates the second, which can then be neglected—this is called the electric dipole approximation and corresponds to ignoring the spatial variation of $\vec{A}$ near the atomic system. Radiative transitions thus coupled are called E1 transitions. In this approximation, the matrix element of the interaction Hamiltonian is

$$\langle b|H'_{\text{E1}}|a\rangle = -\frac{e\mathcal{E}}{2i\omega}\left[e^{-i\omega t}\hat{\varepsilon}^* + e^{+i\omega t}\hat{\varepsilon}\right]\cdot\langle b|\dot{\vec{r}}|a\rangle \tag{6.6}$$

This is called the **velocity** form of the dipole matrix element. More frequently a length form is used, which is obtained by using Schrödinger's equation

$$\begin{aligned}i\hbar\,\langle b|\dot{\vec{r}}|a\rangle &\equiv i\hbar\frac{d}{dt}\,\langle b|\vec{r}|a\rangle \\ &= \langle b|\vec{r}H|a\rangle - \langle b|H\vec{r}|a\rangle \\ &= (E_a - E_b)\,\langle b|\vec{r}|a\rangle\end{aligned}$$

where $E_a = \hbar\omega_a$ and $E_b = \hbar\omega_b$ are the respective energy eigenvalues of H in the states $|a\rangle$ and $|b\rangle$.

Putting this in Eq. (6.6) gives

$$\langle b|H'_{\text{E1}}|a\rangle = \frac{e\mathcal{E}}{2}\frac{\omega_a - \omega_b}{\omega}\left[e^{-i\omega t}\hat{\varepsilon}^* + e^{+i\omega t}\hat{\varepsilon}\right]\cdot\langle b|\vec{r}|a\rangle \tag{6.7}$$

which is the **length** form of the dipole matrix element.

Note that Eq. (6.7) could be derived by treating the electric field as static, interacting through a potential energy term

$$V(\vec{r}) = -e\phi(\vec{r}) = +e\vec{E}\cdot\vec{r}$$

The matrix elements for V are then

$$\langle b|V|a\rangle = \frac{e\mathcal{E}}{2}\left[e^{-i\omega t}\hat{\varepsilon}^* + e^{+i\omega t}\hat{\varepsilon}\right]\cdot\langle b|\vec{r}|a\rangle \tag{6.8}$$

which agrees with Eq. (6.7) for $\omega = (\omega_a - \omega_b)$, i.e. in the region where the oscillating field produces significant effects. Surprisingly, it is possible to show that Eq. (6.8) is a better approximation than Eq. (6.7)—at least in the long wavelength limit (essential for the dipole approximation to be valid)—because it contains the contribution from $H'^{(2)}$, which was neglected in obtaining Eq. (6.7).

3. Higher approximations

The development of the higher terms in H^{int} is carried out fully in a number of textbooks and won't be treated in detail here. It may be shown that the second term in Eq. (6.5) gives rise to electric quadrupole radiation as well as magnetic dipole radiation from the orbital motion of the electron. The three strongest forms of radiative interactions [which all come from H' and $H'^{(s)}$ in Eq. (6.4)] involve

Type	Symbol	Operator	Parity
Electric dipole	E1	$\vec{P} = -e\vec{r}$	$-$
Magnetic dipole	M1	$\vec{M} = -\mu_B(\vec{L} + g_s\vec{S})$	$+$
Electric quadrupole	E2	$\bar{\bar{\eta}} = -e\vec{r}\vec{r}$	$+$

where the parity is negative or positive depending on whether the operator does or does not change sign when $\vec{r} \to -\vec{r}$. The E1 operator has negative parity because it corresponds to a polar vector, while the M1 operator has positive parity because it corresponds to an axial vector. Similarly, the E2 operator has positive parity because it is a dyadic of two polar vectors. In a multi-electron atom the one-electron operators above must be summed over all the electrons (but there are no cross terms in $\bar{\bar{\eta}}$ or $\vec{M}$).

Using the above operators, the matrix elements for the various types of radiation become

$$\text{E1} \qquad \langle b|H^{\text{int}}_{\text{E1}}|a\rangle = -\frac{\mathcal{E}}{2}\left[e^{-i\omega t}\hat{\varepsilon}^* + e^{+i\omega t}\hat{\varepsilon}\right]\cdot\langle b|\vec{P}|a\rangle$$

$$\text{M1} \qquad \langle b|H^{\text{int}}_{\text{M1}}|a\rangle = -\frac{\mathcal{E}}{2}\left[e^{-i\omega t}(\hat{k}\times\hat{\varepsilon}^*) + e^{+i\omega t}(\hat{k}\times\hat{\varepsilon})\right]\cdot\langle b|\vec{M}|a\rangle \tag{6.9}$$

$$\text{E2} \qquad \langle b|H^{\text{int}}_{\text{E2}}|a\rangle = -\frac{\mathcal{E}}{2}\left[e^{-i\omega t}\hat{\varepsilon}^* + e^{+i\omega t}\hat{\varepsilon}\right]\cdot\langle b|\bar{\bar{\eta}}|a\rangle\cdot\vec{k}$$

From the expressions for the E and B fields in Eqs. (6.2) and (6.3), one sees that the E1 operator corresponds to a $-\vec{d}\cdot\vec{E}$ interaction, while the M1 operator corresponds to a $-\vec{\mu}\cdot\vec{B}$ interaction.

B. Selection rules and angular distribution

1. General

The matrix elements for the various types of radiation in (6.9) possess certain symmetry properties (parity for example) that forbid transitions between states with the wrong relative symmetry. As a consequence, a number of conditions must be satisfied in order to ensure the possibility of a particular type of radiation—these are known as **selection rules**.

Frequently the term "selection rules" is taken to mean the selection rules for electric dipole transitions. This has some merit because the matrix elements for other types of transition are smaller by α^n (where $n \geq 1$), and the selection rules may be broken to this order by interactions overlooked in the derivation of the matrix element for E1 radiation. Viewed from this perspective, there are the following reasons why "forbidden transitions" may actually occur:

1. Higher-order radiative processes (e.g. E2, M1, etc.)
2. Multiphoton processes
3. Relativistic effects
4. Interactions with the nucleus (especially hyperfine)
5. Interactions with external fields
6. Collision-induced absorption or emission

From a theoretical perspective the angular distribution of spontaneous radiation emitted by an excited atomic system is closely related to the topic of selection rules because both are grounded on the symmetry properties of the radiation and the system. The semiclassical approach adopted here does not allow a detailed treatment of spontaneous emission, but all properties of the emitted radiation are the same as expected classically from a charge distribution whose moments are

$$2\,\mathrm{Re}\left\{\langle a|O|b\rangle\, e^{i\omega t}\right\}$$

where O is the appropriate radiation operator from (6.9). (This statement is based on the correspondence principle.)

In general, the matrix elements for the various radiative processes can be simplified by using the Wigner–Eckart theorem for irreducible tensor operators $T_q^{(k)}$, where k is the rank of the tensor and q is the projection on the quantization axis

$$\langle njm|T_q^{(k)}|n'j'm'\rangle = \frac{\langle nj||T^{(k)}||n'j'\rangle}{\sqrt{2j+1}}\,\langle j'km'q|jm\rangle \tag{6.10}$$

Here $\langle nj||T^{(k)}||n'j'\rangle$ is called the reduced matrix element, and $\langle j'km'q|jm\rangle$ is a Clebsch–Gordan (C-G) coefficient for the addition of $j_1 = j'$ and $j_2 = k$ to get j. The C-G coefficient vanishes unless the following are satisfied

$$\begin{aligned} &m = q + m' && m\text{-selection rule} \\ &|j' - k| \le j \le j' + k && \text{triangular relation} \end{aligned} \tag{6.11}$$

The various operators in (6.9) are not irreducible tensors, but can be broken up into such tensors. Many of the selection rules and all of the angular distribution information come from the properties of the C-G coefficients. When Eq. (6.10) is used in practice, the reduced matrix element is generally determined by evaluating the left hand side for a particular case—generally the **stretch state** with $m = j$, $m' = j'$, and $q = k$, for which $\langle j'kj'k|jj\rangle = 1$.

2. Electric dipole radiation

Now we shall discuss the interaction of the EM wave with an atomic system in the electric dipole approximation. The E1 operator is easily made into an irreducible tensor of rank 1 [$k = 1$ in Eq. (6.10)] with three components. If we choose the quantization axis z to be along the direction of the light propagation vector $\hat{k}$, then the components are

$$P_{-1} = -e(x - iy)/\sqrt{2} \qquad P_0 = -ez \qquad P_{+1} = e(x + iy)/\sqrt{2}$$

These can be written in terms of the spherical harmonics Y_ℓ^m (ignoring an unimportant normalization constant) as

$$P_{-1} = -erY_1^{-1} \qquad P_0 = -erY_1^0 \qquad P_{+1} = -erY_1^{+1}$$

This form is particularly convenient for discussing the interaction of circularly polarized light since only the scalar products

$$\hat{\varepsilon}_\pm \cdot \langle b|P_{\pm 1}|a\rangle \qquad \text{and} \qquad \hat{\varepsilon}_z \cdot \langle b|P_0|a\rangle$$

are not necessarily equal to zero. Thus σ^- light couples state $|b\rangle$ to state $|a\rangle$ only if $\langle b|P_{-1}|a\rangle \ne 0$.

Imagine that state $|a\rangle$ has a total angular momentum j_a and projection m_a and $|b\rangle$ is a state with j_b and m_b. Then, using the Wigner–Eckart theorem we get

$$\langle b|P_{-1}|a\rangle = \langle n'j_bm_b|P_{-1}|nj_am_a\rangle = \frac{\langle b||P_{-1}||a\rangle}{\sqrt{2j_b+1}} \langle j_a1m_a - 1|j_bm_b\rangle$$

which shows that the C-G coefficient $\langle j_a1m_a - 1|j_bm_b\rangle \ne 0$ is required to prevent $\langle b|P_{-1}|a\rangle$ from vanishing. From (6.11) we see that this requires

$$\begin{aligned} m_b &= m_a - 1 &&\Longrightarrow && \Delta m = -1 \\ |j_a + 1| \ge j_b &\ge |j_a - 1| &&\Longrightarrow && \Delta j = 0, \pm 1 \end{aligned}$$

which are the selection rules for σ^- light in absorption.

Similarly, we can show that the selection rules for σ^+ light in absorption are

$$\Delta m = +1 \qquad \text{and} \qquad \Delta j = 0, \pm 1$$

When LS coupling is appropriate, the J-selection rule becomes an L-selection rule because S does not change in an electric dipole transition; then $\Delta\ell = 0$ is forbidden by the requirement that the parity of states $|a\rangle$ and $|b\rangle$ be different for E1 radiation (because the E1 operator has odd parity).

Linear polarization appears messier at first because

$$\hat{\varepsilon}_x \cdot \langle b|P_{\pm 1}|a\rangle = \pm\frac{1}{\sqrt{2}} \qquad \text{and} \qquad \hat{\varepsilon}_x \cdot \langle b|P_0|a\rangle = 0$$

Thus the selection rules for linear polarization give $\Delta m = \pm 1$ (remember the quantization axis is along $\hat{k}$). In fact the situation is even more complicated because an x polarized incident wave will excite states with $m_b = m_a + 1$ and $m_b = m_a - 1$ coherently. This fact is often overcome by using a quantization axis z' along the x axis, i.e. oriented along $\vec{E}$ but perpendicular to $\hat{k}$. In this coordinate system

$$\hat{\varepsilon}_x \cdot \langle b|P'_{\pm 1}|a\rangle = 0 \qquad \text{and} \qquad \hat{\varepsilon}_x \cdot \langle b|P'_0|a\rangle = 1$$

So the selection rules on m become

Linear polarization	$\Delta m' = 0$	
Circular polarization	$\Delta m' = \pm 1$	(as before)

The primes are a reminder that the unprimed quantization axis has been chosen to be along the propagation direction of the field. It is important to remember that EM radiation is transverse, that it must be a superposition of states with helicity ± 1 (i.e. $m_{\text{photon}} = \pm 1$ along $\hat{k}$) and that it **cannot** produce a transition between states with $\Delta m = 0$ along $\hat{k}$, but only $\Delta m' = 0$ where the prime reminds us that a different quantization axis was necessary to get this simple result.

3. Higher-order processes

For higher-order radiation processes, we just give the selection rules because they can be derived as was done for electric dipole radiation in the previous section.

The M1 operator for magnetic dipole radiation corresponds to an irreducible tensor of rank 1. Therefore we can show that the selection rules (for both circular and linear polarizations) are with **parity unchanged**

$$\begin{aligned} \Delta m &= 0, \pm 1 \\ \Delta j &= 0, \pm 1 \qquad \text{with } 0 \to 0 \text{ forbidden} \end{aligned}$$

Similarly the E2 operator for electric quadrupole radiation corresponds to an irreducible tensor of rank 2. Therefore the selection rules for both polarizations are with **parity unchanged**

$$\begin{aligned} \Delta m &= 0, \pm 1, \pm 2 \\ \Delta j &= 0, \pm 1, \pm 2 \qquad \text{with } 0 \to 0 \text{ forbidden} \end{aligned}$$

C. Transition rates

In this section we find transition rates for closed two-state systems. A closed system has no sources that transfer population into it or decay out of it; the only damping mechanism is the spontaneous decay of population in the upper state into the lower state. Two types of rates are considered—saturated and unsaturated. The **unsaturated rate** is the microscopic rate of transfer induced by the applied field per atom in the lower state to the upper state and also from the upper state to the lower state. The **saturated rate** is the net rate of transfer to the upper state by the applied field per atom in the system. Under steady conditions the saturated rate equals the total loss rate from the excited state by mechanisms other than stimulated emission. We will consider two types of applied field—**monochromatic** and **broadband**.

1. Saturated and unsaturated rates

The rates for two states $|a\rangle$ and $|b\rangle$ may be understood by referring to Fig. 6.1 below.

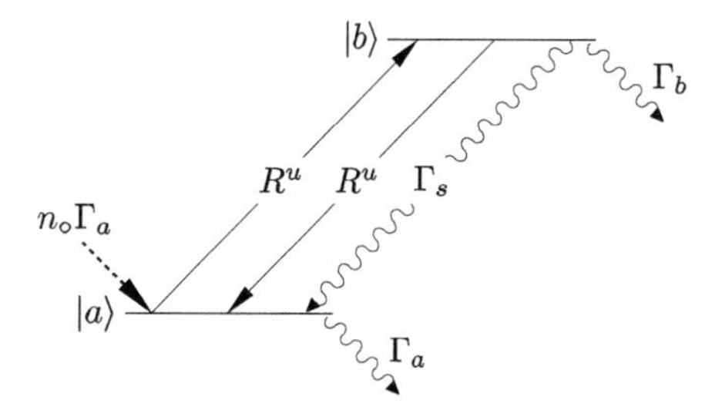

Figure 6.1: Absorption, stimulated emission, and decay rates for two coupled states $|a\rangle$ and $|b\rangle$.

The straight lines represent the unsaturated rate (per sec) of transfer per population in the initial state by the applied field R^u, with the arrows indicating the direction of transfer. This rate is equal in both directions because the absorption and stimulated rates involve the same factors (except for the matrix elements $|\langle a|\vec{P}|b\rangle|^2$ and $|\langle b|\vec{P}|a\rangle|^2$ which are equal). The wavy lines represent the spontaneous decay rates—the decay rate from state $|b\rangle$ to state $|a\rangle$ is Γ_s; we also include transfer out of state $|a\rangle$ at Γ_a and state $|b\rangle$ at Γ_b, and add a counter-balancing source term (dotted line) $n_\circ\Gamma_a$, with the intention of eventually taking these to zero to obtain a closed system.

The differential equations for population transfer to/from $|a\rangle$ and $|b\rangle$ involve the respective populations n_a and n_b. Adding up the transfer and decay rates (except for the source term they are all population in the initial state) gives

$$\begin{aligned}\dot{n}_b &= -n_b[\Gamma_b + \Gamma_s + R^u] + R^u n_a \\ \dot{n}_a &= n_\circ \Gamma_a + n_b(\Gamma_s + R^u) - (R^u + \Gamma_a)n_a\end{aligned} \tag{6.12}$$

In steady state $\dot{n}_a$ and $\dot{n}_b$ are zero, and these equations may be solved for R^u in terms of the ratio $n_b/n_\circ$ by eliminating n_a

$$-n_b\left[(\Gamma_b + \Gamma_s + R^u)(\Gamma_a + R^u) - (\Gamma_s + R^u)R^u\right] + n_\circ \Gamma_a R^u = 0$$

which implies

$$R^u = \frac{(\Gamma_b + \Gamma_s)\Gamma_a(n_b/n_\circ)}{\Gamma_a - (\Gamma_b + \Gamma_a)(n_b/n_\circ)} \tag{6.13}$$

The saturated rate is defined to be the net rate of transfer to state $|b\rangle$ per atom in the system by the applied field. It may be found from

$$R^s = R^u(n_a - n_b)/(n_a + n_b)$$

or, since $\dot{n}_b = 0$, it is also equal to the total loss rate from state $|b\rangle$ excepting field-induced transitions

$$R^s = n_b(\Gamma_s + \Gamma_b)/(n_a + n_b).$$

The saturated rate can be expressed simply in terms of the unsaturated rate by noting that the rate equations in (6.12) yield (in steady state)

$$n_b/n_a = R^u/(\Gamma_b + \Gamma_s + R^u)$$

so we have

$$\begin{aligned}R^s &= R^u \frac{1 - n_b/n_a}{1 + n_b/n_a} \\ &= R^u \frac{(\Gamma_b + \Gamma_s)}{(\Gamma_b + \Gamma_s) + 2R^u} \\ &= \frac{(\Gamma_b + \Gamma_s)}{2} \frac{S}{1 + S}\end{aligned} \tag{6.14}$$

where we have introduced the **saturation parameter**

$$S \equiv 2R^u/(\Gamma_b + \Gamma_s) \tag{6.15}$$

which equals unity when the unsaturated rate equals half the decay rate.

Eq. (6.14) shows that $R^s = R^u$ at low intensities for which $S \ll 1$; and $R^s \to (\Gamma_s + \Gamma_b)/2$ at high intensities for which $S \gg 1$, i.e. R^s saturates and no longer increases with applied power. The fraction $S/[2(1 + S)]$ is the fraction of population in the excited state, and approaches $1/2$ at high intensities.

2. Rates for monochromatic excitation

Now we shall find the saturated and unsaturated rates for monochromatic excitation. Our starting point is the steady-state population of state $|b\rangle$ for a system which is driven by an applied oscillating field with a Rabi frequency of ω_R and a detuning from resonance of δ derived in Eq. (5.26) of Chapter 5, along with the rates described in the preceding section.

The only thing we need additionally is Γ_2 the decay rate of the off-diagonal density matrix elements. Since these decay like the coefficients $c_a^* c_b$ in the wave function, they decay at the average rate for population in the two-states, which gives

$$\Gamma_2 = (\Gamma_a + \Gamma_b + \Gamma_s)/2$$

where the factor of 2 appears because $c_a(t)$ must decay at a rate $\Gamma_a/2$ in order to make the population $\rho_{aa} \sim a^* a$ decay at rate Γ_a.

Thus the steady population is

$$n_b = n_\circ \frac{2\Gamma_2}{\Gamma_s + \Gamma_b}\left(\frac{\omega_R}{2}\right)^2 \left[(\delta^2 + \Gamma_2^2) + \left(\frac{\Gamma_a + \Gamma_b}{\Gamma_a}\right)\left(\frac{2\Gamma_2}{\Gamma_s + \Gamma_b}\right)\left(\frac{\omega_R}{2}\right)^2\right]^{-1}$$

Substituting for $n_b/n_\circ$ in Eq. (6.13) yields a messy expression that simplifies to

$$R^{Mu} = \frac{2\Gamma_2(\omega_R/2)^2}{\delta^2 + \Gamma_2^2} = \frac{\Gamma_s(\omega_R/2)^2}{\delta^2 + (\Gamma_s/2)^2} = \frac{\omega_R^2}{\Gamma_s}\left[\frac{1}{1 + (2\delta/\Gamma_s)^2}\right] \tag{6.16}$$

where Γ_s is the total spontaneous emission rate out of $|b\rangle$, and we have used $\Gamma_2 = \Gamma_s/2$ which is appropriate only when Γ_a and Γ_b are negligible compared to Γ_s. We have replaced R^u with R^{Mu} because this is the **unsaturated rate for monochromatic radiation.** The above expression also shows that the unsaturated rate has a Lorentzian lineshape with FWHM of Γ_s.

The **saturated rate for monochromatic radiation** R^{Ms} is easily calculated from the unsaturated rate by applying the general connection developed between these two quantities in the previous section. Choosing $\Gamma_b = \Gamma_a = 0$ for a closed system, Eq. (6.14) gives

$$R^{Ms} = \frac{\Gamma_s R^{Mu}}{\Gamma_s + 2R^{Mu}} = \frac{\Gamma_s^2(\omega_R/2)^2}{\Gamma_s[\delta^2 + (\Gamma_s/2)^2] + 2\Gamma_s(\omega_R/2)^2}$$

which yields

$$R^{Ms} = \frac{\Gamma_s(\omega_R/2)^2}{\delta^2 + \omega_R^2/2 + (\Gamma_s/2)^2} = \frac{\omega_R^2}{\Gamma_s}\left[\frac{1}{1 + (2\delta/\Gamma_s)^2 + 2(\omega_R/\Gamma_s)^2}\right] \tag{6.17}$$

The expression shows that the lineshape of the saturated rate is Lorentzian, just like the unsaturated one. It also exhibits two features characteristic of saturation:

(i) The rate asymptotes to a finite value of $\Gamma_s/2$ when the field strength increases without limit, i.e. $\omega_R \to \infty$.

(ii) The width of the resonance exhibits power broadening (it increases when ω_R^2 increases), and is equal to $\Gamma_s\sqrt{1+2\omega_R^2/\Gamma_s^2}$.

This formula is probably one of most important in atomic physics.

The fact that the result does not depend critically on Γ_a and Γ_b indicates the validity of the underlying assumption that the effect of monochromatic radiation on a system with several decay rates is simply to transfer population between the states at the rate (per atom in each state) given above.

Often it is desirable to express R^{Mu} in terms of the incident intensity. For monochromatic radiation polarized along z the intensity is

$$I = \frac{c}{8\pi}\mathcal{E}^2$$

Recall that the Rabi frequency [Eq. (5.15) in Chapter 5] is given by

$$\omega_R = \frac{1}{\hbar}\langle b|e\mathcal{E}z|a\rangle \qquad \Longrightarrow \qquad \omega_R^2 = \frac{8\pi I}{c\hbar^2}|\langle b|ez|a\rangle|^2 \tag{6.18}$$

Thus the unsaturated rate of excitation due to a monochromatic beam of radiation of frequency ω and intensity I from Eq. (6.16) is (for a closed system with $\Gamma_a = \Gamma_b = 0$)

$$R_{ab}^{Mu} = R_{ba}^{Mu} = \frac{8\pi I}{c\hbar^2\Gamma_s}|\langle b|ez|a\rangle|^2 \left[\frac{1}{1+(2\delta/\Gamma_s)^2}\right] \tag{6.19}$$

The Lorentzian in square brackets is equal to 1 at exact resonance. Therefore the rate on resonance is

$$R_{ab}^{Mu}\Big|_{\delta=0} = \frac{8\pi I}{c\hbar^2\Gamma_s}|\langle b|ez|a\rangle|^2 \tag{6.20}$$

3. Cross-section for absorption

It is often convenient, particularly for monochromatic radiation, to define a total cross-section for scattering. The cross-section is the area per absorber which scatters the incident beam of photons (at low intensity, so we use the **unsaturated** rate)

$$\sigma_{ab}(\omega) = \frac{\text{Absorption rate}}{\text{Incident photon flux}} = \frac{R_{ab}^{Mu}(\omega)}{I/\hbar\omega}$$

Thus from Eq. (6.19) we get

$$\sigma_{ab}(\omega) = \frac{8\pi\omega}{c\hbar\Gamma_s}|\langle b|ez|a\rangle|^2 \left[\frac{1}{1+(2\delta/\Gamma_s)^2}\right]$$

Written in this form $\sigma_{ab}(\omega)$ (and the corresponding R_{ab}^{Mu}) depends both on the matrix element of z between $|a\rangle$ and $|b\rangle$, and the spontaneous emission rate from $|b\rangle$. In the event that $j_a = 0$ and that $|a\rangle$ is the only state to which $|b\rangle$ decays at a significant rate then

$$\Gamma_s = \frac{4k^3}{3\hbar}|\langle b|ez|a\rangle|^2$$

as we shall show later.

Thus for a spin zero ($j_a = 0$) ground state, we have

$$\begin{aligned}\sigma_{ab}(\omega) &= \frac{8\pi\omega/(c\hbar)\,|\langle b|ez|a\rangle|^2}{4k^3/(3\hbar)\,|\langle b|ez|a\rangle|^2}\left[\frac{1}{1+(2\delta/\Gamma_s)^2}\right] \\ &= 6\pi\lambda\!\!\!^{-}{}^2\left[\frac{1}{1+(2\delta/\Gamma_s)^2}\right] \qquad \text{with } \lambda\!\!\!^{-} = k^{-1}\end{aligned} \tag{6.21}$$

Hence the cross-section at resonance is $6\pi\lambda\!\!\!^{-}{}^2$ **independent** of the matrix element and the polarization of the light. It is a very large cross-section ($\sim 10^{-9}$ cm^2 for visible light, larger for infrared).

The above cross-section represents an upper bound on reality. In practice the cross-sections can be reduced by any factors which reduce the net transition rate—likely possibilities are additional sources of dephasing including spontaneous decay routes of $|b\rangle$ or $|a\rangle$ to other states and non-radiative channels, saturation at high ω_R^2, and Doppler broadening.

4. Rates for broadband excitation

We define broadband excitation to be excitation whose spectral profile is flat across the absorption profile of the system under study, and which has no spectral sub-structure or correlations between components at different frequencies. Thus a mode locked short pulse laser or a multi-mode laser will not qualify. The dynamics of obtaining the saturated rate for broadband excitation R^{Bs} from the unsaturated rate for broadband excitation R^{Bu} are the same as those for monochromatic excitation, and are already covered. Therefore, we need present only a calculation of the unsaturated rate for broadband excitation. We do this simply by integrating the unsaturated rate for monochromatic excitation over frequency. This procedure is justified by our assumption of no correlations between field components at different frequencies—without this we would have to integrate these amplitude for excitation, square it, and divide by the time to get a rate.

Thus our expression for R^{Bu} follows from Eq. (6.16) for R^{Mu}. We integrate over the entire frequency range even though only the region $\omega \approx \omega_\circ$ contributes significantly

$$
\begin{aligned}
R^{Bu} &= \int_{-\infty}^{\infty} R^{Mu}(\omega, \omega_R)\, d\omega \\
&= \frac{\pi \omega_R^2}{2} \int_{-\infty}^{\infty} \frac{\Gamma_s/2\pi}{\delta^2 + (\Gamma_s/2)^2}\, d\delta \\
&= \frac{\pi}{2} \overline{\omega_R^2}
\end{aligned}
$$

We have written the bar in $\overline{\omega_R^2}$ to indicate that it is the transition frequency per radian of frequency. Therefore it has dimensions of t^{-1} not t^{-2}.

If we wish to express R^{Bu} in terms of intensity per radian of frequency $\bar{I}(\omega)$, we can integrate Eq. (6.19) to obtain

$$
R^{Bu} = \frac{4\pi^2}{c\hbar^2} |\langle b|ez|a\rangle|^2\, \bar{I}(\omega) \tag{6.22}
$$

Here $\bar{I}(\omega)$ has units of energy per area per sec per $\sec^{-1}$, i.e. energy per area.

D. Spontaneous emission

1. Thermal equilibrium — Einstein A (emission) and B (absorption) coefficients

Historically the processes of emission and absorption of radiation were known long before stimulated emission was predicted by Einstein.* He deduced the existence of stimulated radiation by considering a thought experiment in which atoms were brought into thermal equilibrium solely by interaction with blackbody radiation. This was a very fruitful juxtaposition at the time because blackbody radiation had a quantum explanation (the Planck radiation spectrum) whereas quantum mechanics was not yet discovered (i.e. second quantization was known but not first quantization). It is a useful combination for us because the Planck spectrum is derived from a quantized radiation field, allowing us to obtain an expression for the spontaneous emission rate Γ_s, which we would otherwise have to obtain by redoing the interaction of the atom and field using a quantized (rather than classical) radiation field.

Consider states $|b\rangle$ and $|a\rangle$ with $E_b - E_a = \hbar\omega$. In thermal equilibrium at temperature T, the numbers of atoms in each state must be related by

$$n_b/n_a = e^{-\hbar\omega/k_B T} \tag{6.23}$$

If the only mechanism of interaction of the atoms is radiation, then these populations must be established by steady-state blackbody radiation. (Einstein realized that stimulated emission is necessary to counterbalance absorption so that the population of state $|b\rangle$ doesn't exceed that of state $|a\rangle$ in the high temperature limit.) The blackbody density of states $\rho_E(\omega, T)$ is: $\hbar\omega$ times the Bose occupation number per mode $n_B(\omega, T)$ times the density of vacuum modes $\rho_m(\omega) = \omega^2/(\pi^2 c^3)$ [from density of states times two polarizations]. Thus

$$\rho_E(\omega, T) = \frac{\hbar\omega^3}{\pi^2 c^3} n_B(\omega, T) \qquad \text{with} \qquad n_B(\omega, T) = \frac{1}{e^{\hbar\omega/k_B T} - 1}$$

Now consider the steady-state population of states $|a\rangle$ and $|b\rangle$ as was done earlier. The rate equations with $\Gamma_a = \Gamma_b = 0$ are

$$R^u = R^{Bu} = \beta\rho_E(\omega) \tag{6.24}$$

where β is a constant related to the absorption coefficient which we will find from our previous expression for the unsaturated rate for excitation by broadband radiation.

*English translation of the paper is available in *Sources of Quantum Mechanics*, B. L. Van Der Waerden, ed. (Dover, New York, 1968). For my personal take on this paper, see essay in Appendix C, "Einstein as Armchair Detective: The Case of Stimulated Radiation."

The rate equations are easily solved for the ratio n_a/n_b

$$\frac{n_a}{n_b} = \frac{\beta\rho_E(\omega) + \Gamma_s}{\beta\rho_E(\omega)} = e^{\hbar\omega/k_B T} \tag{6.25}$$

where we have substituted the temperature dependence from Eq. (6.23). Solving for Γ_s yields

$$\Gamma^s_{ba} = \beta\rho_E(\omega, T)[e^{\hbar\omega/k_B T} - 1] = \frac{\hbar\omega^3}{\pi^2 c^3}\,\beta = \frac{\hbar k^3}{\pi^2}\,\beta \tag{6.26}$$

The Boltzmann factor and the equilibrium conditions have combined to cancel out the denominator of the Bose population number, leaving us with a relationship between the spontaneous decay rate and the absorption coefficient which is independent of temperature. This relationship was first derived by Einstein using different notation: the spontaneous decay rate Γ_s was A, and the absorption coefficient β was $2\pi cB$ (different because the spectral energy density was per wavenumber not per radian per second, and state $|a\rangle$ was above state $|b\rangle$). So his relationship (between the **Einstein A and B coefficients**) was

$$A = 8\pi hc\nu^3\, B$$

A very simple relationship exists between the unsaturated broadband rate and the spontaneous decay rate—their ratio is simply $n_B(\omega, T)$ the average number of photons per mode. To see this, note that Eq. (6.25) implies

$$\frac{\Gamma_s}{\beta\rho_E(\omega)} = \frac{\Gamma_s}{R^{Bu}} = e^{\hbar\omega/k_B T} - 1$$

Hence

$$\frac{R^{Bu}}{\Gamma^s_{ba}} = \frac{1}{e^{\hbar\omega/k_B T} - 1} = n_B(\omega, T)$$

This suggests that spontaneous emission may be regarded as stimulated emission due to zero-point fluctuations.

2. Quantum mechanical expression for spontaneous decay rate

Our objective here is to express the spontaneous decay rate in terms of the transition matrix element. We first calculate the absorption coefficient β from the results of the previous section for the unsaturated broadband absorption rate.

We have from Eq. (6.24)

$$R^{Bu} = \beta\rho_E(\omega)$$

and from Eq. (6.22)

$$R^{Bu} = \frac{4\pi^2}{c\hbar^2} |\langle b|\, ez\, |a\rangle|^2 \bar{I}(\omega)$$

Clearly we must express $\bar{I}(\omega)$ in terms of $\rho_E(\omega)$ in order to find a quantum expression for β.

$\bar{I}(\omega)$ is the intensity of radiation polarized along z per unit area per sec per radian per sec, whereas $\rho_E(\omega)$ is the energy density of unpolarized radiation per unit volume per radian per sec. Hence,

$$\bar{I}(\omega) = \frac{c}{3}\, \rho_E(\omega, T) \tag{6.27}$$

with c to account for the propagation of the energy density (this also makes the dimensions jibe), and 1/3 because only 1/3 of the blackbody radiation has the proper z polarization. Thus

$$\beta = \frac{4\pi^2}{3\hbar^2}\, |\langle b|ez|a\rangle|^2 \qquad \Longrightarrow \qquad B = \frac{\beta}{2\pi c} = \frac{2\pi}{3c\hbar^2}\, |\langle b|ez|a\rangle|^2 \tag{6.28}$$

which expresses β in terms of the transition matrix element.

We can now express the spontaneous decay rate in terms of the matrix element using Eq. (6.26)

$$\begin{aligned} \Gamma^s_{ba} &= \frac{4k^3}{3\hbar} |\langle b|\vec{P}|a\rangle|^2 \\ &= \frac{64\pi^4\nu^3}{3h} |\langle b|\vec{P}|a\rangle|^2 \qquad \text{(traditional notation)} \end{aligned} \tag{6.29}$$

In this expression we have replaced ez by the dipole operator $\vec{P} = -e\vec{r}$ to give a more general expression which applies to the decay of state $|b\rangle$ to state $|a\rangle$ by emission of light of any allowed polarization; it gives the same result since $|a\rangle$ and $|b\rangle$ are coupled by only ez. The second expression is the traditional one for the Einstein A coefficient in which the frequency is measured in wavenumbers ($\nu = k/2\pi = \omega/2\pi c$).

E. Order-of-magnitude of spontaneous emission

Now let us put numbers into some of the expressions for radiative processes to estimate the order-of-magnitude of spontaneous emission rates for different types of radiative processes.

1. Electric dipole radiation

The preceding expression for the spontaneous decay rate uses the dipole operator $\vec{P}$ and is the appropriate one for E1 radiation. Hence

$$\Gamma_{ba}^{\mathrm{E1}} = \frac{4k^3}{3\hbar}|\langle b|\vec{P}|a\rangle|^2$$

The spontaneous emission rate from level $|B\rangle$ to level $|A\rangle$ is the sum of the decay rates to all the lower m_A states of $|A\rangle$ (with this sum and using the operator $\vec{P}$ instead of ez, all states of level $|B\rangle$ decay at an equal rate)

$$\Gamma_{BA}^{\mathrm{E1}} = \frac{4k^3}{3\hbar}\sum_{m_A}|\langle B|\vec{P}|A, m_A\rangle|^2 \tag{6.30}$$

We can express this in terms of oscillator strengths using the relation

$$\sum_{m_A}|\langle B|\vec{P}|A, m_A\rangle|^2 = \frac{3}{2}\frac{e^2\hbar^2}{m}\frac{-f_{BA}}{E_B - E_A}$$

so that

$$\Gamma_{BA}^{\mathrm{E1}} = -2\frac{e^2}{mc^3}\omega_{BA}^2 f_{BA} = -2\alpha^2(ka_\circ)\omega_{BA}f_{BA}$$

If we express the transition energy in atomic units

$$E_B - E_A = \varepsilon_{BA}\left[\frac{me^4}{\hbar^2}\right] \qquad \Longrightarrow \qquad k = \frac{E_B - E_A}{\hbar c} = \alpha\varepsilon_{BA}a_\circ^{-1}$$

then

$$\begin{aligned}\Gamma_{BA}^{\mathrm{E1}} &= -2\alpha^3\left[\frac{me^4}{\hbar^3}\right] f_{BA}\varepsilon_{BA}^2 \\ &= -3.2130\times 10^{10}\, f_{BA}\varepsilon_{BA}^2 \ \mathrm{s}^{-1}\end{aligned}$$

Note that this rate is proportional to ω^2—this reflects the cancellation of one power of the ω^3 term in the spontaneous emission [k^3 in Eq. (6.30)] by the dipole matrix element.

2. Electric quadrupole radiation

The spontaneous emission rate for a quadrupole transition is (after a sum and integral over polarization and direction, respectively)

$$\Gamma_{ca}^{\mathrm{E2}} = \frac{k^5}{10\hbar}|\langle c|e\vec{r}\,\vec{r}|a\rangle|^2$$

Since one does not usually have ready access to a table of quadrupole oscillator strengths, it is often helpful to insert a complete set of states $|b\rangle$ in the middle of the dyadic resulting in an expression containing the familiar dipole matrix elements,

$$\Gamma_{ca}^{\mathrm{E2}} = \frac{k^5}{10\hbar}\left|\sum_b \langle c|\vec{r}|b\rangle \langle b|e\vec{r}|a\rangle\right|^2$$

If the sum above is dominated by one intermediate state $|b\rangle$, then it is easy to obtain the ratio of the electric quadrupole spontaneous decay rate from $|c\rangle$ to $|a\rangle$ to the electric dipole spontaneous decay rate from $|b\rangle$ to $|a\rangle$. It is

$$\frac{\Gamma_{ca}^{E2}}{\Gamma_{ba}^{E1}} = \frac{k_{ca}^5/10\hbar}{4k_{ba}^3/3\hbar}\,\frac{|\langle c|\vec{r}|b\rangle|^2|\langle b|e\vec{r}|a\rangle|^2}{|\langle b|e\vec{r}|a\rangle|^2}$$

$$= \frac{3}{40}\left|k_{ca}\langle c|\vec{r}|b\rangle\right|^2 (k_{ca}/k_{ba})^3$$

This is small because $k_{ca}\langle c|\vec{r}|b\rangle$ is $\sim \alpha(k_{cb}/k_{ca})$. Since the various k's are typically roughly equal, we see that

$$\frac{\Gamma_{ca}^{E2}}{\Gamma_{ba}^{E1}} \sim \frac{\alpha^2}{10} \lesssim 10^{-5}$$

3. Magnetic dipole radiation

The expression for Γ^{M1} is the same as the one for for Γ^{E1} in Eq. (6.30), except that $\vec{M}$ replaces $\vec{P}$. The typical value for the matrix element is

$$\langle b|\vec{M}|a\rangle \equiv \langle b|\mu_B(\vec{L}+g_s\vec{S})|a\rangle \approx \frac{1}{2}\,\alpha e a_\circ$$

Therefore the decay rate is

$$\Gamma_{BA}^{\mathrm{M1}} = \frac{4k^3}{3\hbar}\sum_{m_A}|\langle B|\vec{M}|A,m_A\rangle|^2$$

$$\approx \frac{4k^3}{3\hbar}\left[\frac{1}{2}\alpha e a_\circ\right]^2 = \frac{\alpha^5}{3}\varepsilon_{BA}^3\left[\frac{e^2}{\hbar a_\circ}\right] = 2.85\times 10^5\,\varepsilon_{BA}^3\ \mathrm{s}^{-1}$$

Thus the ratio of magnetic to electric dipole decay rates is

$$\frac{\Gamma_{BA}^{\mathrm{M1}}}{\Gamma_{BA}^{\mathrm{E1}}} \approx \frac{1}{6}\alpha^2 \frac{\varepsilon_{BA}}{-f_{BA}}$$

While the preceding calculations of ratios of transition rates for the various one photon radiation processes give useful order-of-magnitude estimates, this comparison obscures the important fact that these processes obey different selection rules and hence do not occur simultaneously between the same two levels. Indeed, this is the only reason that E2 or M1 transitions are observed—where E1 is allowed it proceeds $\sim 10^5$ times faster than the others. Thus M1 and E2 transitions become important only where E1 is forbidden. For example, the D states in two electron atoms with the configuration $\ldots (n-1)p^6\, ns\, nd$ tend to lie above $\ldots ns^2$ but below $\ldots ns\, np$, and are forced to decay by E2 to $\ldots ns^2$. M1 transitions are usually important when the ground state is split by hyperfine structure (e.g. in H or alkalis) or fine structure (atoms with p or d valence electrons) since several low-lying levels will then originate from the lowest configuration and will have the same parity.

F. Saturation intensities

The discussion so far has shown that the unsaturated rates for absorption and stimulated emission of a transition are proportional to the applied intensity of light. It is therefore convenient to define the **saturation intensity** I_s, because it sets an experimental scale for the intensity required to drive a particular transition. The saturation intensity for monochromatic radiation is defined as the intensity for which the unsaturated rate on resonance is equal to half the decay rate, i.e.

$$R_{ab}^{Mu}\Big|_{\delta=0} = \Gamma_{ba}^{s}/2 \qquad \text{for } I = I_s$$

From the definition of the saturation parameter S in (6.15), we see that the saturation intensity is also the intensity at which $S = 1$.

We similarly define a saturation intensity per cm^{-1} for broadband radiation $\bar{I}_s$ as the intensity per cm^{-1} at which the unsaturated rate for broadband radiation is equal to half the decay rate, i.e.

$$R_{ab}^{Bu} = \Gamma_{ba}^{s}/2 \qquad \text{for } \bar{I} = \bar{I}_s$$

The above definitions allows us to write the unsaturated rates in the following form

$$R_{ab}^{Bu} = \frac{\Gamma_{ba}^{s}}{2}\frac{\bar{I}(\omega)}{\bar{I}_s(\omega_{ba})} \qquad \leftarrow \textbf{broadband}$$

$$R_{ab}^{Mu} = F_{ba}\frac{\Gamma_{ba}^{s}}{2}\frac{I(\omega)}{I_s(\omega_{ba})}\left[\frac{1}{1+(2F_{ba}\delta/\Gamma_{ba}^{s})^2}\right] \qquad \leftarrow \textbf{monochromatic}$$

where F_{ba} is a numerical factor $\lesssim 1$, and is equal to the ratio of the sum of all (radiative and non-radiative) dephasing processes for the $|a\rangle \to |b\rangle$ transition to Γ_{ba}^{s}. Thus $F_{ba} = 1$ for an ideal closed two-state system in which state $|b\rangle$ decays only by spontaneous emission and only to state $|a\rangle$.

1. Closed systems

(i) Broadband radiation

For broadband radiation in a closed system the saturation spectral density $\bar{I}_s(\omega)$ depends only on the frequency (not on the type of transition). This may be seen from the unsaturated rate in Eq. (6.24) and the above definition of $\bar{I}_s(\omega)$, which gives that when $\bar{I} = \bar{I}_s$

$$R_{ab}^{Bu} = \rho_s(\omega)\beta_{ab} = \Gamma_{ba}^{s}/2$$

where $\rho_s(\omega)$ is the energy density per cm^{-1} corresponding to $\bar{I}_s(\omega)$.

Combining this with Eq. (6.26) gives

$$\rho_s(\omega) = \frac{\Gamma^s_{ba}}{2\beta_{ba}} = \frac{\hbar\omega^3}{2\pi^2 c^3}$$

from which using Eq. (6.27)

$$\bar{I}_s(\omega) = \frac{c}{3}\rho_s(\omega) = \frac{\hbar\omega^3}{6\pi^2 c^2}$$

This shows that $\bar{I}_s(\omega)$ is independent of any details of the system or the transition, provided only that it is a two-state system.

The broadband unsaturated rate for a closed system is

$$R^{Bu}_{ab} = \frac{\Gamma^s_{ba}}{2}\frac{\bar{I}(\omega)}{\bar{I}_s(\omega_{ba})}$$

which follows from the fact that R^{Mu}_{ab} is proportional to intensity and equals $\Gamma^s_{ba}/2$ when $I = I_s$.

(ii) Monochromatic radiation

For monochromatic radiation the saturation intensity $I_s(\omega)$ depends linearly on the spontaneous decay rate in contrast to broadband excitation when $\bar{I}_s$ was independent of the system. This follows from the fact that, on resonance, both the monochromatic absorption rate and the cross-section are independent of the dipole matrix element between $|a\rangle$ and $|b\rangle$, as we have seen before in Eq. (6.21). Since the saturation intensity is the intensity which produces a rate of absorption equal to $\Gamma^s_{ba}/2$, a larger Γ^s_{ba} implies a larger I_s.

To find the monochromatic saturation intensity we substitute the expression for Γ^s in terms of the matrix element [Eq. (6.29)] into the rate of absorption at resonance

$$R^{Mu}_{ab}\Big|_{\delta=0} = \frac{8\pi|\langle b|ez|a\rangle|^2}{c\hbar^2\Gamma^s_{ba}}I(\omega) = \frac{6\pi}{c\hbar k^3}I(\omega)$$

Equating the above rate to $\Gamma^s_{ba}/2$ when $I = I_s$ yields

$$I_s(\omega_{ba}) = \frac{c\hbar k^3}{12\pi}\Gamma^s_{ba} = \frac{\hbar\omega^3}{12\pi c^2}\Gamma^s_{ba}$$

Thus the monochromatic unsaturated rate for a closed system is

$$R^{Mu}_{ab} = \frac{\Gamma^s_{ba}}{2}\frac{I(\omega)}{I_s(\omega_{ba})}\left[\frac{1}{1+(2\delta/\Gamma^s_{ba})^2}\right]$$

It is also convenient to express ω_R in terms of I and I_s, which is done as follows

$$\begin{aligned}\omega_R^2 &= \frac{8\pi I}{c\hbar^2}|\langle b|ez|a\rangle|^2 = \Gamma_{ba}^s R_{ab}^{Mu}\Big|_{\delta=0} \\ &= [\Gamma_{ba}^s]^2 \frac{I(\omega_{ba})}{2I_s(\omega_{ba})}\end{aligned} \tag{6.31}$$

Therefore, the monochromatic saturated rate given in Eq. (6.17) can be written in terms of I as

$$R_{ab}^{Ms} = \frac{\Gamma_{ba}^s}{2}\left[\frac{I/I_s}{1+(2\delta/\Gamma_{ba}^s)^2 + I/I_s}\right] \tag{6.32}$$

which shows that the saturation intensity I_s is the intensity at which the transition gets power broadened by a factor of $\sqrt{2}$.

2. Open systems

The results in the preceding subsection are applicable only to ideal closed two-state systems with no decay or dephasing except spontaneous decay of state $|b\rangle$ to state $|a\rangle$. Often it is necessary to deal with systems in which the upper state has additional forms of spontaneous decay (e.g. because there are intermediate levels or other levels near $|a\rangle$), or because there is re-population of the state (e.g. due to collisions or Doppler broadening).

All of these mechanisms can be accounted for by using the density matrix treatment of a general system with decay Γ_a and Γ_b from states $|a\rangle$ and $|b\rangle$ to states outside the system. In the solution for ρ_{bb} [Eq. (5.26) of Chapter 5] we take

$$\Gamma_2 = (\Gamma_{ba}^s + \Gamma_a + \Gamma_b)/2 + \Gamma_{ba}^{\text{nr}} \tag{6.33}$$

where

$$\Gamma_a = \left(\sum_i \Gamma_{ai}^s\right) + \Gamma_a^{\text{nr}} \qquad \text{and} \qquad \Gamma_b = \left(\sum_i \Gamma_{bi}^s\right) - \Gamma_{ba}^s + \Gamma_b^{\text{nr}} \tag{6.34}$$

Thus Γ_2 is augmented by the addition of non-radiative dephasing particular to the $|b\rangle \to |a\rangle$ transition Γ_{ba}^{nr}; and Γ_a and Γ_b include spontaneous decays to all other levels [except from $|b\rangle \to |a\rangle$ which is accounted for explicitly in (6.33) and subtracted out in (6.34)] as well as non-radiative decays (again excepting $|b\rangle \to |a\rangle$).

We define F_{ba} as the ratio of spontaneous decay of $|b\rangle$ only to $|a\rangle$ to the total dephasing, i.e.

$$F_{ba} = \Gamma_{ba}^s/(2\Gamma_2) \lesssim 1$$

where the factor of 2 ensures $F_{ba} = 1$ for an ideal two-state system for which $\Gamma_2 = \Gamma^s_{ba}/2$. Using F_{ba}, we can write the monochromatic unsaturated rate for an open system from Eq. (6.16) as

$$R^{Mu}_{ba} = \frac{2(\omega_R/2)^2}{\Gamma_2}\left[\frac{1}{1+(\delta/\Gamma_2)^2}\right]$$
$$= F_{ba}\frac{\omega_R^2}{\Gamma^s_{ba}}\left[\frac{1}{1+(2\delta F_{ba}/\Gamma^s_{ba})^2}\right]$$

Expressing this in terms of the saturation intensity $I_s(\omega)$, we get

$$R^{Mu}_{ba} = F_{ba}\frac{\Gamma^s_{ba}}{2}\frac{I(\omega)}{I_s(\omega)}\left[\frac{1}{1+(2\delta F_{ba}/\Gamma^s_{ba})^2}\right]$$

which shows that the line profile is lowered by the factor F_{ba} and broadened by the factor F^{-1}_{ba}. Thus, this procedure allows the continued use of Eq. (6.31) for ω_R^2 at the expense of lowering R^{Mu}_{ba} at $\delta = 0$ by a factor of F_{ba}.

The broadband unsaturated rate for an open system is the same as that for a closed system—the saturation intensity (1 photon per mode) is the same and so is the absorption rate. This follows from R^{Mu}_{ba}: although it is lowered it is widened by the same factor, so its integral (the broadband absorption rate) remains constant. Thus an open system also obeys

$$R^{Bu}_{ba} = \frac{\Gamma^s_{ba}}{2}\frac{\bar{I}(\omega)}{\bar{I}_s(\omega_{ba})}$$

To find the saturated rate in open systems, one can simply use the expressions for R^s in terms of R^u discussed earlier. This introduces no computational problems, but gives rise to a semantic one—the saturation intensities are now no longer the intensities at which the saturation parameter S equals unity. Therefore the saturation intensity should more properly be regarded as a reference intensity.

Saturation in open systems is complicated by another phenomenon—optical pumping. If $\Gamma_b \gg \Gamma_a$ (i.e. if alternative spontaneous decay rates exist for $|b\rangle$ but not for $|a\rangle$) then even weak irradiation can dramatically decrease the total number of atoms in the system. R_s will be given correctly, but the net excitation rate can be orders of magnitude below $n_\circ R_s$ due to optical pumping.

G. Problems

1. Classical scattering in 2D

Consider the collisions of small particles with a hockey puck of radius R. (This is a "hard sphere" in two, rather than three, dimensions.)

(a) What is the natural expression for $\sigma_2(\theta)$, the 2D differential cross-section?

(b) What is $\sigma_2(\theta)$ for this case?

(c) Show that the total cross-section is what you would expect.

Solution

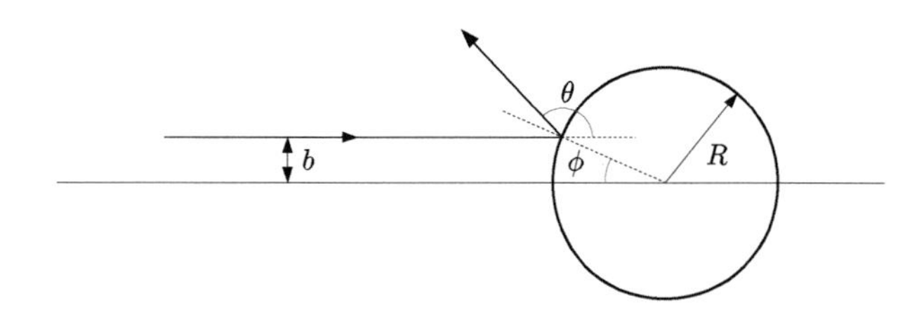

(a) Let the flux of particles crossing unit length in unit time be I.

The number of particles with impact parameter in the range b to $b+db$ $= I\,db \times 2$ (for above and below the axis).

By definition of the scattering cross-section

$$\sigma_2(\theta)\,I\,d\theta = \text{number of particles scattered into angle } \theta \text{ and } \theta + d\theta$$

Since the impact parameter uniquely determines the scattering angle, the two quantities are equal, i.e.

$$I\,db \times 2 = \sigma_2\,\theta\,I\,d\theta \qquad \Longrightarrow \qquad \sigma_2(\theta) = 2\frac{db}{d\theta}$$

(b) For scattering from puck of radius R, the figure shows that

$$\theta + 2\phi = \pi \qquad \Longrightarrow \qquad \phi = \frac{\pi}{2} - \frac{\theta}{2}$$

Using

$$\sin\phi = \frac{b}{R}$$

we get

$$b = R\sin\phi = R\sin\left(\frac{\pi}{2} - \frac{\theta}{2}\right) = R\cos\frac{\theta}{2}$$

Therefore the differential cross-section is ($-$ve sign not important)

$$\sigma_2(\theta) = 2\frac{db}{d\theta} = R\sin\frac{\theta}{2}$$

(c) The total cross-section is

$$\sigma = \int_0^{\pi} R\,\sin\frac{\theta}{2}\,d\theta = -2R\cos\frac{\theta}{2}\bigg|_0^{\pi} = 2R$$

which is expected for a "hard circle" scattering.

2. Rabi frequency from matrix element

Consider the two electric fields

$$\vec{E}_1(\vec{r},t) = \mathcal{E}_\circ\hat{z}\,\mathrm{Re}\left\{e^{i(kx-\omega t)}\right\} \qquad \text{and} \qquad \vec{E}_2(\vec{r},t) = \mathcal{E}_\circ\hat{\varepsilon}_+\,\mathrm{Re}\left\{e^{i(kz-\omega t)}\right\}$$

These interact with a two-level system in which level $|a\rangle$ has $\ell = 0$ and level $|b\rangle$ has $\ell = 1$ (ignore $\vec{S}$). The relevant radial integral is

$$\int_0^{\infty} y_1^*\, r\, y_0\, dr = d \qquad \text{where } y_\ell\text{'s are the radial wavefunctions.}$$

(a) Find the intensity of each wave.

(b) Find the interaction Hamiltonian (in the electric dipole approximation) for the two fields.

(c) Find the Rabi frequency for both cases. ω_R is defined in terms of the interaction matrix element. Check that on resonance $P_2 = \sin^2(\omega_R t/2)$.

(d) What are the Δm selection rules for both fields for absorption and stimulated emission?

(e) Discuss the necessity of using the rotating wave approximation in the two cases.

Solution

The energy transported by an electromagnetic field is given by the Poynting vector. Its average value is related to the intensity as

$$\langle \vec{S} \rangle = \frac{1}{2}\frac{c}{4\pi}\,\vec{E}\times\vec{B}^* = I\hat{k}$$

(a) For both E fields

$$\vec{E}\times\vec{B}^* = \mathcal{E}_\circ^2\hat{\varepsilon}\times\left(\hat{k}\times\hat{\varepsilon}\right) = \mathcal{E}_\circ^2\hat{k} \qquad \Longrightarrow \qquad I = \frac{c}{8\pi}\mathcal{E}_\circ^2$$

(b) The interaction matrix element between two states is

$$\langle b|H'|a\rangle = \frac{e\mathcal{E}_\circ}{2}\,\frac{\omega_a-\omega_b}{\omega}\left[\langle b|\hat{\varepsilon}^*\cdot\vec{r}|a\rangle\, e^{-i\omega t} + \langle b|\hat{\varepsilon}\cdot\vec{r}|a\rangle\, e^{i\omega t}\right]$$

(i) Field $\vec{E}_1$

This field has $\hat{\varepsilon} = \hat{z}$ and $\hat{\varepsilon}^* = \hat{z}$. Therefore the off-diagonal matrix element is

$$\langle b|H_1'|a\rangle = \frac{e\mathcal{E}_\circ}{2}\,\frac{\omega_a-\omega_b}{\omega}\left(e^{-i\omega t}+e^{i\omega t}\right)\langle b|\hat{z}\cdot\vec{r}|a\rangle$$

In terms of spherical harmonics this is

$$\langle b|z|a\rangle = \langle b|r\cos\theta|a\rangle = \sqrt{\frac{4\pi}{3}}\,\langle b|rY_{10}(\theta,\phi)|a\rangle$$

Evaluating this out gives

$$\begin{aligned}
\langle b|z|a\rangle &= \sqrt{\frac{4\pi}{3}}\int dV\,\psi_b^*(\vec{r})\,r\,Y_{10}(\theta,\phi)\,\psi_a(\vec{r}) \\
&= \sqrt{\frac{4\pi}{3}}\int_0^{2\pi} d\phi\int_{-\pi}^{\pi} d\theta\, Y_{1m}^*\,Y_{10}\,Y_{00}\int_0^{\infty} dr\, y_\ell^*\, r\, y_0 \\
&= \sqrt{\frac{4\pi}{3}}\left[\frac{1}{\sqrt{4\pi}}\delta_{m,0}\right] d \\
&= \frac{d}{\sqrt{3}}\delta_{m,0}
\end{aligned}$$

The other matrix elements are

$$\begin{aligned}
&\langle a|H_1'|b\rangle = \langle b|H_1'|a\rangle^* \\
&\langle a|H_1'|a\rangle = \langle b|H_1'|b\rangle = 0 \qquad \text{(by parity)}
\end{aligned}$$

Therefore the interaction Hamiltonian is

$$H_1'(t) = \frac{e\mathcal{E}_\circ d}{2\sqrt{3}}\,\frac{\omega_a-\omega_b}{\omega}\,\delta_{m,0}\begin{bmatrix} 0 & e^{i\omega t}-e^{-i\omega t} \\ e^{i\omega t}-e^{-i\omega t} & 0 \end{bmatrix}$$

(ii) Field $\vec{E}_2$

This field has $\hat{\varepsilon} = \hat{\varepsilon}_+$ and $\hat{\varepsilon}^* = \hat{\varepsilon}_-$. Therefore the off-diagonal element is

$$\langle b|H_2'(t)|a\rangle = \frac{e\mathcal{E}_\circ}{2}\frac{\omega_a - \omega_b}{\omega}\left[\langle b|\hat{\varepsilon}_- \cdot \vec{r}|a\rangle\, e^{-i\omega t} + \langle b|\hat{\varepsilon}_+ \cdot \vec{r}|a\rangle\, e^{i\omega t}\right]$$

Using spherical harmonics, this is evaluated as

$$\begin{aligned}\langle b|\hat{\varepsilon}_- \cdot \vec{r}|a\rangle &= \langle b|r\, Y_{1-1}(\theta,\phi)|a\rangle \\ &= \sqrt{\frac{4\pi}{3}}\int d\Omega\, Y_{1m}^* Y_{1-1}\, Y_{00}\int_0^\infty dr\, y_1^*\, r\, y_0 \\ &= \frac{d}{\sqrt{3}}\delta_{m,-1}\end{aligned}$$

and

$$\begin{aligned}\langle b|\hat{\varepsilon}_+ \cdot \vec{r}|a\rangle &= \langle b|r\, Y_{11}(\theta,\phi)|a\rangle \\ &= \sqrt{\frac{4\pi}{3}}\int d\Omega\, Y_{1m}^* Y_{11}\, Y_{00}\int_0^\infty dr\, y_1^*\, r\, y_0 \\ &= \frac{d}{\sqrt{3}}\delta_{m,1}\end{aligned}$$

Therefore the interaction Hamiltonian is

$$H_2'(t) = \frac{e\mathcal{E}_\circ d}{2\sqrt{3}}\frac{\omega_a - \omega_b}{\omega}\left(\delta_{m,-1}\begin{bmatrix} 0 & e^{i\omega t} \\ e^{-i\omega t} & 0\end{bmatrix} + \delta_{m,1}\begin{bmatrix} 0 & e^{-i\omega t} \\ e^{i\omega t} & 0\end{bmatrix}\right)$$

(c) The interaction Hamiltonian using the Rabi frequency (in electric dipole approximation) is written as

$$H'(t) = \frac{\hbar}{2}\frac{\omega_a - \omega_b}{\omega}\begin{pmatrix} 0 & \omega_R e^{i\omega t} \\ \omega_R e^{-i\omega t} & 0\end{pmatrix}$$

Comparing with (b), the Rabi frequencies in both cases is

$$\omega_R = \frac{e\mathcal{E}_\circ d}{\sqrt{3}}$$

To check the probability of transition on resonance $\omega = \omega_{ba}$, we make the rotating wave approximation which is valid near resonance. Therefore the interaction Hamiltonian simplifies to

$$H'(t) = -\frac{\hbar}{2}\begin{bmatrix} 0 & \omega_R e^{i\omega t} \\ \omega_R e^{-i\omega t} & 0\end{bmatrix}$$

Writing the solution as

$$\psi(t) = \begin{bmatrix} a(t) \\ b(t) \end{bmatrix}$$

and solving Schrödinger's equation

$$H'(t)\psi(t) = i\hbar \frac{d\psi}{dt}$$

we get

$$\dot{a}(t) = \frac{i\omega_R}{2}\, b(t) \qquad \text{and} \qquad \dot{b}(t) = \frac{i\omega_R}{2}\, a(t)$$

which implies

$$\ddot{b}(t) = -\frac{\omega_R^2}{4}\, b(t)$$

Using initial conditions of $a(0) = 1$ and $b(0) = 0$, this gives the solution as

$$b(t) = \sin\left(\frac{\omega_R t}{2}\right) \qquad \Longrightarrow \qquad P_2 = |b(t)|^2 = \sin^2\left(\frac{\omega_R t}{2}\right)$$

(d) Δm selection rules

(i) **$\mathbf{E}_1$ field**

Absorption: $\Delta m = 0$
Stimulated emission: $\Delta m = 0$

(ii) **$\mathbf{E}_2$ field**

Absorption: $\Delta m = +1$
Stimulated emission: $\Delta m = -1$

(e) The rotating wave approximation is necessary in order to get rid of the counter-rotating term, which gets averaged during the interaction.

3. Transitions in Na and H

1. The experimental lifetime of the $3p$ level in Na (which radiates at 589 nm) is 16.3 ns, and for the $2p$ level in H it is 1.60 ns. Find the oscillator strengths $f_{3s,3p}$ for Na, and $f_{2p,1s}$ for H.

2. Estimate the lifetime of H in the $F = 1$ hyperfine level of the $1s$ state. The decay of this level to the $F = 0$ hyperfine level gives rise to the famous 21 cm line of radio astronomy. (You may assume that the matrix element is μ_B in your estimate).

Solution

(a) The discussion in the chapter shows that the oscillator strength for a $|B\rangle \to |A\rangle$ transition is related to the spontaneous decay rate as

$$f_{BA} = -\frac{\Gamma_{BA}^{\mathrm{E1}}}{2\alpha^2(ka_\circ)\omega_{BA}}$$

(i) The $3p \to 3s$ transition in Na

For this transition

$$\lambda = 589 \text{ nm} \qquad \text{and} \qquad \tau = 16.3 \times 10^{-9} \text{ s}$$

Therefore the parameters are

$$\begin{aligned} \Gamma &= \frac{1}{\tau} = 6.13 \times 10^7 \text{ s}^{-1} \\ k &= \frac{2\pi}{\lambda} = 1.07 \times 10^7 \text{ m}^{-1} \\ \omega_{BA} &= kc = 3.2 \times 10^5 \text{ rad/s} \end{aligned}$$

Thus we get (using ω_{AB})

$$f_{3s,3p} = 0.32$$

which has a positive sign because it is for absorption.

(ii) The $2p \to 1s$ transition in H

For this transition

$$\Delta E = R_{\mathrm{H}}\left(1 - \frac{1}{2^2}\right) = 10.2 \text{ eV} \qquad \text{and} \qquad \tau = 1.6 \times 10^{-9} \text{ s}$$

Therefore the parameters are

$$\Gamma = \frac{1}{\tau} = 6.25 \times 10^8 \ \mathrm{s}^{-1}$$

$$\omega_{BA} = \frac{\Delta E}{\hbar} = 1.55 \times 10^{16} \ \mathrm{rad/s}$$

$$k = \frac{\omega}{c} = 5.17 \times 10^7 \ \mathrm{m}^{-1}$$

Thus we get

$$f_{2p,1s} = -0.14$$

which is negative because it is for emission.

(b) The 21 cm line is an M1 transition because it is between two hyperfine levels of the ground state which have the same parity. From the discussion in the chapter

$$\Gamma^{\mathrm{M1}} = \frac{4k^3}{3\hbar}|\langle b|\vec{M}|a\rangle|^2 \approx \frac{4k^3}{3\hbar} \ \mu_B^2$$

Therefore the decay rate is estimated as

$$\Gamma^{\mathrm{M1}} = \frac{4k^3}{3\hbar}\left(\frac{e\hbar}{2mc}\right)^2 = \frac{1}{3}\frac{e^2\hbar k^3}{m^2c^2} = 2.9 \times 10^{-15} \ \mathrm{s}^{-1}$$

which gives the lifetime as

$$\tau = \frac{1}{\Gamma} = 3.4 \times 10^{14} \ \mathrm{s}$$

Chapter 7

Multiphoton

THE advent of the tunable laser has made possible the study of processes in which more than one photon is absorbed so that the atomic system gains several $\hbar\omega$ of energy—this chapter deals with such multiphoton processes. Highly excited states may be populated by these mechanisms creating population inversions with respect to intermediate states so that lasing or superradiant emission (a form of coherent spontaneous emission with short lifetime) occurs. It is even possible to observe lasing action from virtual states (dressed atoms) as in coherent anti-Stokes Raman scattering (CARS).

A. Two-photon absorption

1. General considerations

We shall now calculate the transition rate for a two-photon transition. We expect that this will require a second order perturbation treatment involving $H'_{\rm E1}$ rather than a first order treatment involving $H'^{(2)} = e^2A^2/c^2$ because the $H'^{(2)}$ matrix elements are so small. In consequence of using second order perturbation theory, we expect to start from a state of one parity and wind up in a state of the same parity, with a transition amplitude involving a sum over intermediate states which differ in parity (and in ℓ by 1) from both of these other states. Thus two-photon processes could occur from S $\rightarrow$ S or S $\rightarrow$ D with a sum over intermediate P states; or from P $\rightarrow$ P with a sum over both S and D intermediate states.

For simplicity we assume temporarily that only one intermediate state contributes. The level structure therefore looks like what is shown in Fig. 7.1 below.

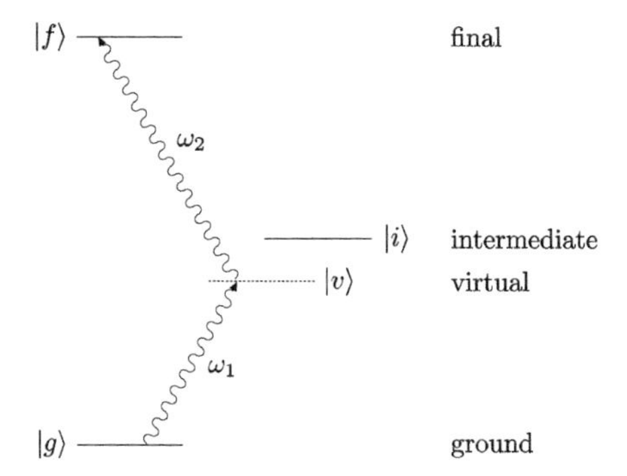

Figure 7.1: Energy level scheme for two-photon transition coupled through an intermediate state. The two photons have frequencies ω_1 and ω_2, and are detuned from the intermediate state.

The states are separated by energies $\hbar\omega_{ig}$ and $\hbar\omega_{fi}$, and in order to conserve energy for the two-photon process we expect resonant behavior when the sum of the two-photons' energy equals $E_f - E_g$, i.e.

$$\omega_{fg} = \omega_1 + \omega_2$$

The two E fields are (in dipole approximation, so no spatial variation)

$$\vec{E}_1(t) = \mathcal{E}_1\hat{\varepsilon}_1 \cos\omega_1 t$$
$$\vec{E}_2(t) = \mathcal{E}_2\hat{\varepsilon}_2 \cos\omega_2 t$$

2. Final-state amplitude from perturbation theory

Now we calculate the amplitude $a_i(t)$ of state $|i\rangle$ assuming boundary conditions of

$$a_i(0) = a_f(0) = 0 \qquad \text{at } t = 0$$
$$a_g = 1 \qquad \text{at all times}$$

Using first order perturbation theory

$$i\hbar\,\dot{a}_i^{(1)} = H_{ig}\,a_g^{(0)}$$

we get

$$\begin{aligned} a_i^{(1)}(t) &= (i\hbar)^{-1}\int_0^t \langle i|H'(t')|g\rangle\, e^{i\omega_{ig}t'}\,dt' \\ &= (i\hbar)^{-1}\left[\langle i|\hat{\varepsilon}_1\cdot\vec{P}|g\rangle\,\frac{\mathcal{E}_1}{2}\int_0^t \left[e^{i(\omega_{ig}+\omega_1)t'} + e^{i(\omega_{ig}-\omega_1)t'}\right]dt'\right] \\ &\quad + [\text{same for field 2}] \\ &= -\frac{\mathcal{E}_1}{2\hbar}\,\langle i|\,\hat{\varepsilon}_1\cdot\vec{P}\,|g\rangle\left[\frac{e^{i(\omega_{ig}+\omega_1)t}-1}{\omega_{ig}+\omega_1} + \frac{e^{i(\omega_{ig}-\omega_1)t}-1}{\omega_{ig}-\omega_1}\right] \\ &\quad + [\text{same for field 2}] \end{aligned}$$

If we define the matrix elements as

$$M_{ig}^{(1)} \equiv \langle i|\hat{\varepsilon}_1\cdot\vec{P}|g\rangle \qquad \text{and} \qquad M_{ig}^{(2)} \equiv \langle i|\hat{\varepsilon}_2\cdot\vec{P}|g\rangle \tag{7.1}$$

and neglect transients and counter-rotating terms, we get

$$a_i^{(1)}(t) = -\frac{1}{2\hbar}\left[\frac{\mathcal{E}_1 M_{ig}^{(1)}\left[e^{i(\omega_{ig}-\omega_1)t}-1\right]}{\omega_{ig}-\omega_1} + \frac{\mathcal{E}_2 M_{ig}^{(2)}\left[e^{i(\omega_{ig}-\omega_2)t}-1\right]}{\omega_{ig}-\omega_2}\right]$$

where the first term is for field 1 and the second for field 2.

In order to find a non-zero amplitude $a_f(t)$ we proceed to second order in perturbation theory

$$i\hbar\,\dot{a}_f^{(2)} = \sum_i H_{fi}\,a_i^{(1)}(t)$$

but we suppress the $\sum_i$ because we are considering only one intermediate state.

Then

$$a_f^{(2)}(t) = (i\hbar)^{-1} \int_0^t \langle f|H'i\rangle\, a_i^{(1)}(t')\, dt'$$

$$= \frac{1}{4\hbar^2} \left[\frac{\mathcal{E}_1^2 M_{fi}^{(1)} M_{ig}^{(1)}}{\omega_{ig} - \omega_1} \frac{e^{i(\omega_{fg}-2\omega_1)t} - 1}{\omega_{fg} - 2\omega_1} \right.$$

$$+ \frac{\mathcal{E}_2^2 M_{fi}^{(2)} M_{ig}^{(2)}}{\omega_{ig} - \omega_2} \frac{e^{i(\omega_{fg}-2\omega_2)t} - 1}{\omega_{fg} - 2\omega_2} \tag{7.2}$$

$$+ \frac{\mathcal{E}_2 \mathcal{E}_1 M_{fi}^{(2)} M_{ig}^{(1)}}{\omega_{ig} - \omega_1} \frac{e^{i(\omega_{fg}-\omega_1-\omega_2)t} - 1}{\omega_{fg} - \omega_1 - \omega_2}$$

$$\left. + \frac{\mathcal{E}_1 \mathcal{E}_2 M_{fi}^{(1)} M_{ig}^{(2)}}{\omega_{ig} - \omega_2} \frac{e^{i(\omega_{fg}-\omega_2-\omega_1)t} - 1}{\omega_{fg} - \omega_1 - \omega_2} \right]$$

where the matrix elements are defined in an analogous manner to what was done in (7.1), and again we have neglected the non-resonant counter-rotating and transient terms. The first two terms in this expression are for two-photon excitation by $\vec{E}_1$ and $\vec{E}_2$ separately, while the third and fourth terms are due to two-photon excitation by $\vec{E}_1$ and $\vec{E}_2$ together.

3. Transition rate

In practice the previous expression must be generalized by summing over all available intermediate states. (Note that there is no restriction that the intermediate states have energy between $\hbar\omega_g$ and $\hbar\omega_f$.) This can produce interesting interference effects, particularly when the intermediate state is split (say by spin-orbit effects) and one laser frequency is adjusted to be in between the components. Further interference can arise between the various terms in Eq. (7.2), especially the second and third.

Often only one of the terms in Eq. (7.2) will contribute significantly because only one of the denominators, e.g. $(\omega_{ig} - \omega_1)(\omega_{fg} - \omega_1 - \omega_2)$, is close to zero. [Obviously only one term contributes if only one field, say $\vec{E}_1$, is present and $(\omega_{fg} - 2\omega_1) \approx 0$.] So we shall look at the probability that the system is in the upper state assuming that only the third term in Eq. (7.2) contributes. This we denote $P^{(2)}$ to emphasize that it is the 2nd order perturbation theory result for a two-photon process.

$$P_f^{(2)}(t) = |a_f^{(2)}(t)|^2$$

$$= \frac{\mathcal{E}_1^2 \mathcal{E}_2^2}{4\hbar^4} \left[\frac{M_{fi}^{(2)} M_{ig}^{(1)}}{\omega_{ig} - \omega_1} \right]^2 \frac{\sin^2 [(\omega_{fg} - \omega_1 - \omega_2)t/2]}{(\omega_{fg} - \omega_1 - \omega_2)^2} \tag{7.3}$$

since $|e^{ix} - 1|^2 = |e^{ix/2} - e^{-ix/2}|^2 = 4\sin^2(x/2)$

The last term is not exact and leads to problems in interpreting this expression. For one thing it lacks the power broadening term present in analogous expressions for a two-state system—this is because we assumed $a_g(t) = 1$, so did not allow the system to "cycle" between states $|g\rangle$ and $|f\rangle$. Furthermore, we have not allowed for the finite lifetime of state $|f\rangle$.

We remedy these difficulties by arguing that Eq. (7.3) must be generalized so that it is similar to the Rabi transition probability equation found previously for monochromatic excitation of an undamped two-state system

$$P_f^{(2)}(t) = \frac{\omega_R^2}{\omega_R'^2} \sin^2\left(\frac{\omega_R' t}{2}\right)$$

This may be achieved by defining

$$\omega_R^2 = \frac{\mathcal{E}_1^2 \mathcal{E}_2^2}{4\hbar^4} \left[\frac{M_{fi}^{(2)} M_{ig}^{(1)}}{\omega_{ig} - \omega_1}\right]^2$$

$$\delta_2^2 = (\omega_1 + \omega_2 - \omega_{fg})^2$$

$$\omega_R'^2 = \omega_R^2 + \delta_2^2$$

Pursuing this argument, we conclude that the introduction of a Γ_2 type coherence dephasing rate Γ_{fg} for the final state will lead to an (unsaturated) rate of excitation given by

$$R_{gf}^{(2)} = \frac{2\Gamma_{fg}(\omega_R/2)^2}{\delta^2 + \Gamma_{fg}^2} = \frac{\omega_R^2}{2\Gamma_{fg}}\left[\frac{1}{1 + (\delta/\Gamma_{fg})^2}\right]$$

Thus in general

$$R_{gf}^{(2)} = \frac{\mathcal{E}_1^2 \mathcal{E}_2^2}{8\hbar^4 \Gamma_{fg}} \left[\sum_i \left(\frac{M_{fi}^{(2)} M_{ig}^{(1)}}{\omega_{ig} - \omega_1 - i\Gamma_{ig}} + \frac{M_{fi}^{(1)} M_{ig}^{(2)}}{\omega_{ig} - \omega_2 - i\Gamma_{ig}}\right)\right]^2 \times \left[\frac{1}{1 + (\delta/\Gamma_{fg})^2}\right] \tag{7.4}$$

where we added the term $-i\Gamma_{ig}$ to ω_{ig} to account for dephasing of the $|g\rangle \to |i\rangle$ transition, a complication of concern only when the intermediate state $|i\rangle$ is almost resonant with ω_1. We also added back the term previously omitted from Eq. (7.2) which accounts for resonance arising when ω_2 is near one of the ω_{ig}'s. The sum over the intermediate states $|i\rangle$'s is put in to emphasize that it is the amplitudes of the various paths from $|i\rangle \to |f\rangle$ which add and not the probabilities.

Note: If $|g\rangle$ decays at Γ_g and $|i\rangle$ decays at Γ_i, then $\Gamma_{ig} = (\Gamma_i + \Gamma_g)/2$ and the coherences of $|i\rangle$ and $|g\rangle$ decay at this rate. The denominator of Eq. (7.4) becomes (upon squaring the modulus) $[(\omega_{ig} - \omega_1)^2 + (\Gamma_{ig})^2]$, giving a FWHM of $2\Gamma_{ig}$.

B. Two-photon de-excitation processes

1. Two-photon stimulated emission

Any absorption process implies the existence of a corresponding stimulated emission process. While thermodynamic arguments like the ones used by Einstein in his discussion of the A and B coefficients provide the most rigorous justification of this statement, it is more easily seen in the case of a two-photon processes by using time dependent perturbation theory. One simply considers the three-state system we have been discussing with the initial conditions of

$$a_f(0) = 1 \qquad \text{and} \qquad a_g(0) = a_i(0) = 0$$

Using the counter-rotating terms in the Hamiltonian and applying second order perturbation theory will result in an expression for $R^{(2)}_{fg}$ which contains the same terms as Eq. (7.4) except that the first resonance denominator will be $(\omega_{fi} - \omega_2 - i\Gamma_{fi}/2)$ instead of $(\omega_{ig} - \omega_1 - i\Gamma_{ig}/2)$. Since the two-photon rate is negligible unless $\omega_{fi} + \omega_{ig} \approx \omega_2 + \omega_1$, these denominators will be close to zero only when the intermediate state $|i\rangle$ is nearly on resonance (in which case stepwise two-photon transitions will become important and the preceding calculation of $R^{(2)}_{fg}$ may no longer be valid). We term the transfer of population from state $|f\rangle$ to state $|g\rangle$ as **two-photon stimulated emission.**

We can describe the two-photon stimulated emission process by saying that the field at ω_1 produces a virtual state $|v\rangle$, and that the field at ω_2 stimulates the $|f\rangle \to |v\rangle$ transition. Thus we are led to define a B coefficient for the virtual state $|v\rangle$ such that (remember B is for broadband radiation)

$$B_{fv}\rho(\omega) = R^{(2)}_{fg}$$

where $R^{(2)}_{fg}$ is the rate at which the system goes from $|f\rangle \to |i\rangle$ under the influence of monochromatic radiation at ω_1 and broadband radiation at ω_2.

The easiest way to find B is to note that Eq. (6.18) for ω_R in the presence of monochromatic radiation and Eq. (6.28) for B (from Chapter 6), both contain the term $\pi|\langle b|ez|a\rangle|^2/(c\hbar^2)$. They combine to yield

$$B = \frac{\omega_R^2}{12I} = \frac{2\pi\omega_R^2}{3c\mathcal{E}_\circ^2}$$

Replacing ω_R^2 we get

$$B_{fv} = \frac{\pi}{6c}\left[\frac{\mathcal{E}_1 M^{(1)}_{ig}}{\hbar(\omega_{ig} - \omega_1)}\right]^2 \left[\frac{M^{(2)}_{fi}}{\hbar}\right]^2$$

We could perform a similar calculation in which we regarded state $|v\rangle$ as being produced by ω_2 and calculated B_{vg}, the rate of stimulated emission from this virtual state to the ground state.

2. One stimulated and one spontaneous photon

Pursuing this analogy further, we would expect that, if the final state were populated, it could decay spontaneously to the virtual state, emitting a photon of frequency $\omega_2 = \omega_{fg} - \omega_1$. This process would have an A coefficient for decay to state $|i\rangle$ of

$$A_{f\to i} = 8\pi hc\nu^3\, B = \frac{16\pi^4\nu^3}{3h}\left[\frac{\mathcal{E}_1 M_{ig}^{(1)}}{\hbar(\omega_{ig}-\omega_1)}\right]^2 |\langle f|\vec{P}|i\rangle|^2 \tag{7.5}$$

where ν is the wavenumber measured in cm^{-1}. Note that $\hat{\varepsilon}_2 \cdot \vec{P}$ has been replaced by $\vec{P}$ since we must sum over the polarization states of the spontaneously emitted photon. This is identical to the usual expression for the A coefficient [Eq. (6.29)] if we replace

$$|\langle b|\vec{P}|a\rangle|^2 \qquad \to \qquad |\langle f|\vec{P}|i\rangle|^2 \left[\frac{\mathcal{E}_1 M_{ig}^{(1)}}{2\hbar(\omega_{ig}-\omega_1)}\right]^2$$

In other words, spontaneous emission proceeds to virtual state $|v\rangle$ at the rate it proceeds to the real state $|i\rangle$ times a factor which is the square of the ratio of the Rabi frequency produced by $\mathcal{E}_1$ to the detuning of field 1 from state $|i\rangle$.

The situation discussed above is rather artificial because it requires population in state $|f\rangle$ which could (and would) decay to state $|i\rangle$. [It could be realized in H where 2s metastable states could have their natural two-photon decay rate enhanced by a strong field, or possibly in Ne atoms in He-Ne discharge since appropriate states are known to be populated.] But its utility is mainly to obtain Eq. (7.5) which will be used to relate stimulated and spontaneous Raman processes in the next section.

C. Raman processes

1. General

We have so far been considering a system in which $E_f > E_i > E_g$. Now we consider a system in which E_i is the highest energy and E_f and E_g can lie in either order. This is a situation which can lead to Raman scattering. We consider for the moment a system with energy levels that look as shown in Fig. 7.2 below.

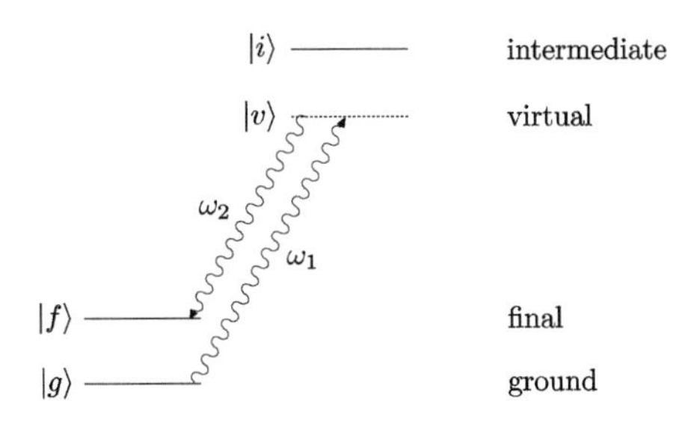

Figure 7.2: Energy level scheme for two-photon transition similar to that in Fig. 7.1 except that the intermediate state is above both the ground and final states.

As the figure shows, the levels are similar to those involved in two-photon absorption except that $E_f < E_i$ so that $E_{fi} < 0$. It is also possible that $E_f < E_g$—this changes nothing fundamental.

The phenomenon involved in Raman process is that radiation at frequency ω_1 is incident on the molecules in state $|g\rangle$. These are virtually excited to state $|v\rangle$ and then they decay to state $|f\rangle$, emitting photons whose frequency ω_2 satisfies

$$\omega_2 = \omega_1 - \omega_{fg}$$

where ω_2 is called the

Stokes component if $\quad \omega_2 < \omega_1$

anti-Stokes component if $\quad \omega_2 > \omega_1$

2. Stimulated Raman scattering

Now we consider the case in which oscillating fields are present at both ω_1 and ω_2 so that the system is driven from $|g\rangle$ to $|f\rangle$ by a simultaneous two-photon process. This is a process in which a photon is absorbed at ω_1 and another is stimulated at ω_2 (so that the beam at ω_2 grows in intensity). The preceding treatment of two-photon absorption is applicable except that

$\omega_{fi} < 0$. This introduces only one charge—in the derivation of Eq. (7.2) counter-rotating terms with denominators $\omega_{ig} + \omega_2$ and $\omega_{fg} - \omega_1 + \omega_2$ were discarded and those with $\omega_{ig} - \omega_2$ and $\omega_{fg} - \omega_1 - \omega_2$ were kept. Now we must keep the former terms, since they can produce resonances when that equation is satisfied. Thus we find instead of Eq. (7.2)

$$a_f^{(2)}(t) = \frac{1}{4\hbar^2} \left[\frac{\mathcal{E}_2 \mathcal{E}_1 M_{fi}^{(2)} M_{ig}^{(1)}}{\omega_{ig} - \omega_1 - i\Gamma_{ig}/2} \frac{e^{i(\omega_{fg} - \omega_1 + \omega_2)t} - 1}{\omega_{fg} - \omega_1 + \omega_2} \right. \\ \left. + \frac{\mathcal{E}_1 \mathcal{E}_2 M_{fi}^{(1)} M_{ig}^{(2)}}{\omega_{ig} + \omega_2 - i\Gamma_{ig}/2} \frac{e^{i(\omega_{fg} - \omega_1 + \omega_2)t} - 1}{\omega_{fg} - \omega_1 + \omega_2} \right]$$

(The terms with $\omega_{fg} - \omega_1 + \omega_1$ or $\omega_{fg} - \omega_2 + \omega_2$ in the denominator produce effects only for $E_f = E_g$; this is called Rayleigh scattering and is neglected here.)

This amplitude may be used to calculate a transition rate for the overall two-photon process. The result is the same as Eq. (7.4) except one must modify the two-photon detuning parameter to

$$\delta_2' = \omega_1 - \omega_2 - \omega_{fg}$$

3. Spontaneous Raman scattering

The first type of Raman process to be predicted or observed involved the absorption of photon at ω_1 and the spontaneous emission of one at ω_2. The rate for this may be found from the results of the preceding section; just as we found the rates for this stimulated processes above we find [from Eq. (7.5)]

$$A_{g\to f}^j = \frac{64\pi^4 \nu_2^3}{3h} \left[\sum_i \left(\frac{\mathcal{E}_1 \langle f|\hat{\varepsilon}_j \cdot \vec{P}|i\rangle M_{ig}^{(1)}}{2\hbar(\omega_{ig} - \omega_1)} + \frac{\mathcal{E}_1 M_{fi}^{(1)} \langle i|\hat{\varepsilon}_j \cdot \vec{P}|g\rangle}{2\hbar(\omega_{if} + \omega_1)} \right) \right]^2 \tag{7.6}$$

This expression has been simplified by noting that

$$\omega_{ig} + \omega_2 = \omega_{ig} + \omega_1 - \omega_{fg} = \omega_{if} + \omega_1$$

Since in Raman scattering there are frequently no near-resonant intermediate states the entire sum over intermediate states has been retained. We have also kept the term with denominator $\omega_{if} + \omega_1$; this term corresponds to emission of a photon at ω_2 and subsequent absorption at ω_1—it can give a large contribution if $|g\rangle$ is an excited state so that some $|i\rangle$ states lie $\approx \hbar\omega_1$ below E_f. It should be noted that we have defined A so that it has a vector component j, which corresponds to the polarization $\hat{\varepsilon}_j$ of the Raman scattered light. The intensity associated with $A_{g\to f}^j$ is

$$I_{g\to f}^j = hc\nu A_{g\to f}^j$$

The quantity inside the square brackets of Eq. (7.6) is a second rank tensor (if $\mathcal{E}_1\hat{\varepsilon}_1$ is brought outside) which is related to the polarizability tensor. This expression agrees with the standard expression, although our derivation is unconventional. The usual derivation is to calculate the polarizability tensor of the atom or molecule and then to find the spontaneous radiation from the radiating dipole

$$\vec{P}(t) = \overleftrightarrow{\alpha}\,\vec{\mathcal{E}}_1 \cos\omega_1 t$$

The possibility of radiating at a frequency different from ω_1 arises from the time-dependent nature of some components of the $\overleftrightarrow{\alpha}$ which appear in the quantum-mechanical calculation.

Frequently Raman scattering is done in molecules, and the final levels $|f\rangle$ are rotational satellites. This introduces the problem of transforming the polarizability tensor calculated in the body frame of the molecule back into the lab frame—a complicated, but standard, procedure in molecular spectroscopy. Detailed discussion of the selection rules and intensities that result is in J. A. Koningstein's book.* Koningstein discusses several variations on regular Raman scattering—two-photon absorption with emission of a single Raman-shifted photon, near-resonant Raman scattering, and stimulated Raman emission and absorption.

4. Raman scattering cross-section

Since the spontaneous Raman process is linear in the applied field, it is possible to define a cross-section for it

$$\begin{aligned}\sigma^R_{gf}(\omega_1) &= \sum_{m_f} \hbar\omega_1 \frac{A_{gf}}{I_1} \\ &= \frac{512\pi^5\nu_1\nu_2^3}{3}\sum_{m_f}\left[\sum_i \left(\frac{\langle f, m_f|\vec{P}|i\rangle\, M^{(1)}_{ig}}{2\hbar(\omega_{ig}-\omega_1)} + \frac{M^{(1)}_{fi}\,\langle i|\vec{P}|g\rangle}{2\hbar(\omega_{if}+\omega_1)}\right)\right]^2\end{aligned}$$

In spite of the very large numerical factor outside the brackets, this is a tiny cross-section if there is no resonance enhancement—typical Raman cross-section for atoms are $\sim 10^{-29}$ cm^2 in the absence of such enhancement. For molecules, the cross-section is typically 10 to 100 times smaller for cases in which $|f\rangle$ and $|g\rangle$ have different vibrational quantum numbers due to imperfect vibrational overlap (e.g. Franck–Condon factors).

* *Introduction to the Theory of the Raman Effect*, D. Reidel Publishing Co. Dordrecht-Holland (1972).

D. Dressed atom for multiphoton processes

1. Two-photon absorption

The idea of this treatment is that E field 1 dresses atomic states $|g\rangle$ and $|i\rangle$, and then E field 2 causes transition to the final state $|f\rangle$.

E field 1 has strength $\mathcal{E}_1$, polarization $\hat{\varepsilon}_1$, and frequency ω_1, so its Rabi frequency is defined in the usual manner as

$$\omega_{R1} = \frac{\mathcal{E}_1}{\hbar} \langle i|\hat{\varepsilon}_1 \cdot e\vec{r}|g\rangle$$

We define $\delta_1 \equiv \omega_1 - \omega_{ig}$; and choose $\delta_1 < 0$ so that dressed atom state $|t^-\rangle \to |g\rangle$ as $\omega_{R1} \to 0$. From Eq. (3.26) of Chapter 3, we know that

$$|t^-\rangle = e^{-i(2\omega_i - \omega'_{R1} + \delta_1)t/2} \left[-e^{i\omega_1 t} \sin\theta \, |g\rangle + \cos\theta \, |i\rangle\right]$$

Approximating the effective Rabi frequency for $|\delta_1| \gg \omega_R$ (and since $\delta_1 < 0$) as

$$\omega'_{R1} \equiv \sqrt{\delta_1^2 + \omega_{R1}^2} \approx -\delta_1 - \frac{\omega_{R1}^2}{2\delta_1}$$

we get the frequency of $|t^-\rangle$ as

$$\begin{aligned}\omega^- &= \frac{2\omega_i - \omega'_{R1} + \delta_1}{2} \\ &\approx \omega_i + \delta_1 + \frac{\omega_{R1}^2}{4\delta_1} \\ &= \omega_i + \omega_1 - \omega_i + \omega_g + \frac{\omega_{R1}^2}{4\delta_1} \\ &= \omega_g + \omega_1 + \frac{\omega_{R1}^2}{4\delta_1}\end{aligned}$$

[The term $\omega_{R1}^2/(4\delta_1)$ is the AC Stark shift of $|g\rangle$, which is negative because $\delta_1 < 0$.]

The cosine factor in $|t^-\rangle$ can be approximated for $|\delta_1| \gg \omega_R$ as

$$\cos\theta = \left[\frac{\omega'_{R1} + \delta_1}{2\omega'_{R1}}\right]^{\frac{1}{2}} \approx \left[\frac{-\delta_1 - \omega_{R1}^2/(2\delta_1) + \delta_1}{-2\delta_1}\right]^{\frac{1}{2}} = \left[\frac{\omega_{R1}^2}{4\delta_1^2}\right]^{\frac{1}{2}} = \frac{\omega_{R1}}{2\delta_1}$$

Thus we get

$$|t^-\rangle \approx \frac{\omega_{R1}}{2\delta_1} e^{-i\left(\omega_g + \omega_1 + \omega_{R1}^2/4\delta_1\right)t} \, |i\rangle \tag{7.7}$$

We now want to find ω_{2R}, the Rabi frequency for the two-photon transition to $|f\rangle$ due to the application of the second laser. E field 2 of the second laser has strength $\mathcal{E}_2$, polarization $\hat{\varepsilon}_2$, and frequency ω_2, so its Rabi frequency is

$$\omega_{R2} = \frac{\mathcal{E}_2}{\hbar} \langle f|\hat{\varepsilon}_2 \cdot e\vec{r}|i\rangle$$

The transition to $|f\rangle$ is from $|t^-\rangle$, and therefore the two-photon transition frequency is given by

$$\omega_2 = \frac{\mathcal{E}_2 e^{-i\omega_2 t}}{\hbar} \langle f|\hat{\varepsilon}_2 \cdot e\vec{r}|t^-\rangle$$

Substituting for $|t^-\rangle$ from Eq. (7.7), we get

$$\begin{aligned}
\omega_2 &= \frac{\omega_{R1}}{2\delta_1} \frac{\mathcal{E}_2}{\hbar} \langle f|\hat{\varepsilon}_2 \cdot e\vec{r}|i\rangle \, e^{+i\omega_f t} e^{-i\omega_2 t} e^{-i\left(\omega_g + \omega_1 + \omega_{R1}^2/4\delta_1\right)t} \\
&= \frac{\omega_{R1}}{2\delta_1} \omega_{R2} \, e^{-i\left[(\omega_1+\omega_2)-(\omega_f-\omega_2)+\omega_{R1}^2/4\delta_1\right]t} \\
&= \omega_{2R} \, e^{-i\delta_2 t}
\end{aligned}$$

with the definitions

$$\omega_{2R} \equiv \frac{\omega_{R1}\omega_{R2}}{2\delta_1} \qquad \text{and} \qquad \delta_2 \equiv (\omega_1 + \omega_2) - (\omega_f - \omega_g) + \frac{\omega_{R1}^2}{4\delta_1}$$

Resonance (when $\delta_2 = 0$) occurs at the frequency

$$\omega_1 + \omega_2 = \omega_f - \omega_g - \frac{(\omega_{R1}/2)^2}{\delta_1} > \omega_f - \omega_g \quad (\text{for } \delta_1 < 0)$$

The (unsaturated) transition rate is

$$R^u_{2,g\to f} = \frac{2\Gamma_{2gf}\,(\omega_{2R}/2)^2}{\delta_2^2 + \Gamma_{2gf}^2}$$

This treatment augments the usual two-photon treatment by inclusion of the AC Stark shift for state $|g\rangle$, but it misses the Stark shift of $|f\rangle$. This can be comparable magnitude.

2. Raman scattering

Our aim here is to find the spontaneous decay rate for a Raman process using the dressed atom picture. For this we want to find the dressed atom state which correlates with atomic state $|g\rangle$ as $\omega_R \to 0$ (for $\delta < 0$). We denote the final state as $|g'\rangle$ to emphasize the fact that it has the same parity as state $|g\rangle$. From the discussion in the previous section, the appropriate dressed atom state is $|t^-\rangle$. Spontaneous decay from this state is called **spontaneous Raman scattering** and occurs at a rate

$$\Gamma^R_{gg'} = \frac{4k'^3}{3\hbar}|\langle g'|\vec{P}|t^-\rangle|^2$$

Using the expression for $|t^-\rangle$, and noting that $\langle g'|\vec{P}|g\rangle = 0$ (same parity), we have

$$\begin{aligned}
\Gamma^R_{gg'} &= \frac{4k'^3}{3\hbar}\left|\langle g'|\vec{P}\cos\theta|i\rangle\ e^{i\left[\omega_{g'}-\left(2\omega_i+\delta-\omega'_R\right)/2\right]t}\right|^2 \\
&= \frac{4k'^3}{3\hbar}\left[\frac{\omega'_R+\delta}{2\omega'_R}\right]\left|\langle g'|\vec{P}\cos\theta|i\rangle\ e^{i\left[\omega_{g'}-\left(2\omega_i+\delta-\omega'_R\right)/2\right]t}\right|^2
\end{aligned}$$

The emission is at the frequency in the exponent in the above expression

$$\begin{aligned}
\omega_{\text{Ram}} &= -\left[\omega_{g'} - \left(\omega_i + \frac{\delta}{2} - \frac{\omega'_R}{2}\right)\right] \\
&= -\omega_{g'} + \omega_i + \delta - \left[\frac{\omega'_R+\delta}{2}\right] \\
&= \left[\omega - (\omega_{g'} - \omega_g)\right] - \left[\frac{\omega'_R+\delta}{2}\right]
\end{aligned}$$

This expression for ω_{Ram} differs from conventional expositions on Raman scattering because it contains the AC Stark shift of the state $|g\rangle$ due to the laser at ω (but it ignores the equally important shift of state $|g'\rangle$).

We have assumed $\delta < 0$, so that when $|\delta| \gg \omega_R$ we can expand

$$\frac{\omega'_R+\delta}{2} \approx \frac{|\delta|}{2}\left[1 + \frac{\omega_R^2}{2\delta^2} - 1\right] = \frac{\omega_R^2}{4\delta}$$

Thus the transition rate in this limit becomes

$$\Gamma^R_{gg'} \approx \frac{k'^3}{3\hbar}\frac{\omega_R^2}{(\omega_{ig}-\omega)^2}|\langle g'|\vec{P}|i\rangle|^2$$

This dressed atom approach assumes that there is only one intermediate state $|i\rangle$, and neglects the counter-rotating term. A more complete expression is

$$\Gamma^{R}_{gg'} = \frac{k'^3}{3\hbar} \sum_{m_{g'}} \left[\sum_{i} \left(\frac{1}{\omega_{ig} - \omega} + \frac{1}{\omega_{ig} + \omega} \right) \frac{\langle g'|\vec{P}|i\rangle \langle i|\mathcal{E}_1 \hat{\varepsilon}_1 \cdot \vec{P}|g\rangle}{\hbar} \right]^2$$

It is necessary to sum over the final $m_{g'}$ states if $j_{g'} \neq 0$. Note that the counter-rotating term gives a comparable contribution when $\omega \ll \omega_{ig}$ (as is the case for visible and invisible gases), and strongly affects the $\omega \to 0$ limit.

E. Problems

1. Magnitude of Raman scattering cross-section

(a) If $|g\rangle$ is the ground state from which the atom (molecule) is excited, $|i\rangle$ is the intermediate state, and $|f\rangle$ is the final state, find the spontaneous Raman cross-section in the presence of an electric field of the form

$$\vec{E}(t) = \mathcal{E}\hat{\varepsilon}\cos\omega t$$

exciting the $|g\rangle \to |i\rangle$ transition. Assume that all the matrix elements are equal

$$\langle i|\vec{P}|g\rangle = \langle i|\vec{P}|f\rangle = dea_\circ$$

and that $\omega_{ig} \gg \omega$, the frequency of the exciting light.

(b) Estimate the cross-section σ_{gf}^R for light at 5000 Å being Raman shifted by 300 cm^{-1} from a molecule whose first electronic excitation energy is ≈ 1 Ry. Take $d = 1$.

Solution

(a) The spontaneous Raman scattering cross-section is

$$\sigma_{gf}^R(\omega) = \frac{\text{Scattering rate}}{\text{Incident photon flux}} = \frac{A_{gf}}{I/\hbar\omega} = \hbar\omega A_{gf}\frac{8\pi}{c\mathcal{E}^2}$$

From the chapter, we have (for one intermediate state)

$$A_{gf} = \frac{64\pi^4\nu_2^3}{3h}\left[\frac{\mathcal{E}\langle f|\vec{P}|i\rangle M_{ig}}{2\hbar(\omega_{ig}-\omega)} + \frac{\mathcal{E}M_{fi}\langle i|\vec{P}|g\rangle}{2\hbar(\omega_{if}+\omega)}\right]^2$$

where ν_2 is the wavenumber $(= k_2/2\pi)$ of the spontaneously emitted light, and M's are matrix elements of $\hat{\varepsilon}\cdot\vec{P}$ between the respective states. Assuming that $\omega_{ig} \gg \omega$, and hence that ω_{if} is also $\gg \omega$ for a small Raman shift, this gives

$$\sigma_{gf}^R(\omega) = \frac{512\pi^5\nu\nu_2^3}{3}\left[\frac{\langle f|\vec{P}|i\rangle M_{ig}}{2\hbar\omega_{ig}} + \frac{M_{fi}\langle i|\vec{P}|g\rangle}{2\hbar\omega_{if}}\right]^2$$

Taking all matrix elements to be $dea_\circ$, we get

$$\begin{aligned}\sigma_{gf}^R(\omega) &= \frac{512\pi^5\nu\nu_2^3}{3}\left[\frac{d^2e^2a_\circ^2}{2\hbar\omega_{ig}} + \frac{d^2e^2a_\circ^2}{2\hbar\omega_{if}}\right]^2\\ &= \frac{512\pi^7\nu\nu_2^3d^4e^4a_\circ^4}{3h^2}\left[\frac{\omega_{if}+\omega_{ig}}{\omega_{ig}\omega_{if}}\right]^2\end{aligned}$$

(b) The exciting light wavelength is 5000 Å; therefore the corresponding wavenumber is $\nu = 20\,000 \text{ cm}^{-1}$.

For an excited state energy of ≈ 1 Ry, the energy of the intermediate state $|1\rangle$ is 1 Ry. Therefore

$$\hbar\omega_{ig} = 13.6 \text{ eV} \quad \Longrightarrow \quad \omega_{ig} = 20.6334 \times 10^{15} \text{ rad/s}$$

The Raman shift is 300 cm^{-1}; hence

$$\omega_{if} = \omega_{ig} - 2\pi c \times 300 \text{ cm}^{-1} \quad \Longrightarrow \quad \omega_{if} = 20.6278 \times 10^{15} \text{ rad/s}$$

and

$$\nu_2 = 20000 - 300 = 19700 \text{ cm}^{-1}$$

Taking $d = 1$, this gives the scattering cross-section as

$$\sigma_{gf}^{R} = 7.03 \times 10^{-28} \text{ cm}^2$$

Chapter 8

Coherence

COHERENCE arises when amplitudes add with a definite relative phase and one observes a physical quantity proportional to the square of the total amplitude. A familiar example is the addition of E fields to produce an interference pattern in classical optics. In quantum optics a coherent (linear) superposition of states with definite numbers of quanta is required to produce a classical oscillating E field in one mode of the quantized field. Coherence is also responsible for a number of interesting interference effects present in scattering of atoms and molecules. In this case it is the quantum mechanical amplitudes associated with different "trajectories" that scatter to the same angle which add to give interference in the probability distribution of the scattered particles.

In atomic physics, the term *coherence* is used for at least two distinctly different physical phenomena—(i) those that occur in a single atom when two states have a definite relative phase (which varies with time if the states are not degenerate), and (ii) those that occur when there is more than one atom radiating so that interference effects and coupling of the atoms become important. Radiative coupling dominates when the atoms are localized in a spatial region smaller than a wavelength of the radiation, whereas interference and concomitant directional specificity become important for larger samples.

In this chapter, we will study the effects of coherence in single atoms followed by effects in ensembles—both localized and extended. In addition, we will see different facets of **coherent control**—phenomena where the properties of atoms can be modified using control lasers on auxiliary transitions, made possible by the fact that atoms are multilevel systems. To be consistent with terminology in this field, we will use the terms "states" and "levels" interchangeably—thus atoms are two-level, three-level, or multilevel.

A. Coherence in single atoms

In this section, we treat coherence in single atoms. Here the coherence results from the fact that the atom has been placed in a superposition of its eigenstates so that the expectation values of certain operators (e.g. the x component of polarization P_x) exhibit interference structure. Such structure may manifest itself as a time dependence in the expectation value of these operators, or as difference in the steady-state processes (e.g. anisotropy of radiation). We shall first consider quantum beats—a phenomenon in which the radiation rate of an atom oscillates in time; and then level crossing—a situation in which interference between two nearly degenerate states can modify the polarization and intensity of fluorescent light from the system.

1. Quantum beats

Let us now consider what happens when two neighboring states of an atomic systems are excited coherently by a short pulse. "Short" pulse in this context means shorter than $\pi/\Delta E$ of the two states, and since we want to discuss the case when ΔE is larger than the natural linewidth, "short" also means much less than the lifetime of the excited states (both presumed to have lifetimes $\tau = 1/\Gamma$). What happens will be monitored by the temporal behavior of the fluorescent radiation of polarization $\hat{\varepsilon}_2$, given that the pulse arrived at $t = 0$ with polarization $\hat{\varepsilon}_1$.

For a very short pulse $\Delta t \ll 1/(\omega_{ig} - \omega_1)$ the system has a state vector at time t (using first-order perturbation theory, the dipole interaction, the rotating wave approximation, and limiting E_1 appropriately)

$$\begin{aligned} |t\rangle^{(1)} &= \frac{\Delta t}{2\hbar} E_1 \sum_i M_{ig}^{(1)} |i\rangle \\ &\propto E_1 \sum_i M_{ig}^{(1)} e^{-i\omega_i t} |i\rangle_\circ \end{aligned}$$

where $\omega_i = E_i/\hbar - i\Gamma/2$ and $|i\rangle_\circ$ is $|i\rangle$ at $t = 0$. The rate of radiation from $|t\rangle^{(1)}$ to a state $|g'\rangle$ with the same parity as the ground state is proportional to the square of the matrix element

$$\begin{aligned} R(t) &\propto \left| \langle g' | \hat{\varepsilon}_2 \cdot \vec{P} | t \rangle^{(1)} \right|^2 \\ &\propto E_1^2 \left| \sum_i M_{g'i}^{(2)} M_{ig}^{(1)} e^{-i\omega_i t} \right|^2 \\ &\propto I_1 \left| M_{g'a}^{(2)} M_{ag}^{(1)} e^{-i\omega_a t} + M_{g'b}^{(2)} M_{bg}^{(1)} e^{-i\omega_b t} \right|^2 \\ &\equiv I_1 \left| A_{g'g}^{21} e^{-i\omega_a t} + B_{g'g}^{21} e^{-i\omega_b t} \right|^2 \end{aligned}$$

where we have assumed that there are only two intermediate states, $|a\rangle$ and $|b\rangle$.

Expanding this out gives

$$\begin{aligned} R(t) \propto\ & I_1 e^{-\Gamma t} \left[\left| M_{g'a}^{(2)} M_{ag}^{(1)} \right|^2 + \left| M_{g'b}^{(2)} M_{bg}^{(1)} \right|^2 \right] && \leftarrow \text{normal decay} \\ & + I_1 e^{-\Gamma t} \, 2\,\mathrm{Re} \left\{ M_{g'a}^{(2)^*} M_{ag}^{(1)^*} M_{g'b}^{(2)} M_{bg}^{(1)} e^{-i(\omega_b - \omega_a)t} \right\} && \leftarrow \text{interference} \end{aligned}$$

Defining $\Delta\omega$ as $\omega_b - \omega_a$, this can be written as

$$\begin{aligned} R(t) \propto I_1 e^{-\Gamma t} & \left[\left| M^{(2)}_{g'a} M^{(1)}_{ag} \right|^2 + \left| M^{(2)}_{g'b} M^{(1)}_{bg} \right|^2 \right] \\ & + I_1 e^{-\Gamma t} \left[\mathrm{Re} \left\{ M^{(2)^*}_{g'a} M^{(1)^*}_{ag} M^{(2)}_{g'b} M^{(1)}_{bg} \right\} \cos(\Delta\omega\, t) \right. \\ & \qquad \left. - \mathrm{Im} \left\{ M^{(2)^*}_{ag'} M^{(1)^*}_{ga} M^{(2)}_{g'b} M^{(1)}_{bg} \right\} \sin(\Delta\omega\, t) \right] \end{aligned} \tag{8.1}$$

The point is that the observed fluorescence does not decay smoothly—there are oscillations (beats) in the decay whose frequency is $\Delta\omega$. They can be large enough to cause zeros in $R(t)$.

Quantum beats in the decay fluorescence permit measurement of the energy splitting $E_b - E_a$ even though (in fact because) the spectrum of the exciting light is too broad to resolve the states $|a\rangle$ and $|b\rangle$. It is thus well suited to the measurement of closely spaced levels (in fact widely spaced levels give trouble because the excitation and fluorescence electronics must be very fast). Another advantage is that there is no first-order Doppler broadening on ω (only on $\Delta\omega$). Furthermore, the method offers the possibility of making subnatural linewidth measurements of $\Delta\omega$ by following the oscillations for several lifetimes.

It should be clear from the above derivation that the exact nature of the excitation source is immaterial as long as it is short. Thus collisional excitation should produce quantum beats if it can be pulsed (electrons, for instance). The best examples of this (in fact the best examples of quantum beats, period) are obtained using the beam foil techniques where fast-moving atoms are excited by collisions with a thin foil and the quantum beats are observed as a function of distance downstream from the foil. The distance gives a good measurement of t because the velocity distribution can be very narrow even downstream of the foil. A further advantage of beam foil excitation is that the selection rules are not so restrictive as radiative excitation (particularly if the foil is tilted with respect to the beam), allowing more states to be excited coherently.

It should be stressed that the coherence that produces quantum beats is only in the excited states $|a\rangle$ and $|b\rangle$. There is no interference between different radiators, and it is not necessary to observe fluorescence from the system back into its original level. Thus it might be possible to excite $|a\rangle$ and $|b\rangle$ using quadrupole radiation and observe beats from a spontaneous dipole transition. In fact, it is not necessary to observe the beats using fluorescence; quantum beats have been observed in absorption using a second short pulse to excite $|a\rangle$ and $|b\rangle$ to a higher level whose population was monitored.

2. Level crossing

It might seem at first that there would be no interference in the fluorescence of states $|a\rangle$ and $|b\rangle$ in the preceding situation if the exciting source remained on continuously, even though it had a wide enough spectrum to excite both states. This is not the case, however, when $\Delta\omega$ becomes comparable to Γ because then the excited atoms decay before they have a chance to get out of phase. Thus coherence effects may be observable in situations where $\Delta\omega$ becomes small, for example near a value of magnetic field where the energies E_a and E_b become degenerate. This situation is called a level crossing.

To determine the steady-state fluorescence observed near a level crossing, we simply integrate $R(t)$ out to $t = \infty$. [This produces the same result as adding a new randomly distributed variable $t_\circ$ and considering the integral of $R(t - t_\circ)$ back to $t_\circ = -\infty$.] We keep only the oscillating terms in Eq. (8.1). Since the signal is independent of the time of observation—remember the source is a continuous light source—we omit the t as an argument of the fluorescence rate R

$$R^{(2)} \propto I_1 \left[\int_0^\infty e^{-\Gamma t}\, \mathrm{Re}\left\{ M_{g'a}^{(2)^*} M_{ag}^{(1)^*} M_{g'b}^{(2)} M_{bg}^{(1)} \right\} \cos(\Delta\omega\, t)\, dt \right. \\ \left. + \int_0^\infty e^{-\Gamma t}\, \mathrm{Im}\left\{ M_{g'a}^{(2)^*} M_{ag}^{(1)^*} M_{g'b}^{(2)} M_{bg}^{(1)} \right\} \sin(\Delta\omega\, t)\, dt \right]$$

which yields

$$R^{(2)} \propto I_1\, \mathrm{Re}\left\{ M_{g'a}^{(2)^*} M_{ag}^{(1)^*} M_{g'b}^{(2)} M_{bg}^{(1)} \right\} \frac{\Gamma}{\Gamma^2 + (\Delta\omega)^2} \\ + I_1\, \mathrm{Im}\left\{ M_{g'a}^{(2)^*} M_{ag}^{(1)^*} M_{g'b}^{(2)} M_{bg}^{(1)} \right\} \frac{\Delta\omega}{\Gamma^2 + (\Delta\omega)^2}$$

Thus there are two components—a Lorentzian with FWHM of 2Γ, and a dispersive curve. In general the matrix element is neither purely real nor purely imaginary, and as a result the experimentally measured curve will have to be fit with a sum of the two terms in the above equation.

It is worth noting that the FWHM of the curve 2Γ, in contrast to a FWHM of Γ which one normally expects from resonance experiments on systems with an excited state lifetime $1/\Gamma$. This is because coherence in a normal resonance experiment is a superposition of a ground and an excited state whose amplitude decays like $e^{-\Gamma t/2}$, whereas coherence in a level crossing experiment is due to superposition of two excited states and its amplitude decays as $e^{-\Gamma t}$.

The first level crossing experiment was performed by Hanle in 1925, who discovered that the resonance fluorescent light of a system had different properties around zero magnetic field than at other values of the field. The first non-zero field level crossing experiment was not performed until 34 years later by Colegrove et al., who monitored the absorption rather than the fluorescence. This experiment improved knowledge of the fine structure of helium.

3. Double resonance

Looking over our discussion of level crossing, we note that it is basically a resonance technique without an RF resonance field. The lineshape is Lorentzian and occurs at a certain value of magnetic field, just as for an RF-induced resonance if the RF frequency is fixed while the field is tuned (as is generally done in NMR spectroscopy, for example). The method of detection involves the pattern of fluorescence radiation rather than the RF power absorbed, which makes the method much more sensitive (you get an optical photon rather than an RF photon for each system "resonated").

It is possible to use level crossing techniques in conjugation with RF fields—the crossing signal produced by the RF occurs where the excited states are spaced apart by an energy $\hbar\omega_{\mathrm{RF}}$. The technique of using changes in the fluorescent light to detect when two (not necessarily excited) states are coupled together by RF is called double resonance. When applied to the ground state it is very similar to optical pumping in which the light produces and monitors the populations of the ground state. Both optical pumping and level crossing offer (in addition to the great sensitivity mentioned above) a way to "tag" levels of a complicated system, simplifying the spectrum obtained when sweeping the other frequency.

B. Coherence in localized ensembles

In this section, we discuss effects which arise in localized ensembles because the coupling of the radiation to different members of the ensemble is sensitive to their relative phase. **Localized** means closer together than a wavelength, and this implies that retardation effects are unimportant so that the radiation will not be strongly directional (but may, for example, have the distribution characteristic of dipole or quadrupole radiation).

1. Superradiance

(i) Qualitative discussion

The previous section of this chapter dealt with coherence produced in single atomic systems and involved superpositions of eigenstates with a definite relative phase. Although this relative phase must have a non-zero expectation value for the ensemble of systems which are generally required to get a detectable signal, there is no necessity to have a dense system in order to observe the coherent effects. Now we consider systems where several identical radiators are close to one another, and discuss the effects of this neighborliness on the radiation of the system. Even in the absence of external driving fields, spontaneously developed coherence of the radiators can result in a marked departure of the radiation from a simple sum of intensities of the spontaneous radiation patterns of the individual systems—this is called **superradiance** or superfluorescence.

Before discussing a system of specifically quantized radiators in detail, consider a hypothetical pulsed type NMR experiment on a sample containing N spin 1/2 systems. Imagine that a $\pi/2$ pulse of radiation is applied to the system, which initially had all its spins in the lower energy state (say spins along $-z$). This creates an oscillating magnetic moment in the sample

$$M \sim N\mu_B \operatorname{Re}\left\{\hat{\varepsilon}_+ e^{-i\omega_\circ t}\right\} e^{-t/T_{2N}}$$

where T_{2N} is the inverse of the dephasing rate Γ_{2N} for the N-spin system (each with T_2 type dephasing as discussed in Chapter 5, "Resonance").

These spins constitute a coherent ensemble of radiators confined within a region much smaller than a wavelength. They induce a voltage in the pick-up coil of

$$V \sim \frac{dM}{dt} \sim N\mu_B\omega_\circ \operatorname{Re}\left\{\hat{\varepsilon}_+ e^{-i\omega_\circ t}\right\} e^{-t/T_{2N}}$$

where we have assumed that $\omega_\circ \gg 1/T_{2N}$.

Thus the power collected by the detector (which is roughly equal to V^2 divided by its impedance) is proportional to N^2. Nothing seems strange about this—in NMR one realizes that the **voltage** is proportional to the number of spins in the sample, whereas in a fluorescence experiment one "knows" that the **power** is proportional to the number of atoms in the sample. One hardly thinks it unnatural that the spins in the NMR experiment radiate coherently in response to the external stimulation pulse at $t = 0$.

Now consider what happens to the energy in the spin system. Pretend that T_2 arises solely from spontaneous radiative process, so that the total energy radiated must be

$$E_{\text{rad}} = \int_0^\infty R(t)dt = N\hbar\omega_\circ/2 \qquad \text{(proportional to N)}$$

(since $\langle m_z \rangle = 0$ after a $\pi/2$ pulse, the average spin energy is $\hbar\omega_\circ/2$).

On the other hand, it is clear from the above discussion that

$$R_N = N^2 R_1 \qquad \text{(proportional to } N^2\text{)}$$

where R_N is the power radiated by a system with N spins.

It is impossible to satisfy the above two equations simultaneously unless the lifetime of the spontaneous radiation also depends on N. Thus if

$$T_{2N} = T_{21}/N \qquad \text{and} \qquad R_N(t) = N^2 I_\circ e^{-t/T_{2N}}$$

then the total energy radiated will vary as N while the initial rate varies as N^2. This behavior is the hallmark of cooperative radiative processes (although N is often less than the total number of available radiators).

(ii) For 2 two-level systems

Imagine that N identical two-level systems are contained in a region of space much smaller than λ, the wavelength of radiation which they emit. This restriction means that the phase of the field is constant everywhere in the sample so we do not have to worry about retarded times, etc. The systems are not close enough to perturb each other—the only manifestation of the symmetry restrictions on the overall wavefunction arises in the interaction with the radiation field.

Such a situation was considered by Dicke in a seminal paper in 1954.* To dramatize the effects of coherence, he first considered the case of 2 spin 1/2 particles (neutrons) in a magnetic field. One neutron in its higher energy state will decay spontaneously (type M1) with time constant T_{21}, if it is placed in a small box alone (the box is for localizing the particle only and does not interfere with the radiation). If a second neutron in its lower

*R. H. Dicke, "Coherence in spontaneous radiation processes," *Phys. Rev.* **93**, 99–110 (1954).

energy state is already in the box, then when the excited state neutron is added, the system is in an equal superposition of singlet $|s = 0, m_s = 0\rangle$ and triplet $|s = 1, m_s = 0\rangle$ states. Hence there is a probability of 1/2 that the two neutrons are in a subradiant state (the singlet) and a probability of 1/2 that they are in superradiant state (the triplet). The triplet component decays to the ground state $|s = 1, m_s = -1\rangle$ twice as fast as the lone neutron system, but the singlet component cannot decay to the triplet ground state (via M1) and remains there forever. Hence the radiation rate at $t = 0$ is not altered by the presence of the extra spin down neutron. However, the total energy is only 1/2 that for the lone neutron, so the radiation rate must decay twice as fast as for a single neutron. This discussion may be applied to any type of two-level systems (and Dicke gives such a semiclassical treatment).

(iii) For N two-level systems

Dicke then gives a general discussion of the coherent radiative behavior of a system of N two-level systems. We shall consider these systems to be spin 1/2 systems in a magnetic field* since then his cooperation operators become familiar angular momentum operators

$$\begin{aligned} \text{Total angular momentum:} \quad & \vec{R} = \sum \vec{S}_i \\ \text{Eigenvalue of } \vec{R}^2\text{:} \quad & r(r+1) \le \frac{N}{2}\left(\frac{N}{2}+1\right) \\ z \text{ component of } \vec{R}\text{:} \quad & m = \sum m_i = \frac{1}{2}(n_+ - n_-) \le r \end{aligned} \tag{8.2}$$

where n^+ is the number of up spins, and n^- is the number of down spins.

Using this formalism Dicke shows that the dipole matrix element governing interaction with the field (and raising m from $m-1$ to m) is proportional to $\sqrt{(r+m)(r-m+1)}$ so that the intensity is

$$I = I_1(r+m)(r-m+1)$$

where I_1 is the rate for a single system.

This is largest when r is big (e.g. the spins all line up) and m is small (i.e. the total spin precesses in the xy plane). In this case $I \propto N^2 I_1$ in accord with our earlier qualitative discussion. When r is big ($r = N/2$) and m is big ($m = r$) the rate is $I = NI_\circ$, i.e. just what you expect from N independent radiators. When r is small (it can be 1/2 if N is odd, 0 if N is even), then $I \approx I_1$ if N is odd and 0 if N is even—the system is subradiant (radiates less rapidly than N independent systems). Dicke also points out that the rate of stimulated emission/absorption is not enhanced by N^2, but only by N, even in the state with $r = N/2$.

*Remember any two-level system is analogous to a spin 1/2 system in a magnetic field—from the Feynman–Vernon–Hellwarth theorem discussed in Chapter 5, "Resonance."

Dicke suggests two ways to produce a superradiant system with $r \approx N/2$ and $m \approx 0$. One is to put the system in its ground state (say by cooling it so $k_BT \ll E_+ - E_-$) whereupon it is in the state $|r = N/2, m = -r\rangle$, and then to give it a $\pi/2$ pulse, raising m to ≈ 0. The second way is to put the system in a state $|r = N/2, m = +r\rangle$ (this corresponds to a state with every system in its upper level and might be achieved by optical pumping) and then **wait**. At first the system will decay with an intensity $I = NI_1$, i.e. independently. The decay will be via $\Delta m = -1, \Delta r = 0$ transitions, however, so that the system will radiate faster and faster until $m \approx 0$. At this point the intensity will be (for a short time) a factor $\sim N/2$ larger—if N is large this can be dramatic.

In conclusion we stress the simple physical picture involved in superradiance and its relationship to the concept of coherence, illustrating the discussion by considering N spins. Remember that when a single spin is not purely up or down, it has a non-zero expectation value for its x or y spin projection, which is related to the relative phase of the coefficients a and b of the spin up and down states—this can be seen by noting that the S_x and S_y spin matrices have only off-diagonal elements, implying that $\langle S_x \rangle$ involves terms like a^*b or b^*a (i.e. the off-diagonal elements of the density matrix). In a localized ensemble of such spins the radiation field depends on the total S_x and S_y which involve the off-diagonal matrix elements of the (ensemble averaged) density matrix. If the spins are in phase—coherent—then the radiation rate varies as N^2. If they are out of phase there is a reduced total S_x and S_y and the radiation rate is smaller—in fact it can vanish if the spins are suitably arranged. Large S_x and S_y total moments are associated with values of $s \approx N/2$ and $m \approx 0$, and can be produced from states with large s and $m \approx \pm s$ by $\pi/2$ pulses of radiation. Alternatively, they will evolve spontaneously if the system has large s and m corresponding to the higher energy level of the spins. The central idea of superradiance is that spontaneous radiation, which is a spontaneous (i.e. self-induced) and random process for a single atom, can be a coherent process for a localized ensemble of atoms.

2. Spin echoes

In the preceding section we discussed the radiation from N identical systems. In real life the system, while truly identical, may have slightly different resonant frequencies due to physical effects such as local field inhomogeneities or Doppler broadening. Such broadening is referred to as **inhomogeneous** (which we will see more about in Chapter 9, "Lineshapes") because it is not the same for all members of the ensemble—it will cause a coherently radiating ensemble to become incoherent. The reduction of $\langle S_x \rangle$ and $\langle S_y \rangle$ due to this is reflected in a small value of T_2, the phenomenological decay time of the off-diagonal elements of the density matrix. Echo phenomena is general, and spin echoes in particular, arise when the inhomogeneous broadening results in a **rephasing** of the dephased radiators. Obviously this situation will not occur by itself—in general a coherent oscillating moment is produced (e.g. with a $\pi/2$ pulse), which then decays due to rapid inhomogeneous processes. A second pulse (or pulses) is then applied, which reverses the relative phase of the dephased radiators so that the inhomogeneous broadening brings the system back into coherence at a later time.

The easiest spin echo to understand is the one produced in a two-state system whose population vector $\vec{r}$ is initially $\vec{r} = \hat{z}$ (the system is in the lower state). At $t = 0$ this system is subject to a $\pi/2$ pulse of resonance radiation so that $\vec{r}$ (which we shall hereafter call a "spin") is shifted into the xy plane where it decays at a rate $\Gamma_2 = 1/T_2$ due to dephasing of the spins of the individual systems. At $t = T$, a π pulse is applied to the system. This reverses the relative phases of the individual spins and as a result they will rephase (or be in phase again) at $t = 2T$.

The crucial step in the echo process is the π pulse. Say that a particular spin i has a resonance frequency $\omega_i = \omega_\circ + \delta_i$ (δ_i is small; the variation in δ_i is roughly the inverse of the coherence decay time T_2). If the $\pi/2$ pulse is phased so that the angle in the xy plane is $\phi = 0$ at $t = 0$, then the ith spin will be at $\phi_i(T_-) = \omega_i T$ just before the π pulse is applied. Assume that this pulse has its $\vec{\omega}$ vector along the arbitrary angle ϕ_T so that it does not change the direction of a spin with phase $\phi_i(T_-) = \phi_T$. If $\phi_i \neq \phi_T$ then the spin will precess 180° about $\vec{\omega}$ in the rotating frame and will wind up at an angle

$$\phi_i(T_+) = \phi_T - [\phi_i(T_-) - \phi_T] = 2\phi_T - \omega_i T$$

For $t > T$ its phase will be (since it continues to precess at ω_i)

$$\begin{aligned}\phi_i(t > T) &= \phi_i(T_+) + \omega_i(t - T) \\ &= 2\phi_T + \omega_i(t - 2T)\end{aligned}$$

It is clear that at $t = 2T$, the phase $\phi_i = 2\phi_T$ independent of the value of ω_i! All the spins will be in phase again at this time and the system will radiate coherently.

The pulse sequence, radiated field, and the spins are shown in Fig. 8.1 below.

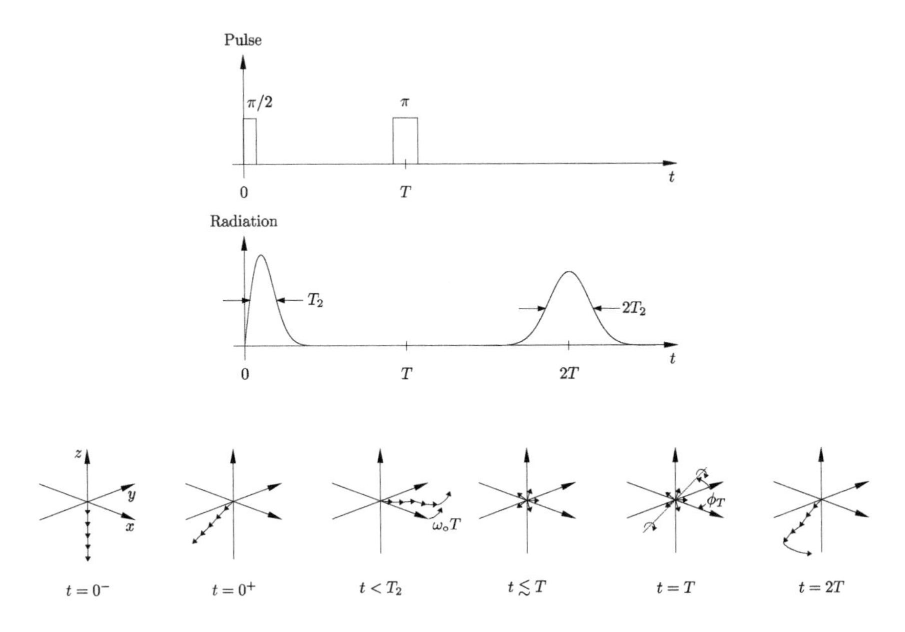

Figure 8.1: Spin echo scheme showing the pulse sequence, radiated field, and the spins at different times.

As seen, the amplitude of the echo signal is reduced because not all of the T_2-type damping is inhomogeneous—some is homogeneous (e.g. spontaneous decay) and some comes from energy loss (T_1) processes. In addition, the figure shows that the width of the pulse affects the time of the echo slightly by taking the $t = 0$ point inside the pulse.

While echos are very dramatic coherent phenomena, the diehard spectroscopist will note that they are not very useful as a spectroscopic tool since all information about the central frequency $\omega_\circ$ is lost. Even if the phase of the echo is somehow measured, it is simply $2\phi_T$ —independent of the frequency.

C. Coherence in extended ensembles

When an ensemble of atomic systems whose dimensions are large compared with a wavelength interacts with radiation, the phase of the radiation (whether emitted or absorbed) varies throughout the ensemble, and the phase of the polarization of the system varies from place to place in consequence. If the radiators interfere to produce radiation propagating in a given direction, then the relative phase of the radiators must reflect the phase of this traveling wave, and the radiators will not be phased correctly to produce radiation traveling in some direction. When the polarizations of the radiators are phased so as to optimally produce a wave, they are said to be **phase matched**. In this section, we shall derive the expression for phase matching in multi-photon processes, discuss the effects of imperfect phase matching, and then give some examples of how it may be achieved in practice.

1. Phase matching

Let us consider an ensemble of radiators that absorb radiation of wave vector $\vec{k}_1$ and emit radiation with wave $\vec{k}_2$. The direction $\hat{k}_1$ is determined by the incident radiation, but the direction $\hat{k}_2$ must be regarded as a variable. In general the length of $\vec{k}_2$ is fixed by the relationship

$$k_2 = n(\omega_2)\, \omega_2/c \tag{8.3}$$

where $n(\omega_2)$ is the index of refraction of the medium at the frequency of the emitted radiation ω_2, which in turn will be determined by the equation of energy conservation

$$\omega_1 \pm \omega_2 \pm \cdots \pm \omega_{n+1} = 0 \tag{8.4}$$

or else by the level scheme of the atoms.

We now seek to calculate the phase of the radiation that arrives at a planar detector perpendicular to $\hat{k}_2$ relative to the phase of the initial wave at the source. (A planar detector might consist of a lens that focuses plane waves traveling along $\hat{k}_2$ onto a small detector.) The detected intensity will, of course, be maximum when the phase ϕ_i of radiation arriving at the detector from an atom at $\vec{r}_i$ is independent of $\vec{r}_i$ so that the effect of all the atoms add coherently. We shall make our calculation of ϕ_i independent of this consideration.

The situation under discussion is shown in Fig. 8.2 below.

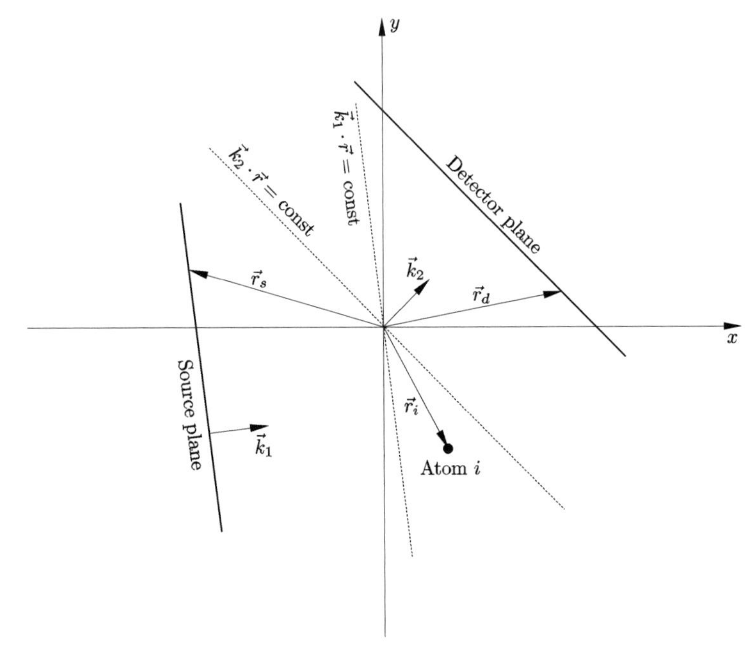

Figure 8.2: Scheme for phase matching, showing source, detector, and atom i.

In general, each traveling wave has the form $e^{i(\vec{k}\cdot\vec{r}-\omega t)}$, with k and ω related as in Eq. (8.3). We now calculate the phase of the radiation received at the detector plane. It is a sum of two parts

$$\begin{array}{ll} \text{From source to atom } i: & \phi_{\text{sa}} = \vec{k}_1 \cdot (\vec{r}_i - \vec{r}_s) \\ \text{From atom to detector:} & \phi_{\text{ad}} = \vec{k}_2 \cdot (\vec{r}_d - \vec{r}_i) \end{array} \tag{8.5}$$

Hence the total phase for the ith atom is

$$\phi_i = \vec{r}_i \cdot \left(\vec{k}_1 - \vec{k}_2\right)$$

(we dropped the terms depending on $\vec{r}_s$ and $\vec{r}_d$ because they are the same for all atoms in the ensemble and it is the relative phase that is most interesting).

For a multi-photon process the phase becomes

$$\phi_i = \vec{r}_i \cdot \left(\vec{k}_1 \pm \vec{k}_2 \pm \cdots \pm \vec{k}_{n+1}\right) \equiv \vec{r}_i \cdot \Delta\vec{k}$$

where the signs are the same as in Eq. (8.4), and $\Delta\vec{k}$ is defined conventionally as

$$\Delta\vec{k} = \vec{k}_1 \pm \vec{k}_2 \pm \cdots \pm \vec{k}_{n+1} \tag{8.6}$$

Perfect phase matching occurs when $\Delta\vec{k} = 0$, in which case all ϕ_i's are the same. Then the radiators interfere constructively, and the detected intensity will be maximum in the direction $\hat{\vec{k}}_{n+1}$ for which $\Delta\vec{k} = 0$.

2. Intensity for finite mismatch

Sometimes it is not possible to achieve perfect phase matching. For instance, if $n(\omega)$ is not constant, then the fact that Eq. (8.4) is satisfied for the ω_i means that $\Delta\vec{k}$ in Eq. (8.6) will not in general be zero for a collinear arrangement of the beams. If an attempt is made to go to a non-collinear arrangement of the beams $\vec{k}_1 \dots \vec{k}_n$, then a slight misalignment will make it impossible for the system to find a direction for $\vec{k}_{n+1}$ in which $\Delta\vec{k} = 0$. (Once a signal is found, the $\vec{k}_1 \dots \vec{k}_n$ beams can be adjusted to minimize $\Delta\vec{k}$).

The length of $\vec{k}_{n+1}$ is determined by ω_{n+1} and the index of refraction of the ensemble. For collinear beams, the minimum for $\Delta\vec{k}$ will therefore be produced for $\vec{k}_{n+1}$ parallel to $\sum_{i=1}^{n} \vec{k}_i$ and $\Delta\vec{k}$ will also be parallel to $\sum_{i=1}^{n} \vec{k}_i$. If the system extends a distance L in this direction the field produced at the detector will be given by

$$E_d = n \,\mathrm{Re}\left\{ \int_0^L dz \iint dx\, dy\, e^{i\phi(\vec{r})} \right\}$$

where n is the number density of radiators in the ensemble, the z axis has been chosen along $\Delta\vec{k}$, and ϕ_i has been replaced by $\phi(\vec{r}) = \vec{r}\cdot\Delta\vec{k} = z\,\Delta k$.

Defining N as the total number of radiators in the system, this yields

$$\begin{aligned} E_d &= N\,\mathrm{Re}\left\{ \frac{1}{L}\int_0^L dz\, e^{i\Delta k\, z} \right\} \\ &= N\,\mathrm{Re}\left\{ \frac{e^{i\Delta k\, L} - 1}{i\Delta k\, L} \right\} \\ &= N\,\frac{\sin(\Delta k\, L)}{\Delta k\, L} \end{aligned}$$

Thus the total intensity will contain the factor

$$I \sim N^2 \left[\frac{\sin \Delta k\, L}{\Delta k\, L} \right]^2 \tag{8.7}$$

which is characteristic of diffraction of radiation from a one-dimensional slit (in which case $\Delta k\, L$ is replaced by $2\pi L \sin\theta/\lambda$).

In this and in the preceding section we have assumed an isotropic medium in which n depends only on ω. In a crystal n can also depend on the relative orientation of the crystal lattice and the polarization of the E field. Thus it may become possible to obtain $\Delta\vec{k} = 0$ by changing the angle of the crystal while holding the directions of $\vec{k}_i$ fixed—this is called **angle tuning**.

3. Examples

(i) Degenerate four-wave mixing

It is possible to observe intense signals generated by four-wave mixing in a two-state system in which all four ω's are the same (degenerate) but their directions differ. The relevant ω_i's and the $\vec{k}_i$'s are shown in Fig. 8.3 below.

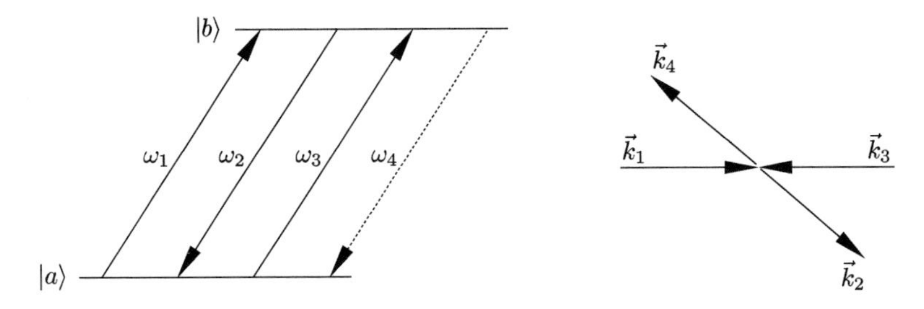

Figure 8.3: Energy level structure for degenerate four-wave mixing, showing the relevant ω_i's and the $\vec{k}_i$'s.

ω_4 is shown dashed to indicate that it is generated by the system rather than imposed from without. The arrows on the $\vec{k}_i$'s are placed to indicate whether the system absorbs or emits the radiation.

The relation

$$\omega_1 - \omega_2 + \omega_3 - \omega_4 = 0$$

is trivially satisfied in this case, and $\Delta\vec{k} = 0$ simply implies that $\vec{k}_4$ will be opposite to $\vec{k}_2$ if $\vec{k}_1$ and $\vec{k}_3$ are anti-parallel. Non-coplanar geometries which satisfy $\Delta\vec{k} = 0$ also exist.

(ii) Generation of UV radiation

Coherent ultraviolet radiation can be generated by four-wave mixing processes in atomic vapors which satisfy

$$\omega_1 + \omega_2 + \omega_3 - \omega_4 = 0$$

The efficiency of the process is obviously enhanced by the presence of near-resonant states, as seen from the level structure shown in Fig. 8.4 below.

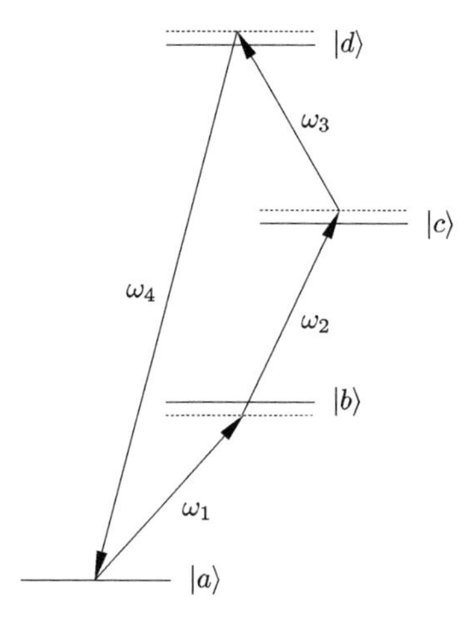

Figure 8.4: Energy level structure for generating ultraviolet radiation.

Frequently $\omega_2 = \omega_{ca} - \omega_1$ and level $|c\rangle$ is excited by a resonant two-photon process. Sometimes level $|d\rangle$ is not a bound level, but is in the one-electron continuum.

Several schemes for phase matching have been tried, since this is obviously essential if highly efficient UV generation is desired. Since collinear arrangement of the beams provides the greatest overlap of the focused beams necessary for high power generation, phase matching by changing the angles of the $\vec{k}_i$ is not desirable. One alternative is to introduce a second atomic species into the system (in addition to the atoms used for the four-wave process) whose refractive index varies in such a way as to make $\Delta\vec{k} = 0$. Another scheme is to exploit the fact that the index of refraction of the primary atomic gas changes rapidly near a resonance and to choose the three frequencies ω_1, ω_2, and ω_3 so that the index matching condition is satisfied.

D. Mixed examples

In this section we consider several examples in that the coherent behavior of an ensemble involves ideas that we have discussed in earlier sections of this chapter. We consider echos in two and three level systems first, and conclude with a discussion of superradiance in samples whose dimensions are $\gg \lambda$, the wavelength of the radiation emitted.

1. Echoes in extended media

(i) Spin echo in an extended sample

Now we consider the possibility of observing a spin echo in an extended inhomogeneously broadened medium such as a crystal. We let $\vec{k}_1$, $\vec{k}_2$, and $\vec{k}_3$ be propagation directions of the $\pi/2$ pulse, the π pulse, and the echo respectively. Generalizing the treatment of Section B2 simply involves adding in the appropriate phase factors $\vec{k}_1 \cdot \vec{r}_i$. Using the same notation for time as before one has

$$\phi_i(T_-) = \vec{k}_1 \cdot \vec{r}_i - (\omega_\circ + \delta_i)T$$

(The $-$ sign arises because we now define ϕ as the phase from $\vec{k} \cdot \vec{r}_i - \omega_i t$ instead of simply ω_i as before).

In an extended sample the phase of the π pulse depends on $\vec{r}_i$—we can take account of this by setting the phase of the π pulse to be $\phi_T = \vec{k}_2 \cdot \vec{r}_i$. Then

$$\begin{aligned}\phi_i(t > T) &= 2\phi_T - \phi_i(T_-) + \omega_i(t - T)\\ &= 2\vec{k}_2 \cdot \vec{r}_i - \phi_i(T_-) - (\omega_\circ + \delta_i)(t - T)\\ &= (2\vec{k}_2 - \vec{k}_1) \cdot \vec{r}_i - (\omega_\circ + \delta_i)(t - 2T)\end{aligned}$$

For an echo pulse emitted in the direction $\vec{r}_3$, the phase is (remember emission means the phase is $-\vec{k}_3 \cdot \vec{r}_i$)

$$\phi_i(t > T) = (2\vec{k}_2 - \vec{k}_1 - \vec{k}_3) \cdot \vec{r}_i + (\omega_\circ + \delta_i)(2T - t)$$

In order to get an echo one must have ϕ_i become independent of i for some time. It is clear that at the time $t_{\text{echo}} = 2T$, the effects of inhomogeneous broadening (i.e. those involving δ_i) will vanish just as they did for the small sample. In an extended sample, however, one must also satisfy the phase matching condition that

$$\Delta\vec{k} = \vec{k}_1 + \vec{k}_3 - 2\vec{k}_2 = 0$$

Since $\omega_1 = \omega_2 = \omega_3$, this phase matching condition is difficult to satisfy except in the case that $\vec{k}_1$ and $\vec{k}_2$ are inclined relative to each other at a small angle α so that

$$|\Delta\vec{k}| \approx \alpha^2|\vec{k}|$$

If α is a few milliradians it is possible to keep the intensity factor in Eq. (8.7) close to unity for sample sizes ~1 cm and still benefit from the spatial separation of the echo pulse from the exciting pulses.

(ii) Echoes in multilevel atoms

The idea of engineering an echo by a suitable sequence of pulses to a system is not limited to a two-state system. Recall that for a two-state system, an echo occurred whenever the radiators rephased themselves at some later time (and with some definite spatial pattern if the ensemble were an extended one). If the atomic systems have several levels, then an echo can result whenever the relative phase of *any* two states connected by an allowed transition is rephased throughout the ensemble. The presence of additional levels therefore offers more flexibility in generating an echo because a coherence produced between two states can be transferred to other levels (where it evolves with a different frequency) and then back to the original states.

2. Strong superradiance in extended samples

Dicke's comment in the superradiance paper* states that "A classical system of simple harmonic oscillators distributed over a large region of space can be so phased relative to each other that coherent radiation is obtained in a particular direction. It might be expected also that the radiating gas under consideration would have energy levels such that spontaneous radiation occurs coherently in one direction." One might also expect some complications since the system would not "know" which way to radiate (unless given a $\pi/2$ pulse), and Doppler broadening may also be a problem.

The restriction to a region $\ll \lambda$ had the subtle effect of suppressing Doppler broadening since a system with large enough speed to cause a frequency shift outside the radiation width $(1/T_{2N})$ would hit the walls and bounce back before radiating. In a real extended sample some form of inhomogeneous broadening will occur and greatly complicate the superradiance problem by making the number of subsystems capable of superradiating dependent on the superradiant decay time. We circumvent this difficulty by considering only **strong** superradiance, which means that the big bang portion

*See footnote on p. 242 for complete reference.

of the superradiant process happens fast enough to interact with all the subsystems (i.e. $1/T_{2N} >$ Doppler width).

Now consider an extended sample of area A and length L. How must we restrict the direction of superradiant emission so that all the subsystems can participate? We must have the angle within the forward 1-slit diffraction pattern characteristic of a slit of width $A^{1/2}$. Such superradiance can occur only in a cone of solid angle

$$\Delta\Sigma \approx \theta_{\text{diff}}^2 \approx \left(\frac{\lambda}{D}\right)^2 = \frac{\lambda^2}{A}$$

Thus instead of finding a superradiant emission power

$$P = N^2 P_\circ \qquad (P_\circ \text{ is for one system})$$

for this extended system, this rate is multiplied by $f \equiv \Delta\Sigma/4\pi$, the geometric fraction of solid angle into which superradiant emission can occur coherently. Thus we get

$$P = f N^2 P_\circ$$

for the radiation rate. This will radiate the total energy ($= NP_\circ T_{21}$) away in a superradiant time

$$T_{\text{SR}} = \frac{NP_\circ T_{21}}{P} = \frac{T_{21}}{Nf} = \frac{T_{\text{SP}}}{N_{\text{eff}}}$$

where T_{SP} is the spontaneous decay time of a single radiator and

$$N_{\text{eff}} = fN = \frac{\lambda^2}{4\pi A} nAL = \frac{\lambda^2 L}{4\pi} n = \pi \lambda\!\!\!^{-2} Ln$$

i.e. is the number of subsystems within a cylinder of radius $\lambda\!\!\!^{-}$ and length L. More detailed calculations show N_{eff} to be 1/2 as big as the above equation indicates.

Superradiance in two atom systems has been known for a long time—a homonuclear molecule has **gerade** and **ungerade** states one of which is subradiant and the other which radiates at ~two times the rate of the free atom if the molecule is weakly bound (so that both the dipole matrix element and the emission frequency are not appreciably changed).

E. Coherent control in multilevel atoms

1. Coherent population trapping — CPT

Coherent population trapping is a phenomenon in which atoms are driven into a coherent superposition state from which they cannot absorb light. Coherence arises because the amplitudes for absorption from the dark state for the two laser beams cancel to create a non-absorbing state. This shows that the two laser beams have to be **phase coherent**.

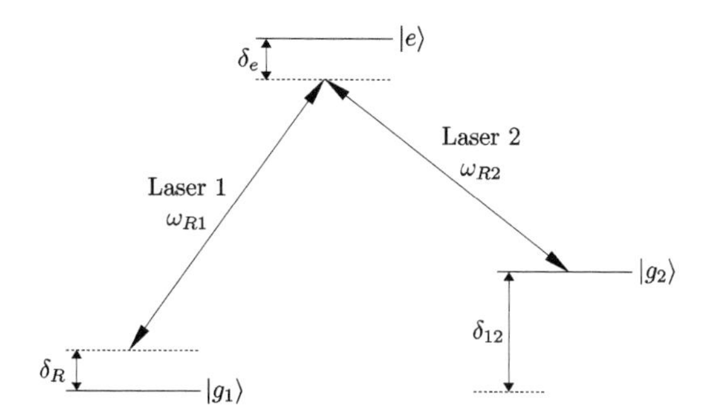

Figure 8.5: Energy level structure for observing CPT.

The creation of the dark state can be understood by considering the three-level Λ system shown in Fig. 8.5. There are two electric dipole allowed transitions, namely $|g_1\rangle \to |e\rangle$ and $|g_2\rangle \to |e\rangle$. These are coupled by two phase coherent lasers with respective Rabi frequencies of ω_{R1} and ω_{R2}. The lasers are detuned by an equal amount δ_e from the upper level $|e\rangle$. They have a relative detuning of $\delta_{12} - \delta_R$, where δ_{12} is the difference in frequency between $|g_1\rangle$ and $|g_2\rangle$, and δ_R is the Raman detuning from the two-photon resonance coupling $|g_1\rangle$ and $|g_2\rangle$ through an intermediate level near $|e\rangle$. Thus the CPT resonance occurs as δ_R is scanned at the point $\delta_R = 0$, while $|e\rangle$ plays the role of a "stepping stone" for the process. In fact, the larger the detuning from this level the better for the coherence of the process since the possibility of spontaneous decay due to real transitions to $|e\rangle$ is reduced, but this comes at the price of reduced signal.

The underlying physics of the CPT phenomenon can be understood most easily when the atomic system is expressed in a different basis. Instead of the energy eigenstates $|g_1\rangle$ and $|g_2\rangle$, we consider the linear combination $|+\rangle$ and $|-\rangle$ given by

$$|+\rangle = \frac{\omega_{R1}^* |g_1\rangle + \omega_{R2}^* |g_2\rangle}{\sqrt{|\omega_{R1}|^2 + |\omega_{R2}|^2}}$$

$$|-\rangle = \frac{\omega_{R2} |g_1\rangle - \omega_{R1} |g_2\rangle}{\sqrt{|\omega_{R1}|^2 + |\omega_{R2}|^2}}$$

Since these states are not energy eigenstates, they evolve with time. In the rotating wave approximation, and when the light fields satisfy the Raman resonance condition ($\delta_R = 0$) with no phase difference, transitions starting from $|-\rangle$ are not allowed, i.e.

$$\langle e|H_{E1}^{\text{int}}|-\rangle = 0$$

$|-\rangle$ is therefore called a coherent dark state. Atomic population is driven into the dark state by optical pumping and gets trapped in the state; thus the process is called coherent population trapping. The zero absorption from this state is due to the destructive interference of the transition amplitudes along the two possible excitation pathways to level $|e\rangle$.

The above analysis shows that the linewidth of the CPT resonance is limited by the decoherence rate between the two ground levels. CPT is usually studied in alkali atoms with the two lower levels formed by the two ground hyperfine levels. Thus the transition between $|g_1\rangle$ and $|g_2\rangle$ is E1 forbidden, and the lifetime-limited linewidth is below 1 Hz. However, when CPT is studied in vapor cells, the ground state coherence can be destroyed by spin-exchange collisions with the walls. To reduce this effect, CPT studies normally use vapor cells with paraffin coating on the walls, or cells that are filled with a buffer gas like N_2 or Ne.

CPT in vapor cells has several applications in precision spectroscopy, such as sensitive magnetometry and atomic clocks. For magnetometry, the basic principle is that the CPT resonance gets split into several components in the presence of a magnetic field. The amount of shift depends on the Zeeman shift of the sublevels involved, and hence can be used to determine the strength of the B field. One gets high sensitivity because the linewidths of the resonances is very small, and even a small field can cause a well resolved shift. For use as atomic clocks, recall that the CPT resonance occurs when the frequency difference between the two laser beams is exactly equal to the ground hyperfine interval. Since the SI unit of 1 second is defined in terms of the ground hyperfine level in ^{133}Cs, the phenomenon naturally lends itself to the realization of an atomic clock using Cs in a vapor cell. Even with Rb where the interval is different, the CPT resonance can be used to realize a secondary time standard.

2. Electromagnetically induced transparency — EIT

Electromagnetically induced transparency is a phenomenon in which a strong control laser is used to modify the properties of the medium for a weak probe laser. In the simplest manifestation of EIT, we take the medium to be composed of three-level atoms, so that the control and probe lasers can act on different transitions. Applications of EIT include slowing and storage of light, lasing without inversion, enhanced non-linear optics, and high resolution spectroscopy.

EIT occurs due to two effects caused by the control laser—(i) creation of dressed states, and (ii) interference between the decay pathways to (or from) these dressed states—both of which will be discussed in the following sections. We will study the phenomenon of EIT in the three canonical types of three-level atoms—lambda (Λ), ladder (Ξ), and vee (V). To facilitate a density matrix analysis consistent in all the systems, we define the three atomic levels as $|1\rangle, |2\rangle$, and $|3\rangle$. The probe laser is on the $|1\rangle \rightarrow |2\rangle$ transition with Rabi frequency ω_{Rp} and detuning $\delta_p \equiv \omega_p - |\omega_{21}|$. The control laser is on the $|2\rangle \rightarrow |3\rangle$ transition with Rabi frequency ω_{Rc} and detuning $\delta_c \equiv \omega_c - |\omega_{32}|$. Both these transitions are electric dipole allowed, which means that the levels involved have opposite parity. Thus the levels $|1\rangle$ and $|3\rangle$ have the same parity, and transition between them is electric dipole forbidden. The respective spontaneous decay rates are Γ_1, Γ_2, and Γ_3. For simplicity, we will take all ground-like levels to have zero decay rate, and all excited states to have the same decay rate of Γ. In the density matrix analysis, control induced susceptibility for the probe transition is given by the density matrix element ρ_{12}.

3. EIT in a Λ system

We first consider a Λ system with level structure as shown in Fig. 8.6(a).

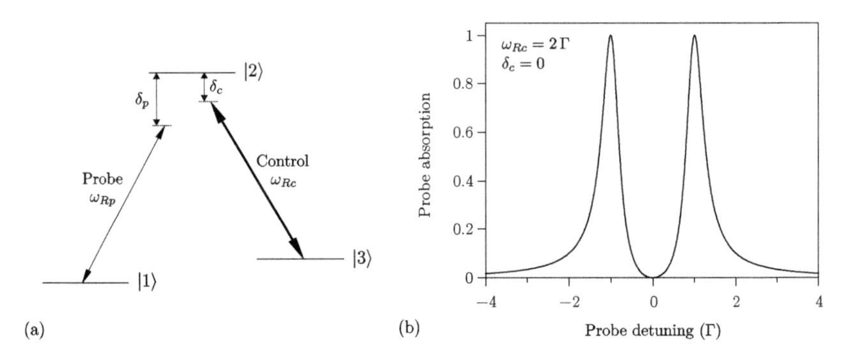

Figure 8.6: EIT in a Λ system. (a) Level structure. (b) Probe absorption calculated from $\mathrm{Im}\{\rho_{12}\Gamma/\omega_{Rp}\}$ as a function of probe detuning, with $\omega_{Rc} = 2\Gamma$ and $\delta_c = 0$.

The density matrix elements (in the rotating wave approximation) obey the following equations:

$$\begin{aligned}
\dot{\rho}_{11} &= \frac{\Gamma_2}{2}\rho_{22} + \frac{i}{2}\left(\omega_{Rp}\rho_{12} - \omega_{Rp}\rho_{21}\right) \\
\dot{\rho}_{22} &= -\Gamma_2\rho_{22} + \frac{i}{2}\left(\omega_{Rp}\rho_{21} - \omega_{Rp}\rho_{12}\right) + \frac{i}{2}\left(\omega_{Rc}\rho_{23} - \omega_{Rc}\rho_{32}\right) \\
\dot{\rho}_{33} &= \frac{\Gamma_2}{2}\rho_{22} + \frac{i}{2}\left(\omega_{Rc}\rho_{32} - \omega_{Rc}\rho_{23}\right) \\
\dot{\rho}_{12} &= \left(-\frac{\Gamma_2}{2} - i\delta_p\right)\rho_{12} + \frac{i}{2}\omega_{Rp}\left(\rho_{11} - \rho_{22}\right) + \frac{i}{2}\omega_{Rc}\rho_{13} \\
\dot{\rho}_{13} &= -i(\delta_p - \delta_c)\rho_{13} + \frac{i}{2}(\omega_{Rc}\rho_{12} - \omega_{Rp}\rho_{23}) \\
\dot{\rho}_{23} &= \left(-\frac{\Gamma_2}{2} + i\delta_c\right)\rho_{23} + \frac{i}{2}\omega_{Rc}\left(\rho_{22} - \rho_{33}\right) - \frac{i}{2}\omega_{Rp}\rho_{13}
\end{aligned}$$

[Note that there are only six equations for the nine elements since the off-diagonal elements are complex conjugates of each other.]

These equations are solved in the weak probe limit (i.e. neglecting higher powers of ω_{Rp}), and in steady state (i.e. all the LHSs are 0). Since probe absorption is given by $\mathrm{Im}\{\rho_{12}\Gamma/\omega_{Rp}\}$, the relevant density matrix element is ρ_{12}. Specializing to the case of $\delta_c = 0$ (control laser on resonance), the steady-state solution yields

$$\rho_{12} = \frac{i\omega_{Rp}/2}{\Gamma/2 + i\delta_p - i\dfrac{\omega_{Rc}^2/4}{\delta_p}} \tag{8.8}$$

The poles of the above equation give the location of the dressed states (in terms of probe detuning) as

$$\delta_{p\pm} = \pm\frac{\omega_{Rc}}{2}$$

Thus probe absorption as a function of probe detuning will show two peaks at these locations—called an **Autler–Townes doublet**. This is seen in Fig. 8.6(b), where we show a plot of $\mathrm{Im}\{\rho_{12}\Gamma/\omega_{Rp}\}$ vs δ_p. The figure shows that there is a minimum in probe absorption at line center because the absorption peaks move to the locations of the dressed states created by the control laser—thus the **transparency** induced by the control laser is called electromagnetically induced transparency.

In order to get the linewidths of the two subpeaks, note that the control laser couples the $|2\rangle$ and $|3\rangle$ levels, so the linewidth of each dressed state is

$$\Gamma_{\pm} = \frac{\Gamma_2 + \Gamma_3}{2}$$

The probe laser measures absorption to these dressed states from level $|1\rangle$, hence the linewidth of each subpeak in the doublet is

$$\Gamma_{p\pm} = \Gamma_1 + \frac{\Gamma_2 + \Gamma_3}{2} \tag{8.9}$$

In the Λ system, $\Gamma_1 = \Gamma_3 = 0$ and $\Gamma_2 = \Gamma$, and the subpeak linewidths are

$$\Gamma_{p\pm}^{\Lambda} = \frac{\Gamma}{2}$$

Thus probe absorption splits into two peaks each with a linewidth of half the original linewidth, exactly as seen in the figure.

4. Role of dressed state interference in EIT

The above discussion of the linewidth ignores any interference between the two dressed states. When we take that into account, it results in probe absorption decreasing due to destructive interference. The effect of the interference is seen from the results of the calculations shown in Fig. 8.7. The solid line represents the complete density matrix calculation, while the dotted lines represent what the spectrum would look like with just the two dressed states located at $\pm\omega_{Rc}$ with linewidth $\Gamma/2$, and *no interference.* The spectra are normalized so that the maximum absorption is unity. The figure shows that interference causes the transparency dip to increase and become zero at line center. The effect is significant up to about $\omega_{Rc} = 3\Gamma$, and then its effect becomes small because of the increased dressed state separation.

Thus EIT in a Λ system is explained by a combination of the AC Stark shift caused by the control laser and the interference of the decay pathways from the dressed states. As we will see later, dressed state interference is not so important in Ξ and V systems.

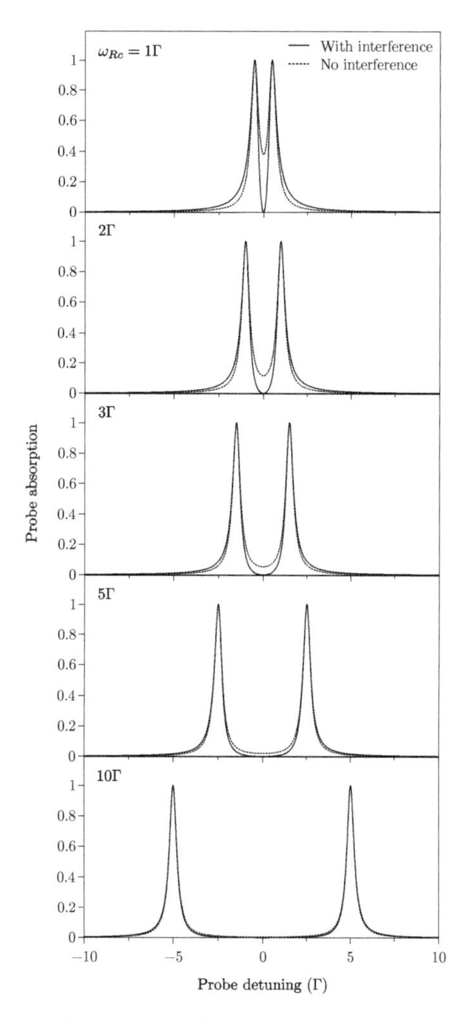

Figure 8.7: Dressed state interference in EIT in a Λ system. "With interference" represents the results of a complete density matrix calculation, whereas "No interference" shows the results of calculation with two Lorentzians at $\pm\omega_{Rc}$ having linewidth $\Gamma/2$ (representing the dressed states).

5. EIT vs CPT

The physics underlying the two phenomena are related, particularly when EIT is studied in a Λ system. For example, the Λ system with two ground levels allows the formation of a perfect dark state, which is required for CPT to occur and shows up in EIT as dressed state interference. But there are also important differences between the two—the aim of this section is to highlight these differences. Many of these points have already been mentioned in the sections discussing CPT and EIT, but are repeated here for ease of comparison. The differences in terms of various experimental parameters are as follows:

(i) **Phase coherence.** The primary requirement for CPT experiments is that the two lasers are phase coherent. When used in clock applications, one way to achieve this is to beat the two outputs on a sufficiently fast photodiode, and then phase lock the beat signal using an RF oscillator. Alternately, the two beams are derived from the same laser (so that they are phase coherent by definition), and the required frequency difference is obtained using an electro-optic modulator (EOM). In contrast, the control and probe lasers in EIT experiments are usually independent since there is no requirement for phase coherence.

(ii) **Scan axis.** The above discussion of phase coherence tells us that the scan axes in the two cases are different. In CPT, the scan axis is the relative detuning between the two beams, while in EIT it is the frequency of the (phase-independent) probe laser. If the probe transition is in the optical regime, this frequency is of order 10^{15} Hz, whereas the CPT frequency is of order 10^9 Hz (6 orders of magnitude smaller).

(iii) **Natural linewidth.** The difference of the scan axes in the two cases brings up the question of what defines the natural linewidth for the resonance. In CPT experiments, the resonance occurs when the relative detuning between the two beams matches the difference frequency between the two ground levels—δ_{12} in Fig. 8.5. The relevant natural linewidth is therefore the linewidth of transition between the two ground levels, which is sub-Hz because the transition is electric-dipole forbidden. The linewidth of the upper level is not relevant. For example, the CPT resonance can be used to create an atomic clock using a Cs vapor cell with the two laser beams tuned to the D_2 line of Cs. But the 5 MHz linewidth of the upper level does not enter the picture—a resonance with a linewidth of 100 Hz (which has been observed) would not be called **subnatural**. This discussion shows that, in general, the CPT phenomenon can only be used for precision spectroscopy on the ground levels.

On the other hand, the EIT phenomenon can be used for high resolution spectroscopy of the upper level. This is because the probe beam couples to the upper level. Its scan axis is the frequency of the probe beam, and the relevant natural linewidth is the linewidth of the upper level. In the case of the Rb D_2 line used in many EIT studies, the upper level has a linewidth of 6 MHz. Therefore any feature that is narrower than 6 MHz would be called subnatural.

(iv) **Power in the two beams.** In CPT both beams play an equally important role, while EIT is studied in the regime where the control is strong and the probe is weak. This leads to an important difference—the powers (and hence intensity) in the two beams are roughly equal for CPT, while the power in the probe beam in EIT is negligibly small. This also has implications for the theoretical density matrix analysis in the two cases. EIT is analyzed under the approximation of neglecting higher-order terms in the probe intensity, whereas this approximation in not valid for CPT. In addition, the quantity of interest in EIT is ρ_{12} which corresponds to the coherence between levels $|1\rangle$ and $|2\rangle$, while in CPT it is ρ_{22} which gives the population in the upper state.

(v) **Fluorescence versus absorption.** One important consequence of the fact that CPT results from the creation of a dark state is that there is a concomitant decrease in the fluorescence from the cell. In fact the first observation of CPT was the appearance of a dark region in the fluorescence in a Na vapor cell, with an inhomogeneous magnetic field applied so that the dark state was only formed in a localized region. By contrast, in EIT the strong laser is always being absorbed (and the atoms are fluorescing), and the induced transparency is seen only in the absorption signal of the weak probe laser. Therefore, (to first order) there is no change in the fluorescence whether the probe laser is on or off.

(vi) **Effect of buffer gas.** Because CPT is a ground-state coherence phenomenon, any technique that increases the coherence time will give a narrower linewidth. One of the common methods is to use a buffer gas in the vapor cell—typically a few torr of a gas like Ne or N_2. For example, when CPT resonances are used for clock applications, the vapor cell is filled with buffer gas. On the other hand, the use of such cells for EIT experiments actually kills the signal. This is because the presence of the buffer gas broadens the probe absorption signal due to collisions and prevents the observation of any modification due to the control laser.

(vii) **Effect of detuning from the upper level.** Real transitions to the upper level in CPT cause decoherence of the dark state because of spontaneous decay. Therefore detuning from the upper level (increasing δ_e in Fig. 8.5) results in the linewidth of the resonance becoming smaller. Of course, there is a simultaneous reduction in signal strength, but this can be compensated by increasing the power in the two laser beams. Note that the CPT resonance still occurs at the same relative detuning between the two beams—$\delta_R = 0$, the point at which the two-photon Raman resonance condition is satisfied. In addition, there is no change in the lineshape. Thus many CPT experiments are done with a detuning of several linewidths from the excited state. On the other hand, detuning the control laser from the upper level in EIT ($\delta_c \neq 0$) causes the resonance to shift within the absorption profile of the probe laser (appear at a different location), and also makes the lineshape very different.

6. EIT in a Ξ system

We now consider the Ξ system with level structure as shown in Fig. 8.8(a).

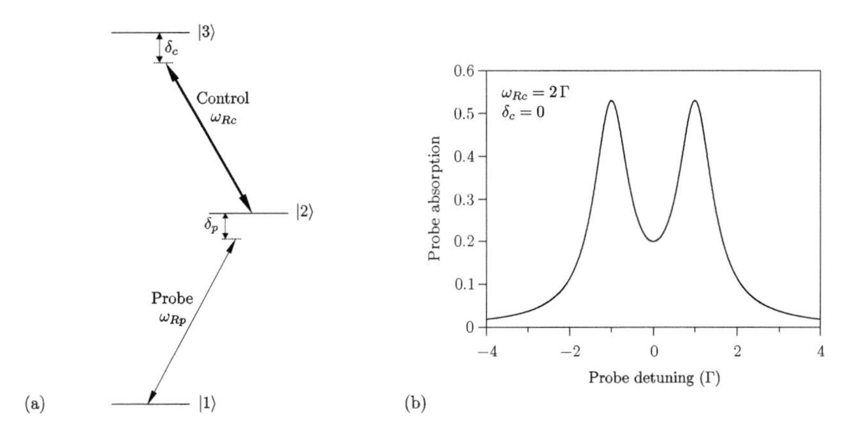

Figure 8.8: EIT in a Ξ system. (a) Level structure. (b) Probe absorption calculated from $\text{Im}\{\rho_{12}\Gamma/\omega_{Rp}\}$ as a function of probe detuning, with $\omega_{Rc} = 2\Gamma$ and $\delta_c = 0$.

The steady-state value of ρ_{12} with $\delta_c = 0$, obtained from a similar density matrix analysis as was done for the Λ system, is

$$\rho_{12} = \frac{i\omega_{Rp}/2}{\Gamma/2 + i\delta_p + \dfrac{\omega_{Rc}^2/4}{\Gamma/2 + i\delta_p}} \tag{8.10}$$

The pole structure of the denominator shows that again the dressed states are located at $\pm\omega_{Rc}/2$. In this case, $\Gamma_1 = 0$ and $\Gamma_2 = \Gamma_3 = \Gamma$. Eq. (8.9) tells us that the subpeak linewidths are

$$\Gamma_{p\pm}^{\Xi} = \Gamma$$

Probe absorption as calculated from $\text{Im}\{\rho_{12}\Gamma/\omega_{Rp}\}$ with $\omega_{Rc} = 2\Gamma$ is shown in Fig. 8.8(b). As seen, the spectrum splits into two with a transparency dip at line center. Each subpeak has a linewidth of Γ as discussed above. But in this case, the absorption at line center does not go to zero as in the Λ case, and the maximum absorption is less than unity. More interestingly, dressed state interference does not play a role, which can be verified by doing calculations with and without interference as done for the Λ system.

7. EIT in a V system

We now consider the V system with level structure as shown in Fig. 8.9(a).

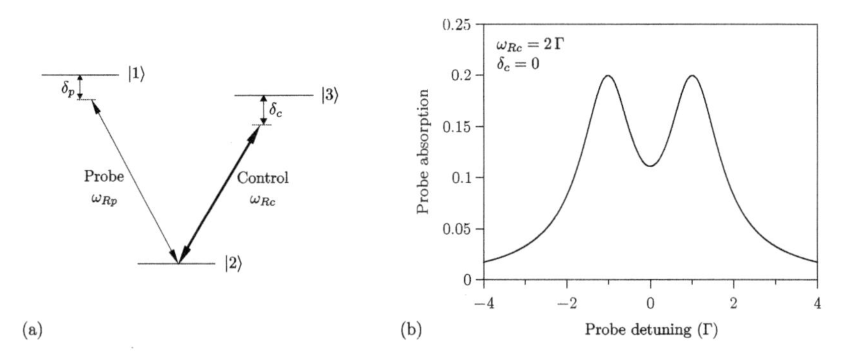

Figure 8.9: EIT in a V system. (a) Level structure. (b) Probe absorption calculated from $\mathrm{Im}\{\rho_{21}\Gamma/\omega_{Rp}\}$ as a function of probe detuning, with $\omega_{Rc} = 2\Gamma$ and $\delta_c = 0$.

In this system, level $|1\rangle$ is above level $|2\rangle$; therefore the relevant matrix element for probe absorption is ρ_{21}. The steady-state value of ρ_{21} with $\delta_c = 0$, obtained from a similar density matrix analysis as was done for the Λ system, is

$$\rho_{21} = \frac{i\omega_{Rp}/2\left[\left(\Gamma^2 + \omega_{Rc}^2\right) - \dfrac{\omega_{Rc}^2\Gamma/2}{\Gamma + i\delta_p}\right]}{\left(\Gamma^2 + 2\omega_{Rc}^2\right)\left[\Gamma/2 + i\delta_p + \dfrac{\omega_{Rc}^2/4}{\Gamma + i\delta_p}\right]} \tag{8.11}$$

The pole structure of the denominator shows that the dressed states are located at $\pm\omega_{Rc}/2$ as for the other two cases. In this case, $\Gamma_1 = \Gamma_3 = \Gamma$ and $\Gamma_2 = 0$. Eq. (8.9) tells us that the subpeak linewidths are

$$\Gamma_{p\pm}^{\mathrm{V}} = \frac{3\Gamma}{2}$$

Probe absorption as calculated from $\mathrm{Im}\{\rho_{21}\Gamma/\omega_{Rp}\}$ with $\omega_{Rc} = 2\Gamma$ is shown in Fig. 8.9(b). As seen, the spectrum splits into a doublet with a transparency dip at line center. Each subpeak has a linewidth of $3\Gamma/2$ as shown above. But in this case also (as for the Ξ system), EIT is only partial with non-zero absorption at line center, and the peak absorption is less than unity. In addition, the effect of dressed state interference is quite small, as can be verified using calculations with and without interference.

F. Other effects in coherent control

1. Optical rotation

Just like the imaginary part of the susceptibility induced by the strong control laser can be used to modify the absorption of the weak probe laser, the real part can be used to modify the refractive index of the medium, and hence the phase change experienced by the probe laser. One consequence of this is the rotation of the plane of polarization of a linearly polarized probe beam as it propagates through the medium—called **optical rotation**.

The theoretical analysis proceeds in the same way as for EIT, except that the quantity of interest is the **real** part of $\rho_{12}\Gamma/\omega_{Rp}$. The relevant density matrix element ρ_{12} for the three types of three-level systems has been derived previously in Eqs. (8.8), (8.10), and (8.11). The quantity $\mathrm{Re}\{\rho_{12}\Gamma/\omega_{Rp}\}$ with the previous control Rabi frequency of $\omega_{Rc} = 2\Gamma$ for the three cases is shown in Fig. 8.10. As expected, the rotation has a dispersive lineshape with two resonances at the location of the dressed states created by the control laser, exactly where the peaks for the EIT signal are located. The separation between the maximum and minimum of each resonance is equal to the linewidth of the probe absorption peak, as derived earlier for the three cases. In addition, the amplitude is different in the three cases, consistent with the different absorption maxima.

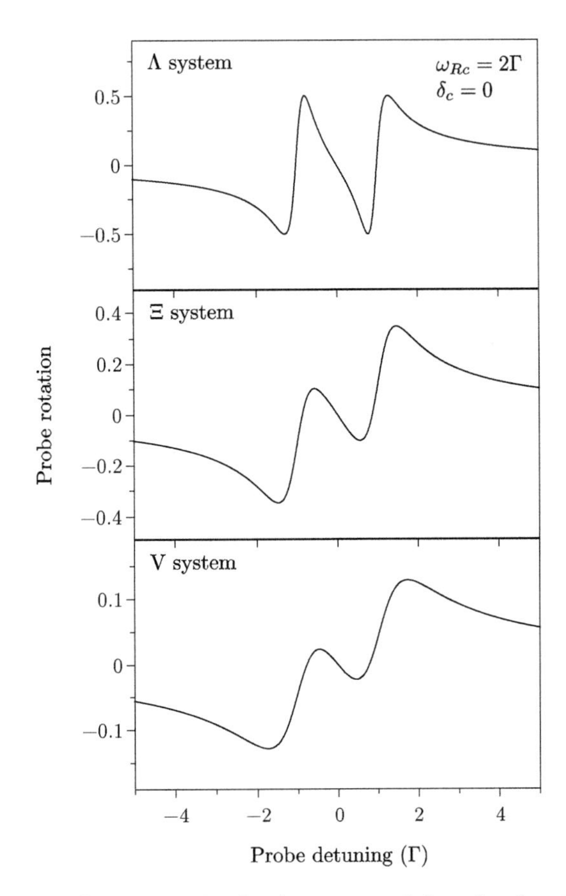

Figure 8.10: Probe rotation in the three types of three-level systems. The quantity plotted is $\mathrm{Re}\{\rho_{12}\Gamma/\omega_{Rp}\}$ as a function of probe detuning, with $\omega_{Rc} = 2\Gamma$ and $\delta_c = 0$.

2. Electromagnetically induced absorption — EIA

Electromagnetically induced absorption is the analog of EIT in the sense that (two) control lasers are used to create increased **absorption** for the probe laser, instead of increased transparency. The phenomenon requires an N-type level structure as shown in Fig. 8.11(a). There are two control lasers—(i) Control 1 coupling the $|1\rangle \rightarrow |2\rangle$ transition with Rabi frequency ω_{Rc1}, and (ii) Control 2 coupling the $|3\rangle \rightarrow |4\rangle$ transition with Rabi frequency ω_{Rc2}. Both control lasers are on resonance with no detuning. The probe laser is on the $|3\rangle \rightarrow |2\rangle$ transition. For simplicity we assume that $\Gamma_1 = \Gamma_3 = 0$ and $\Gamma_2 = \Gamma_4 = \Gamma$.

Before solving the density matrix equations for this system, we note that the strong control 1 laser optically pumps all the population into the $|3\rangle$ level. Control 2 then cycles the population between levels $|3\rangle$ and $|4\rangle$, so that in steady state the populations are

$$\rho_{44} = \frac{(\omega_{Rc2}/\Gamma)^2}{1 + 2(\omega_{Rc2}/\Gamma)^2} \qquad \text{and} \qquad \rho_{33} = 1 - \rho_{44}$$

The relevant matrix element for probe absorption is ρ_{32}. Density matrix analysis (similar to what was done for the Λ system) yields in steady state

$$\rho_{32} = -\frac{i\omega_{Rp}\rho_{33}}{2\beta_{32}\alpha_2} + \frac{i\omega_{Rp}\omega_{Rc2}^2(\rho_{33} - \rho_{44})}{8\beta_{42}\beta_{32}\beta_{43}\alpha_2} - \frac{i\omega_{Rp}\omega_{Rc1}^2\omega_{Rc2}^2(\rho_{33} - \rho_{44})}{32\beta_{42}\beta_{32}\beta_{41}\beta_{43}\alpha_1\alpha_2}\left(\frac{1}{\beta_{31}} + \frac{1}{\beta_{42}}\right)$$

where

$$\alpha_1 = 1 + \frac{\omega_{Rc1}^2}{4\beta_{41}\beta_{42}} + \frac{\omega_{Rc2}^2}{4\beta_{41}\beta_{31}}$$

$$\alpha_2 = 1 + \frac{\omega_{Rc1}^2}{4\beta_{32}\beta_{31}} + \frac{\omega_{Rc2}^2}{4\beta_{32}\beta_{42}} - \frac{1}{16}\frac{\omega_{Rc1}^2\omega_{Rc2}^2\left(\frac{1}{\beta_{31}} + \frac{1}{\beta_{42}}\right)^2}{\beta_{32}\beta_{41}\alpha_1}$$

$$\beta_{31} = -i\delta_p$$

$$\beta_{32} = -\frac{\Gamma}{2} - i\delta_p$$

$$\beta_{41} = -\frac{\Gamma}{2} - i\delta_p$$

$$\beta_{42} = -\Gamma - i\delta_p$$

$$\beta_{43} = -\frac{\Gamma}{2}$$

A plot of $\mathrm{Im}\{\rho_{32}\Gamma/\omega_{Rp}\}$ is shown in Fig. 8.11(b) for the case of $\omega_{Rc1} = \omega_{Rc2} = 2\Gamma$. The figure shows that there is a peak in absorption at line center—**electromagnetically induced absorption**.

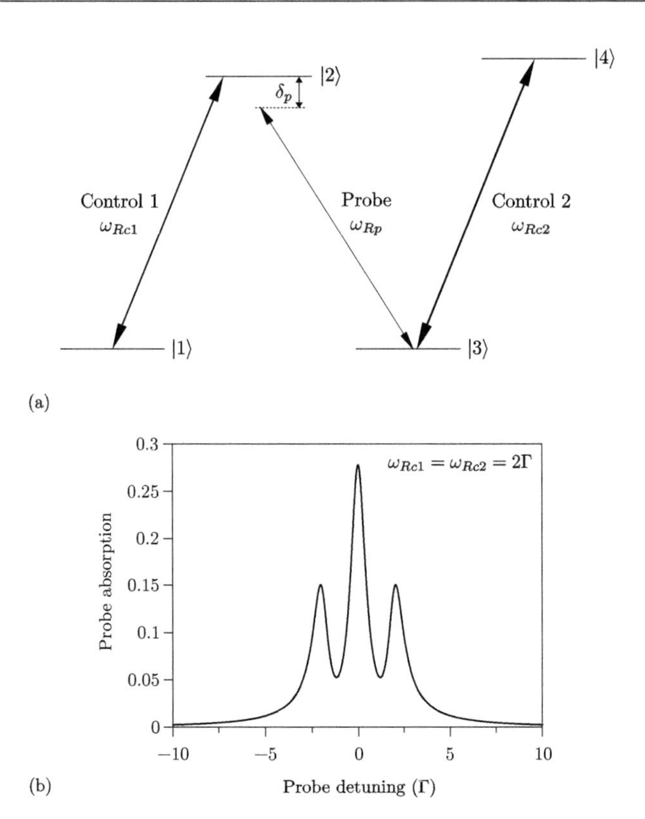

Figure 8.11: Electromagnetically induced absorption in an N-type system. (a) Level structure. (b) Probe absorption calculated from $\text{Im}\{\rho_{12}\Gamma/\omega_{Rp}\}$ as a function of probe detuning, with $\omega_{Rc1} = \omega_{Rc2} = 2\Gamma$.

G. Problems

1. Quantum beats

Consider a spinless electron that has an $\ell = 0$ ground state and $\ell = 1$ excited state. In a magnetic field B along the z axis, the excited state energy is

$$E^*_m = \hbar\omega_\circ - g\mu_B mB$$

Assume that the system is exposed to a short pulse of radiation of polarization $\hat{\varepsilon}_1$ at $t = 0$.

(a) Which polarizations $\hat{\varepsilon}_1 = \hat{\varepsilon}_x, \hat{\varepsilon}_y, \hat{\varepsilon}_+, \hat{\varepsilon}_-$ (where $\hat{\varepsilon}_\pm$ are circular polarizations referred to the z axis) can possibly produce a situation where quantum beats can be observed in the subsequent fluorescence?

(b) Assume that $\hat{\varepsilon}_1 = \hat{\varepsilon}_x$. Which two excited states can produce beats? At what value of B do they cross?

(c) If the excited state has a spontaneous lifetime of Γ_s^{-1}, calculate (within a constant) the rate of emission of photons of polarization (i) $\hat{\varepsilon}_2 = \hat{\varepsilon}_x$ and (ii) $\hat{\varepsilon}_2 = \hat{\varepsilon}_y$, at time t after the pulse.

(d) Can you give a classical explanation for the above results?

(e) Can you suggest a polarization that could produce beats between the $m = 0$ and $m = 1$ states?

Solution

(a) In order to produce quantum beats, the incident radiation should excite more than one level. Since

$$\hat{\varepsilon}_x = \frac{\hat{\varepsilon}_+ + \hat{\varepsilon}_-}{\sqrt{2}} \qquad \text{and} \qquad \hat{\varepsilon}_y = \frac{\hat{\varepsilon}_+ - \hat{\varepsilon}_-}{i\sqrt{2}}$$

the Δm selection rules tell us that $\hat{\varepsilon}_x$ and $\hat{\varepsilon}_y$ are the only polarizations that can produce quantum beats.

The other two $\hat{\varepsilon}_+$ and $\hat{\varepsilon}_-$ only excite one Zeeman level, namely $|1,1\rangle$ and $|1,-1\rangle$ respectively.

(b) If $\hat{\varepsilon}_1 = \hat{\varepsilon}_x$, quantum beats are produced by the Zeeman levels $|1,1\rangle$ and $|1,-1\rangle$. These levels cross for $B = 0$.

(c) As discussed in the chapter,

$$R(t) \propto I_1 e^{-\Gamma_s t}\left[\left|M_{ga}^{(2)} M_{ag}^{(1)}\right|^2 + \left|M_{gb}^{(2)} M_{bg}^{(1)}\right|^2\right]$$
$$+ I_1 e^{-\Gamma-st}\, 2\,\mathrm{Re}\left\{M_{ga}^{(2)^*} M_{ag}^{(1)^*} M_{gb}^{(2)} M_{bg}^{(1)} e^{-i\Delta\omega t}\right\}$$

where

$$|g\rangle = |0,0\rangle\,, \quad |a\rangle = |1,1\rangle\,, \quad |b\rangle = |1,-1\rangle\,, \quad \Delta\omega = \omega_b - \omega_a = \frac{2g\mu_B B}{\hbar}$$

(i) **Polarizations** $\hat{\varepsilon}_1 = \hat{\varepsilon}_x$ **and** $\hat{\varepsilon}_2 = \hat{\varepsilon}_x$

Expressing the polarizations in terms of the spherical harmonics $Y_{\ell m}$'s, the matrix elements are

$$\begin{aligned} M_{ag}^{(1)} &= \langle 1,-1|\hat{\varepsilon}_x \cdot \vec{P}|0,0\rangle \\ &\sim \langle 1,-1|\left(Y_{11} + Y_{1-1}\right)|0,0\rangle \\ &\sim \langle 1,-1|Y_{1-1}|0,0\rangle \\ M_{ag}^{(2)} &= M_{ag}^{(1)} \end{aligned}$$

If we denote this matrix element by A, then from a similar analysis

$$\begin{aligned} M_{bg}^{(1)} &= \langle 1,1|\hat{\varepsilon}_x \cdot \vec{P}|0,0\rangle \\ &\sim \langle 1,1|\left(Y_{11} + Y_{1-1}\right)|0,0\rangle \\ &\sim \langle 1,1|Y_{11}|0,0\rangle = A \\ M_{bg}^{(2)} &= M_{bg}^{(1)} = A \end{aligned}$$

Therefore the rate of emission goes as

$$\begin{aligned} R(t) &\sim I_1 e^{-\Gamma_s t}\left[2\left|M_{ag}^{(1)} M_{ga}^{(1)}\right|^2\right. \\ &\qquad \left. +2\,\mathrm{Re}\left\{M_{ga}^{(1)*} M_{ag}^{(1)*} M_{ga}^{(1)} M_{ag}^{(1)} e^{-i\Delta\omega t}\right\}\right] \\ &\sim 2I_1 e^{-\Gamma_s t}|A|^4\left[1 + \cos\Delta\omega\, t\right] \\ &= 4I_1 e^{-\Gamma_s t}|A|^4 \cos^2\left(\frac{g\mu_B B t}{\hbar}\right) \end{aligned}$$

(ii) **Polarizations** $\hat{\varepsilon}_1 = \hat{\varepsilon}_x$ **and** $\hat{\varepsilon}_2 = \hat{\varepsilon}_y$

Since the input polarization is again along x, the matrix elements for (1) are the same as before, namely

$$M_{ag}^{(1)} = M_{bg}^{(1)} = A$$

The matrix elements for (2) with y polarization are

$$M_{ag}^{(2)} \sim i \langle 1,-1| (Y_{11} + Y_{1-1}) |0,0\rangle = -iA$$
$$M_{bg}^{(2)} \sim i \langle 1,+1| (Y_{11} + Y_{1-1}) |0,0\rangle = iA$$

Therefore the rate of emission goes as

$$\begin{aligned} R(t) &\sim I_1 e^{-\Gamma_s t} \left[2 \left| M_{ag}^{(1)} M_{ga}^{(1)} \right|^2 \right. \\ &\qquad \left. + 2\,\mathrm{Re}\left\{ M_{ga}^{(1)*}\, iM_{ag}^{(1)*}\, M_{gb}^{(1)}\, iM_{bg}^{(1)}\, e^{-i\Delta\omega\, t} \right\} \right] \\ &\sim 2I_1 e^{-\Gamma_s t} |A|^4 \left[1 - \cos \Delta\omega\, t\right] \\ &= 4I_1 e^{-\Gamma_s t} |A|^4 \sin^2\left(\frac{g\mu_B B t}{\hbar}\right) \end{aligned}$$

(d) Classically, we can imagine that the exciting pulse creates a set of electric dipoles aligned with the electric field of the radiation. They then precess at the Larmor frequency about the magnetic field—z axis—and the motion is damped by re-radiation. The electric dipole radiation therefore shows a maximum whenever the axis of the dipoles is aligned with respect to the axis of observation.

(e) In order to produce beats between the $m = 0$ and $m = 1$ states we need a polarization that is a linear combination of $\hat{\varepsilon}_+$ and $\hat{\varepsilon}_z$, for example

$$\hat{\varepsilon}_1 = \tfrac{1}{\sqrt{2}} \left(\hat{\varepsilon}_+ + \hat{\varepsilon}_z\right) = \tfrac{1}{2} \left(\hat{\varepsilon}_x + i\hat{\varepsilon}_y\right) + \tfrac{1}{\sqrt{2}} \hat{\varepsilon}_z$$

2. Hanle effect

Consider the above situation except with continuous "white" excitation.

(a) Sketch a curve for $R_y(B)$, the radiation rate with polarization $\hat{\varepsilon}_y$. Sketch an experimental setup which would measure this.

(b) If this system were studied in a level crossing experiment, what would be the half-width-at-half-maximum (HWHM) of the level crossing signal observed with polarization $\hat{\varepsilon}_2 = \hat{\varepsilon}_x$?

(c) Can you suggest a way in which the sign of g could be determined using the Hanle effect?

Solution

(a) In the case of continuous excitation, we integrate $R(t)$ from $t = 0$ to ∞ to obtain

$$R \sim I_1 \operatorname{Re}\left\{M_{ga}^{(2)*} M_{ag}^{(1)*} M_{gb}^{(2)} M_{bg}^{(1)}\right\} \frac{\Gamma_s}{\Gamma_s^2 + (\Delta\omega)^2}$$
$$+ I_1 \operatorname{Im}\left\{M_{ga}^{(2)*} M_{ag}^{(1)*} M_{gb}^{(2)} M_{bg}^{(1)}\right\} \frac{\Delta\omega}{\Gamma_s^2 + (\Delta\omega)^2}$$

For $\hat{\varepsilon}_2 = \hat{\varepsilon}_y$, the term in brackets is pure real, and from the previous problem we have

$$M_{ag}^{(1)} = M_{bg}^{(1)} = A, \qquad M_{bg}^{(2)} = -iM_{ag}^{(2)} = iA$$

Therefore the rate has an inverted Lorentzian lineshape given by (and sketched below)

$$R_y(B) \sim -I_1|A|^4 \frac{\Gamma_s}{\Gamma_s^2 + (\Delta\omega)^2} \qquad \text{where } \Delta\omega = \frac{2g\mu_B B}{\hbar}$$

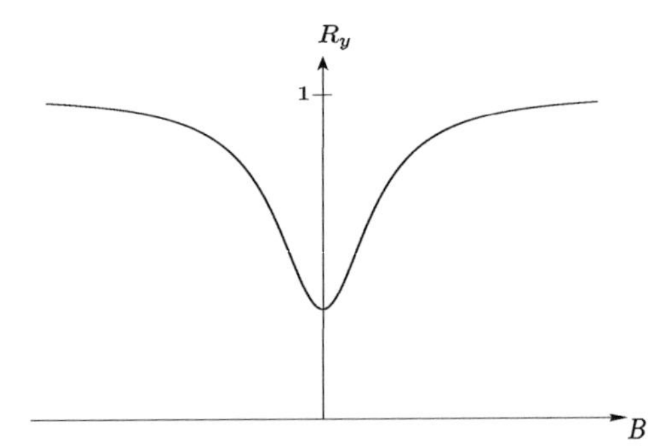

An experimental schematic to measure R_y is shown below.

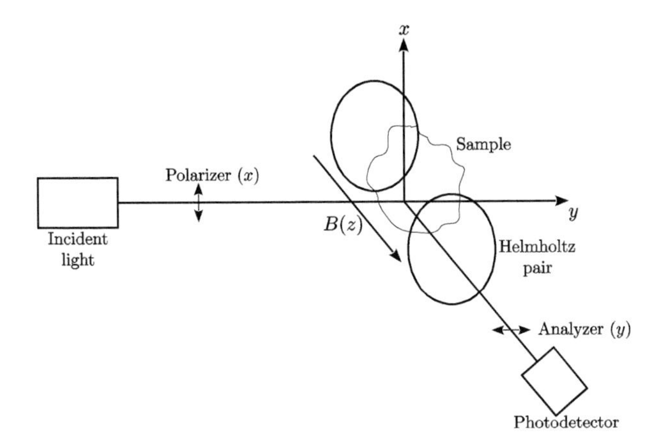

(b) For $\hat{\varepsilon}_2 = \hat{\varepsilon}_x$, the term in the brackets is still real and the curve is a Lorentzian with FWHM $= 2\Gamma_s$. Therefore the HWHM is Γ_s, or in terms of B it is

$$B_{\text{HWHM}} = \frac{\hbar\Gamma_s}{2g\mu_B}$$

(c) If the analyzer is rotated by 45° from the polarizer, the resultant curve obtained has a dispersive lineshape. The sign of the dispersion for a given sign of B field depends on the sign of g. Hence, g can be measured. The idea is shown in the figure below.

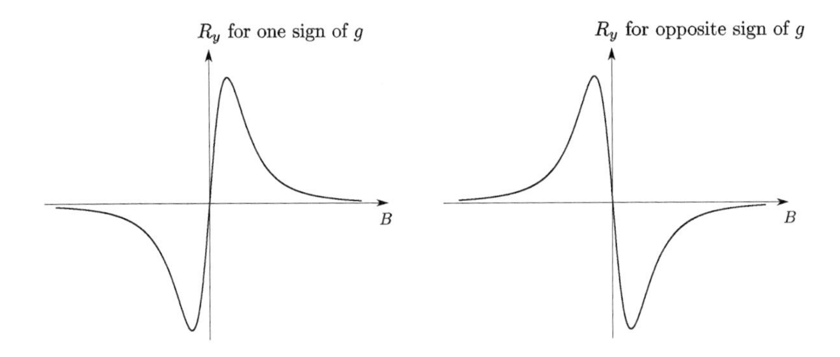

3. Dicke superradiance

Consider the two spin 1/2 magnetic moments mentioned by Dicke as an example of coherence in spontaneous emission. You are given that the spontaneous decay of a single spin proceeds at a rate

$$R_1 = |\langle\uparrow|H'_{\mathrm{M1}}|\downarrow\rangle|^2$$

(a) Find the spontaneous decay rate for the states of a two-spin system

$$|S, m_s\rangle = |1,+1\rangle, |1,0\rangle, |1,-1\rangle, \text{and } |0,0\rangle$$

corresponding to the triplet and singlet states, respectively.

(b) How do these rates compare with the expectation based on the idea that the particles decay independently?

Solution

(a) The states of the two-spin system in terms of the single spins are

(i) **Triplet**

$$\begin{aligned} |1,+1\rangle &= |\uparrow_1\uparrow_2\rangle \\ |1,0\rangle &= \tfrac{1}{\sqrt{2}}\left(|\uparrow_1\downarrow_2\rangle + |\downarrow_1\uparrow_2\rangle\right) \\ |1,-1\rangle &= |\downarrow_1\downarrow_2\rangle \end{aligned}$$

(ii) **Singlet**

$$|0,0\rangle = \tfrac{1}{\sqrt{2}}\left(|\uparrow_1\downarrow_2\rangle - |\downarrow_1\uparrow_2\rangle\right)$$

The perturbing Hamiltonian is defined as

$$H'_{\mathrm{M1}} \equiv H^1_{\mathrm{M1}} + H^2_{\mathrm{M1}}$$

The selection rules for this Hamiltonian are $\Delta S = 0$ and $\Delta m_s = \pm 1$. Therefore the spontaneous decays are as follows.

(i) **Triplet** — the $|1,+1\rangle$ state can spontaneously decay to the $|1,0\rangle$ state which, in turn, can decay to the $|1,-1\rangle$ ground state. The corresponding decay rates are

$$|1,+1\rangle \qquad R = |\langle 1,0|H_{\text{M1}}|1,1\rangle|^2$$
$$= \left| \tfrac{1}{\sqrt{2}} \left(\langle \uparrow_1 \downarrow_2| + \langle \downarrow_1 \uparrow_2| \right) | \left(H_{\text{M1}}^1 + H_{\text{M1}}^2 \right) |\uparrow_1 \uparrow_2\rangle \right|^2$$
$$= \left| \tfrac{1}{\sqrt{2}} \langle \downarrow_1 |H_{\text{M1}}^1| \uparrow_1\rangle + \tfrac{1}{\sqrt{2}} \langle \downarrow_2 |H_{\text{M1}}^2| \uparrow_2\rangle \right|^2$$
$$= 2R_1$$

$$|1,0\rangle \qquad R = |\langle 1,-1|H_{\text{M1}}|1,0\rangle|^2$$
$$= \left| \langle \downarrow_1 \downarrow_2 | \left(H_{\text{M1}} 1 + H_{\text{M1}}^2 \right) |1/\sqrt{2} \left(|\uparrow_1 \downarrow_2\rangle + |\downarrow_1 \uparrow_2\rangle \right) \rangle \right|^2$$
$$= \left| \tfrac{1}{\sqrt{2}} \langle \downarrow_1 |H_{\text{M1}}^1| \uparrow_1\rangle + \tfrac{1}{\sqrt{2}} \langle \downarrow_2 |H_{\text{M1}}^2| \uparrow_2\rangle \right|^2$$
$$= 2R_1$$

$$|1,-1\rangle \qquad R = 0$$

(ii) **Singlet** — the $|0,0\rangle$ state cannot decay, and hence

$$|0,0\rangle \qquad R = 0$$

(b) If the particles were to decay independently the $|1,+1\rangle$ state (or the $|\uparrow_1\uparrow_2\rangle$ state in the two-spin basis) would decay at the same rate $2R_1$, but to the ground state $|1,-1\rangle$ (or $|\downarrow_1\downarrow_2\rangle$). The $|1,0\rangle$ and $|0,0\rangle$ states have one excited spin and both would have decayed at R_1 to the same ground state $|1,-1\rangle$.

Thus it is important to include coherence in spontaneous emission to get the rates correct.

4. Spin echoes in extended media

The idea of this problem is to use lasers to create a coherent spin echo that propagates in a different direction than either of the incident pulses.

Assume a gas of identical two-state systems, each with resonant frequency $\omega_\circ$, and position $\vec{r}_i$ and velocity $\vec{v}_i$ at $t = 0$. Do not consider the Doppler shift explicitly—it is accounted for using $\vec{r}_i(t) = \vec{r}_i + \vec{v}_i t$ in the phase factors $\vec{k} \cdot \vec{r}$.

- At $t = 0$ apply a $\pi/2$ pulse with $\vec{k}_1$ along x.
- At $t = T$ apply a π pulse with $\vec{k}_2$ at an angle α from x.

Due to the fact that $\alpha \neq 0$, the phase matching and Doppler cancellation will not be perfect. Nevertheless, there will be an echo for sufficiently small α.

(a) Find T_e and $\vec{k}_e$, the time and direction of the echo. Find expressions for the phase matching error and for the velocity space dephasing.

(b) By what factor will the intensity be reduced due to failure of perfect phase match? Assume $\lambda = 500$ nm, cell size 1 cm, and $\alpha = 3 \times 10^{-3}$.

(c) Crudely estimate the loss of signal due to velocity space mismatch—assume $v_{\text{rms}} = 10^3$ m/s and $T = 200$ ns.

Solution

Consider the phase of the ith radiator.

- $t = 0$: $\pi/2$ pulse with $\vec{k}_1$ along $\hat{x}$.

$$\phi_i(t \leq T) = \vec{k}_1 \cdot \vec{r}_i - \omega_\circ t$$

- $t = T$: π pulse with $\vec{k}_2$ at an angle α from $\hat{x}$

$$\phi_\pi = \vec{k}_2 \cdot (\vec{r}_i + \vec{v}_i T)$$

Therefore the phases are

$$\phi_i(T_+) = 2\phi_\pi - \phi_i(T_-) = 2\vec{k}_2 \cdot (\vec{r}_i + \vec{v}_i T) - \vec{k}_1 \cdot \vec{r}_i + \omega_\circ T$$
$$\phi_i(t > T) = 2\vec{k}_2 \cdot (\vec{r}_i + \vec{v}_i T) - \vec{k}_1 \cdot \vec{r}_i - \omega_\circ (t - 2T)$$

Echo radiation adds a phase $-\vec{k}_e \cdot (\vec{r}_i + \vec{v}_i t)$; therefore the final phase is

$$\phi_i\,(t > T) = \left(2\vec{k}_2 - \vec{k}_1 - \vec{k}_e\right) \cdot \vec{r}_i + \left(2\vec{k}_2 T - \vec{k}_e T - \vec{k}_e t\right) \cdot \vec{v}_i - \omega_\circ\,(t - 2T)$$

(a) The echo occurs when the above phase in minimized for all i, i.e. it is coherent over the sample. Since the position and velocity are unpolarized (isotropic) in a gas, $\vec{k}_e$ is obtained by minimizing $\left(2\vec{k}_2 - \vec{k}_1 - \vec{k}_e\right)$ and T_e is obtained by minimizing $\left(2\vec{k}_2 T - \vec{k}_e t\right)$. All oscillators are identical so the $\omega_\circ$ term is the same.

By energy conservation

$$|\vec{k}_1| = |\vec{k}_2| = |\vec{k}_e| = k$$

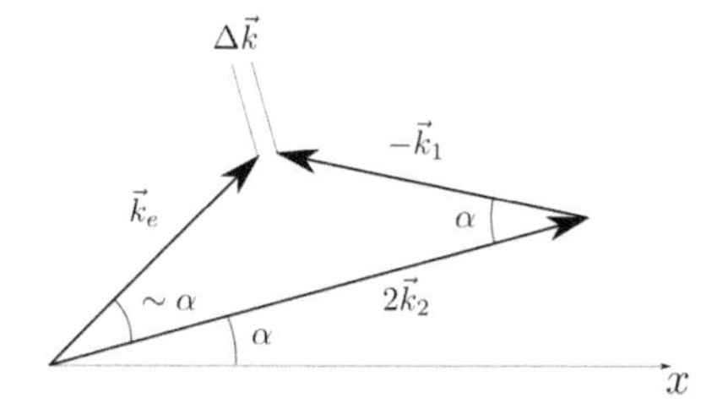

As can be seen from the figure, $\left(2\vec{k}_2 - \vec{k}_1 - \vec{k}_e\right)$ is minimized when $\vec{k}_e$ is at an angle 2α from x, i.e. along $\left(2\vec{k}_2 - \vec{k}_1\right)$.

Therefore the time of the echo is such that

$$|2\vec{k}_2 T - \vec{k}_e T_e| = \text{minimum}$$

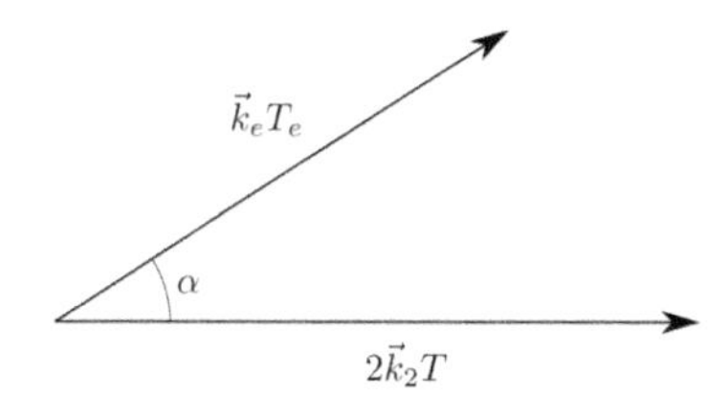

The figure shows that this happens when

$$kT_e = 2kT\cos\alpha \qquad \Longrightarrow \qquad T_e = 2T\cos\alpha \approx 2T \qquad (\alpha \ll 1)$$

(i) **Phase-matching error** is

$$\Delta k = |2\vec{k}_2 - \vec{k}_1| - k$$

Using

$$\vec{k}_2 = k\left(\cos\alpha\hat{x} + \sin\alpha\hat{y}\right) \qquad \text{and} \qquad \vec{k}_1 = k\hat{x}$$

we get

$$\Delta k = k\left[\sqrt{(2\cos\alpha - 1)^2 + (2\sin\alpha)^2} - 1\right] = k\left[\sqrt{5 - 4\cos\alpha} - 1\right]$$

For $\alpha \ll 1$, we have

$$\Delta k \approx k\left[\sqrt{1 + 2\alpha^2} - 1\right] = \alpha^2 k$$

(ii) **Velocity space dephasing** is

$$\Delta\phi_v = \left(2\vec{k}_2 T - 2\vec{k}_e T\right)\cdot\vec{v}_i = 2T\left(\vec{k}_2 - \vec{k}_e\right)\cdot\vec{v}_i$$

Therefore for $\alpha \ll 1$, we get

$$\Delta\phi_v \approx 2T\vec{v}_i\cdot\hat{y}\,k\sin\alpha \approx 2\alpha kT\,\vec{v}_i\cdot\hat{y}$$

(b) The intensity reduction due to phase mismatch is

$$I \sim \frac{\sin^2(\Delta kL/2)}{(\Delta kL/2)^2}\,I_\circ$$

For cell extending from $-L/2$ to $L/2$

$$\frac{\Delta kL}{2} \approx \frac{\alpha^2 kL}{2} = \frac{\alpha^2\pi L}{\lambda}$$

Using the values of $\lambda = 500$ nm, $L = 1$ cm, and $\alpha = 3\times10^{-3}$, we get

$$I = 0.8978\,I_\circ$$

(c) We can estimate the loss in intensity due to $\Delta\phi_v$ by assuming a uniform velocity distribution from $-v_{\text{rms}}$ to $+v_{\text{rms}}$ along y. Then

$$I \sim \frac{\sin^2(2\alpha kTv_{\text{rms}})}{(2\alpha kTv_{\text{rms}})^2}\,I_\circ$$

The term in brackets is

$$2\alpha\,kTv_{\text{rms}} = 2\alpha\frac{2\pi}{\lambda}Tv_{\text{rms}} = 15$$

Therefore

$$I \approx 1.5\times10^{-3}\,I_\circ$$

Chapter 9

Lineshapes

No resonance line is infinitely narrow, even in theory, because spontaneous emission always introduces a finite lifetime to the upper state and therefore (by the uncertainty principle) a natural width to the transition. In practice, a number of other processes such as collisions, Doppler broadening, power broadening, field inhomogeneities, etc., also add width to resonance lines. It is important to study the effect of these processes on the frequency dependence of the transition probability (the **lineshape**) for two important reasons: (i) it is not possible to find the resonance frequency $\omega_\circ$ with high precision unless the line (or especially several overlapping lines) can be fitted accurately, and (ii) it is not possible to estimate either the magnitude of the "signal" or its dependence on the intensity of the driving radiation without understanding the broadening mechanisms.

This chapter is devoted to the study of the different kinds of lineshapes encountered in spectroscopy experiments. We will also see the description of Gaussian beams in optical systems because the Gaussian function represents the intensity profile of lasers used in such experiments.

A. Low intensity and simple collisions

1. Homogeneous vs inhomogeneous broadening

It is customary to divide broadening mechanism into two classes—homogeneous and inhomogeneous.

(i) Homogeneous

This broadening results from a similar widening of the transition probability curve for each atom, e.g. the natural linewidth or collision-induced mechanisms. In homogeneous broadening, all of the atoms exchange energy with the impressed radiation field equally. Since this kind of broadening arises from random interruptions in the coherence of the radiation from the atoms, it results in a mathematical lineshape called a Lorentzian. The normalized (so that $\int_{-\infty}^{\infty} \mathscr{L}\, d\omega = 1$) functional form is

$$\mathscr{L}(\omega) = \frac{1}{2\pi} \frac{\Gamma}{(\omega - \omega_\circ)^2 + (\Gamma/2)^2}$$

where $\omega_\circ$ is the center frequency and Γ is the linewidth (FWHM).

(ii) Inhomogeneous

This broadening results from a small random shift of the resonance frequency for different atoms which widens the transition probability curve for the ensemble of atoms, e.g. Doppler effect or inhomogeneous fields (in solids). In inhomogeneously broadened systems only a fraction of the atoms exchange energy efficiently with the field at a given frequency (and time). It results from random perturbations of the frequency of the atoms which generally follow a normal (Gaussian) distribution. Thus the mathematical form of the lineshape is Gaussian, which has the normalized form

$$G(\omega) = \frac{1}{\sqrt{2\pi}\,\sigma} \exp\left[-\frac{(\omega - \omega_\circ)^2}{2\sigma^2}\right]$$

where $\omega_\circ$ is the center frequency, and σ is the rms value of $\overline{(\omega - \omega_\circ)}$. If we want to express the linewidth as FWHM, then it is $2\sqrt{2\ln 2}\,\sigma = 2.355\,\sigma$.

The distinction between homogeneous and inhomogeneous broadening can often be made only approximately. For example the (homogeneous) broadening produced by collisions may be different for atoms in different parts of the (inhomogeneous) Doppler profile owing to the fact that the atoms in the tail of the Doppler profile are moving faster and suffer collisions more frequently. Interatomic forces between colliding atoms can shift the frequency of the radiation slightly, causing asymmetric lineshapes. Therefore, throughout this chapter, we shall assume that the collisions are simple—they occur at a constant rate, and they are instantaneous. Moreover we shall assume that the radiation intensities are low in order to avoid saturation effects.

2. Lorentzian line

Let us assume that there is some mechanism which **randomly** limits the duration of the interaction of the radiating (or absorbing) subsystems with the oscillating field. This can arise from the natural decay of the excited state, from collisional interruption of the oscillators (T_2 type processes), from loss of atoms, etc.—all that matters is that it is a simple stochastic process characterized by

Γ randomization rate for radiating subsystems

$1/\Gamma$ mean lifetime of uninterrupted interaction

For such a process the probability of uninterrupted interaction for a time between t and $(t+dt)$ is $f(t)dt$ where the distribution function is

$$f(t) = \Gamma e^{-\Gamma t} \qquad \text{with} \qquad \int_0^\infty f(t)dt = 1$$

Now let us investigate the lineshape for an ensemble of two-state subsystems that interact with the field for a random time discussed above. For each subsystem the (Rabi) transition probability is

$$P(\omega, t) = \frac{\omega_R^2}{\omega_R'^2} \sin^2\left(\frac{\omega_R' t}{2}\right) \qquad \text{with} \qquad \omega_R'^2 = (\omega - \omega_\circ)^2 + \omega_R^2$$

when the subsystem has interacted for time t with radiation at ω.

Thus the ensemble average of the transition probability is

$$\begin{aligned}
\langle P \rangle_\Gamma &= \int_0^\infty P(\omega, t) f(t)\, dt \\
&= \frac{\omega_R^2}{\omega_R'^2} \Gamma \int_0^\infty \sin^2\left(\frac{\omega_R' t}{2}\right) e^{-\Gamma t}\, dt \\
&= \frac{\omega_R^2}{2\omega_R'^2} \Gamma \int_0^\infty (1 - \cos\omega_R' t)\, e^{-\Gamma t}\, dt \\
&= \frac{\omega_R^2}{2\omega_R'^2} \left[1 - \frac{\Gamma^2}{\Gamma^2 + \omega_R'^2}\right] \\
&= \frac{\omega_R^2}{2\omega_R'^2} \left[\frac{\omega_R'^2}{\Gamma^2 + \omega_R'^2}\right] \\
&= \frac{1}{2} \left[\frac{\omega_R^2}{(\omega - \omega_\circ)^2 + \omega_R^2 + \Gamma^2}\right]
\end{aligned} \tag{9.1}$$

which shows that the probability distribution has a Lorentzian lineshape. The terms in the denominator are, respectively, the frequency offset, the power broadening factor, and the randomization rate.

This average was carried out using t to represent the duration of interaction —in a real system one imagines that these intervals all extend backward in time from the moment of observation.

As we have seen in Chapter 5, "Resonance," ω_R^2 is proportional to the intensity of the incident wave—in fact ω_R is the natural measure of this intensity because it tells the frequency at which the subsystems are cycled back and forth between the two states (on resonance). It should be noted that the squares of the power broadening and randomization rates add, whereas, if several processes (e.g. collisions, natural decay) act together, the randomization rate Γ is the sum of the Γ's for each of them. The combined FWHM of the resonance curve given above is

$$\Gamma_F = 2\sqrt{\omega_R^2 + \Gamma^2}$$

which shows that the width is 2Γ in the low power limit and becomes $2\omega_R$ at high powers. Finally, note that the ensemble average has removed oscillations both in frequency and in time that were present in the two-state transition probability.

3. Spontaneous decay lineshape

Weisskopf and Wigner* considered the problem of spontaneous emission, showing that to a very good approximation it caused the excited state population to decay exponentially with a rate given by the Einstein A coefficient. (Slight deviations from the exponential law are predicted at very short times and also at very long times; no deviation has yet been seen experimentally.) Phenomenologically speaking this decay can be represented by adding a small imaginary term to the energy "eigenvalue" for each excited state

$$E'_a = E_a - i\hbar\Gamma_a/2 \qquad \Longrightarrow \qquad \omega'_a = \omega_a - i\Gamma_a/2$$

where $\Gamma_a = A$.

Then the decaying behavior appears in the expression for the wavefunction

$$|\psi_a(t)\rangle = e^{-iHt/\hbar}\,|\psi_a(0)\rangle = e^{-\Gamma_a t/2}e^{-iE_a t/\hbar}\,|\psi_a(0)\rangle$$

and the probability of being in state $|a\rangle$

$$P_a(t) \sim \psi_a^*\psi_a = e^{-\Gamma_a t}P_a(0)$$

decays at the rate Γ_a.

Since the decay of the excited state is exponential, the lineshape of a naturally broadened transition is Lorentzian. This works out formally when substituted in the transition for a two-state system (the $\sin^2$ term is replaced by its average value of $1/2$)

$$\begin{aligned}
P(\omega) &= \frac{1}{2}\frac{\omega_R^2}{|\omega'_\circ - \omega|^2 + \omega_R^2} \\
&= \frac{1}{2}\frac{\omega_R^2}{|\omega_a - i\Gamma_a/2 - \omega_b - \omega|^2 + \omega_R^2} \qquad \text{taking } \omega_a > \omega_b \\
&= \frac{1}{2}\frac{\omega_R^2}{|(\omega_\circ - \omega) - i\Gamma_a/2|^2 + \omega_R^2} \\
&= \frac{1}{2}\frac{\omega_R^2}{(\omega_\circ - \omega)^2 + (\Gamma_a/2)^2 + \omega_R^2}
\end{aligned}$$

This agrees with the expression in Eq. (9.1) for two-state atoms in which the probabilities of the two states were taken to decay at a combined rate of Γ, when Γ is identified with $\Gamma_a/2$. This identification is required because only one state decays in the present example. In general

$$\Gamma = (\Gamma_a + \Gamma_b)/2$$

*V. F. Weisskopf and E. P. Wigner, "Calculation of the natural brightness of spectral lines on the basis of Dirac's theory," *Z. Phys.* **63**, 54–73 (1930).

if both states can decay. Note that in this case one must make the replacement

$$\omega'_{ab} = \omega_{ab} - i(\Gamma_a + \Gamma_b)$$

because with a complex eigenvalue for each state and using $\omega'_{ab} = \omega'_a - \omega'_b$ will give (incorrectly) $\Gamma = (\Gamma_a - \Gamma_b)/2$.

B. Relativistic effects in emission and absorption

Doppler broadening arises from the motion of the absorbing or emitting atoms relative to the observer. It is an inhomogeneous form of broadening and is nearly always the limit on attainable resolution in conventional spectroscopy (because it decreases only as $T^{1/2}$ and is therefore not easy to eliminate).

Doppler broadening is but one manifestation of special relativity. Several others are important in high precision resonance physics and we shall give a treatment that is general enough to include all effects observed so far.

1. Photon recoil

The (angular) frequency of a photon given off by an atomic system of mass M is making a transition between two states by an energy E_T is not exactly $\omega_\circ = E_T/\hbar$. The discrepancy arises because of the recoil of the atomic system and is termed **photon recoil** (although atomic recoil might be a better term!).

Consider an atom at rest in its excited state. Its total energy W is

$$W = Mc^2 + E_T$$

The emitted photon has energy $E_e = \hbar\omega_e$ and momentum $p_\gamma = \hbar\omega_e/c = E_e/c$.

By momentum conservation, this momentum gets transferred to the atom and is felt by the atom as a recoil, as shown in Fig. 9.1 below.

p_γ ω_e

Figure 9.1: Recoil momentum felt by atom after it emits a photon.

The energies before and after the decay are

$$W_i = Mc^2 + E_T \qquad \text{and} \qquad W_f = Mc^2 + \frac{p_\gamma^2}{2M} + E_e$$

where $p_\gamma^2/2M$ is the recoil energy of the atom. By energy conservation ($W_i = W_f$) we have

$$E_T = E_e\left(1 + \frac{E_e}{2Mc^2}\right)$$

which gives

$$E_e \approx E_T\left(1-\frac{E_T}{2Mc^2}\right) \qquad \Longrightarrow \qquad \omega_e \approx \omega_\circ\left(1-\frac{\hbar\omega_\circ}{2Mc^2}\right) \tag{9.2}$$

The difference between ω_e and $\omega_\circ$ is quite small since $Mc^2 \approx A \times 10^9$ eV, where A is the atomic number.

In order to be absorbed, a photon must have a frequency

$$\omega_a = \omega_\circ\left(1+\frac{\hbar\omega_\circ}{2Mc^2}\right)$$

which is greater than $\omega_\circ$ because some of the energy must go into kinetic energy of the atom. The recoil of an atom due to the emission or absorption of a photon depends on the direction of the photon and thus produces kinetic splittings of the line in high precision spectroscopy employing photons moving in several directions.

2. Doppler shift

Now consider the transformation of the photon back into the laboratory coordinate system in which the atom is initially moving with velocity $\vec{v}$. Assume that the atom emits a photon at an angle θ with respect to its velocity $\vec{v}$, as shown in Fig. 9.2 below.

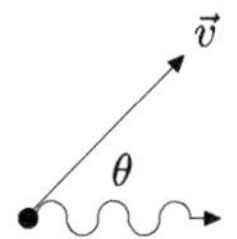

Figure 9.2: Angle θ between atomic velocity and direction of photon emission.

The relativistic Doppler formula gives

$$\omega_{Le} = \frac{\omega_e}{\gamma\,(1-\beta\cos\theta)}$$

where

$$\gamma = \sqrt{\frac{1}{1-\beta^2}} \qquad \text{and} \qquad \beta = \frac{v}{c}$$

Thus the frequency in the lab frame is red shifted or blue shifted from the frequency of the emitted photon depending on whether $\theta < \pi/2$ or $\theta > \pi/2$.

Specifically, the frequency in the lab frame can be written as

$$\omega_{Le} = \omega_e \left[1 - \left(\frac{v}{c}\right)^2\right]^{1/2} \left[1 - \frac{v}{c}\cos\theta\right]^{-1}$$

$$\approx \omega_e \left[1 + \frac{v\cos\theta}{c} - \left(\frac{v}{c}\right)^2 \left(\frac{1}{2} - \cos^2\theta\right)\right]$$

The first term is the first-order Doppler shift ($v\cos\theta$ is the velocity component along the direction of the photon), and the second term is the second-order Doppler shift.

If we now add the recoil term from Eq. (9.2) we get

$$\omega_{Le} \approx \omega_\circ \left[1 + \frac{v\cos\theta}{c} - \left(\frac{v}{c}\right)^2 \left(\frac{1}{2} - \cos^2\theta\right) - \frac{\hbar\omega_\circ}{2Mc^2}\right]$$

In most experimental situations the second-order Doppler term (which is of order $k_B T/Mc^2$) is even smaller than the photon recoil term, which is itself usually unobservable. Hence only the first-order Doppler shift is important, and one finds for the laboratory frequency of the photons

$$\omega_L \approx \omega_\circ \left[1 + \frac{v\cos\theta}{c}\right] = \omega_\circ \left[1 + \frac{v_\ell}{c}\right] = \omega_\circ + \vec{k}\cdot\vec{v}$$

where v_ℓ is the component of the atom's velocity along the emitted photon, and $\vec{k}$ is the photon wavevector. This formula also applies to absorption, since the only difference between laboratory absorption and emission frequencies comes from the recoil term.

C. Lineshape of atoms in a gas

1. Gaussian distribution

When a gas of atoms of mass M is at equilibrium at temperature T, the velocity distribution is a Maxwell–Boltzmann distribution. The most familiar form of this distribution is actually for the distribution of speeds, and is proportional to $v^2 e^{-Mv^2/2k_BT}$. In order to determine the lineshape for emission or absorption of radiation from a gas it is necessary to use the (simpler) one dimensional form of this distribution

$$\begin{aligned} f(v_\ell)\, dv_\ell &= \sqrt{\frac{M}{2\pi k_B T}} \exp\left(-\frac{Mv_\ell^2}{2k_BT}\right) dv_\ell \\ &= \frac{1}{\sqrt{2\pi} v_T} \exp\left(-\frac{v_\ell^2}{2v_T^2}\right) dv_\ell \qquad \text{where} \qquad v_T = \sqrt{\frac{k_BT}{M}} \end{aligned}$$

v_T is the characteristic thermal speed of an atom in the gas.

Combining the above distribution function with the first-order Doppler shift results in the frequency-space distribution of the Doppler-broadened line. Setting

$$\omega - \omega_\circ = \omega_\circ \frac{v_\ell}{c} = 2\pi \nu_\circ v_\ell = k v_\ell$$

one finds

$$\mathscr{D}(\omega)\, d\omega = \frac{1}{\sqrt{2\pi}\, \omega_D} \exp\left[-\frac{(\omega - \omega_\circ)^2}{2\omega_D^2}\right] d\omega \tag{9.3}$$

The function $\mathscr{D}(\omega)$ is the fractional strength of the system at frequency ω and

$$\omega_D = 2\pi\nu_\circ \frac{v_T}{c} = 2\pi \frac{v_T}{\lambda_\circ} = k v_T$$

is the Doppler width. At room temperature an atom of mass 40 amu has $\omega_D/2\pi = 0.5$ GHz at 5000 Å.

The distribution above has the form of the normalized Gaussian function seen at the beginning of this chapter.

2. Voigt profile

The preceding expression for the Doppler profile of a spectral line neglected any other sources of broadening of the transition. When these are included the actual lineshape becomes a convolution of the Gaussian and the Lorentzian functions

$$\mathscr{V}(\omega - \omega_\circ, \Gamma_F, \omega_D) = \frac{1}{2\pi\sqrt{2\pi}\,\omega_D} \times \int_{-\infty}^{\infty} d\omega' \frac{\Gamma_F}{(\omega' - \omega_\circ)^2 + (\Gamma_F/2)^2} \exp\left[-\frac{(\omega' - \omega_\circ)^2}{2\omega_D^2}\right] \tag{9.4}$$

assuming that the other broadening mechanisms are homogeneous—i.e. having Lorentzian lineshapes and all linewidths adding to an effective Γ_F. The above lineshape is known as the **Voigt profile** and cannot be evaluated analytically; but commercial programs are available to use different ratios of Γ_F and ω_D to get the best fit to experimental data.

For real gases at low pressures (<1 torr at visible frequencies) the line is Gaussian near line center, and Eq. (9.3) is a good approximation to Eq. (9.4). This situation arises because the natural linewidth is typically 100 times smaller than ω_D and the integrand in Eq. (9.4) peaks near $\omega' = \omega_\circ$.

However, there are two frequently encountered cases in which Eq. (9.3) is a poor approximation to Eq. (9.4). One occurs when the randomization rate (sum of natural decay, collisional processes, etc.) becomes comparable with ω_D so that many values of ω' contribute to $\mathscr{V}$. The other occurs far from line center where the Gaussian falls off faster than the Lorentzian; the dominant contribution to $\mathscr{V}$ is then a Lorentzian tail which comes from the region $\omega' \approx \omega_\circ$ in the integral in Eq. (9.4) and gives a line amplitude equal to $\Gamma_F/(\omega - \omega_\circ)^2$ for large detunings.

D. Confined particles

Trapped particles offer the possibility of reaching the ultimate in spectroscopic precision: cooling the particle with lasers or electronics can reduce second-order Doppler shifts at least to 10^{-17} (for 1 mK and atomic mass 10 amu), proper design of the cavity can suppress spontaneous emission, and collisions can be virtually eliminated for single trapped particles in cryogenically pumped environments. The first-order Doppler shift can be entirely eliminated also, in spite of the fact that v/c is not particularly small ($\sim 10^{-8}$ in the above example). Suppression of the first-order Doppler shift results from the nature of the spectrum emitted/absorbed by a trapped particle—it consists of an unshifted central line with sidebands spaced apart by multiples of the frequency of oscillation. The amplitude of the sidebands may be reduced by lowering the amplitude of oscillation of the trapped particle, but it is also possible to address spectroscopically the unshifted central line—this approach underlies the Mössbauer effect, as well as the use of buffer gases and specially coated containers to get narrow spectra.

1. Spectrum of oscillating emitter

We now consider the lineshape of the emission spectrum of a harmonically bound particle.* If the particle oscillates with amplitude $x_\circ$ at frequency ω_t, then the phase of radiation emitted by the atom towards a detector situated at large x will contain the term

$$\phi(t) = -kx(t) - \omega_\circ t = -kx_\circ \sin\omega_t t - \omega_\circ t \tag{9.5}$$

A wave with this phase will have an instantaneous frequency

$$\omega(t) = -\dot{\phi}(t) = kx_\circ\omega_t \cos\omega_t t + \omega_\circ = kv(t) + \omega_\circ \tag{9.6}$$

consistent with the usual Doppler shift term into the lab system of $\vec{k}\cdot\vec{v}$.

In the parlance of electrical engineering, signals with the above phase and frequency correspond to phase and frequency modulation, respectively. We shall find the spectrum from the phase expression in Eq. (9.5) since the amplitude of the phase oscillation is the physically important **modulation index** β, which is just the maximum phase shift relative to $\omega_\circ t$

$$\beta = kx_\circ$$

This approach also avoids the common pitfall of assuming that the phase corresponding to frequency modulation is $-\omega(t)t$. Thus we must find the spectrum of a wave whose amplitude is proportional to

$$a(t) = \cos\,\phi(t) = \cos\,(\omega_\circ t + \beta\sin\omega_t t)$$

*The absorption spectrum has the same shape, so we do not need to consider it separately.

The spectrum will contain n^{th} order Bessel functions J_n. Some algebra using the following identities

$$\cos(z \sin\theta) = J_\circ(z) + 2\sum_{k=1}^{\infty} J_{2k}(z)\cos(2k\,\theta)$$

$$\sin(z \sin\theta) = 2\sum_{k=0}^{\infty} J_{2k+1}(z)\sin[(2k+1)\theta]$$

gives

$$a(t) = \sum_{n=-\infty}^{\infty} (-1)^n J_n(\beta)\cos\left[(\omega_\circ + n\omega_t)t\right]$$

Obviously the system does not have a continuous lineshape, rather the emission is either at $\omega_\circ$ or sidebands that differ from $\omega_\circ$ by a multiple of the trapping frequency, $n\omega_t$. Physically this results from the exactly periodic nature of the motion. The probability of emission at frequency $\omega_\circ + n\omega_t$ is simply $J_n^2(\beta)$, hence the intensity spectrum of the emitted light is given by

$$I(\omega) \propto \sum_{n=-\infty}^{\infty} J_n^2(\beta)\,\delta(\omega - \omega_\circ - n\omega_t) \tag{9.7}$$

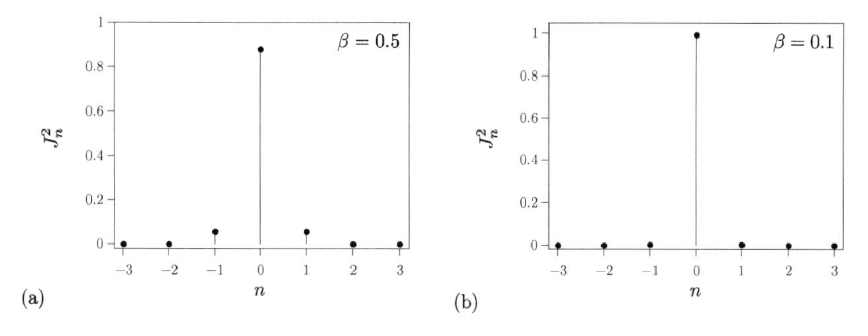

Figure 9.3: Spectral intensity of central peak and sidebands in FM modulation with modulation index β. (a) $\beta = 0.5$. (b) $\beta = 0.1$, showing nearly zero intensity in the sidebands.

The effect of β on the spectral intensity is shown graphically in Fig. 9.3. The height of the central peak and three sidebands are shown in the figure for two values of β. For $\beta = 0.5$ shown in Fig. 9.3(a), only the $n = \pm 1$ sidebands have significant intensity, while the others are practically zero. For $\beta = 0.1$ shown in the Fig. 9.3(b), all the sidebands are almost zero. Thus the sidebands get increasingly suppressed as β becomes small compared to 1, or equivalently (since $\beta = kx_\circ$) as $x_\circ$ becomes small compared to λ.

An alternative and intuitively appealing derivation of these results is to consider the quantum number of the bound oscillating particle explicitly in the calculation. Then the initial state is an atom in state $|b\rangle$ trapped in quantum state n_i of the trap; after emission the atom is in state $|a\rangle$ in quantum state n_f of the trap. The frequency of the emitted photon determined from energy conservation

$$\begin{aligned} \hbar\omega &= \hbar\omega_{ba} + E_i^{\text{trap}} - E_f^{\text{trap}} && \text{in general} \\ &= \hbar(\omega_{ba} + n\omega_t) && \text{for harmonic trap with } n = n_i - n_f \end{aligned}$$

This expression needs no correction for recoil since the initial and final kinetic energies of the atom are explicitly accounted for in E_i^{trap} and E_f^{trap}. The transition rate is

$$R = \frac{4k^3}{3h} |\langle b|\hat{\varepsilon} \cdot \vec{p}|a\rangle|^2 |\langle \psi_f^{\text{trap}}|e^{-i\vec{k}\cdot\vec{r}}|\psi_i^{\text{trap}}\rangle|^2 \tag{9.8}$$

The second term is the confinement factor and depends on the phase variation of the outgoing wave and the trap eigenstates $|\psi_f\rangle$ and $|\psi_i\rangle$. If this matrix element is evaluated in the momentum representation, $e^{-i\vec{k}\cdot\vec{r}}$ is a translation operator (with $-\hbar k$) so this factor becomes

$$|\langle \phi_f(\vec{p} - \hbar k)|\phi_i(\vec{p})\rangle|^2$$

In the case of a harmonic oscillator with $n_i, n_j \gg 1$, the confinement factor will yield the Bessel function expression consistent with Eq. (9.3).

The preceding view bears much similarly to electronic transitions in molecular spectroscopy in that an electronic transition occurs between two states with quantized vibrational motion. Indeed, the matrix element involving the trap states in Eq. (9.8) is analogous to the Franck–Condon factor in molecular spectroscopy (except for $e^{i\vec{k}\cdot\vec{r}} \approx 1$ owing to the small size of atoms, the matrix element of $\vec{p}$ depends on $\vec{r}$ and must be brought inside the spatial matrix element). This association emphasizes the generality of Eq. (9.8)—it applies equally to non-harmonic traps, and even to traps (as for neutral atoms) in which the confinement potential differs for state $|a\rangle$ and $|b\rangle$.

2. Tight confinement

The most dramatic effects associated with tightly confined radiators occur when the particles are confined to dimensions smaller than one wavelength of the emitted light—tight confinement also called the **Lamb–Dicke regime**. This is evident from the suppression of sidebands seen in Fig. 9.3. It can also be derived from the confinement matrix element in Eq. (9.8): if the spatial extent of the wavefunctions associated with ψ^{trap} is $\ll \lambda$, then

$\vec{k} \cdot \vec{r} \ll 1$ and it is reasonable to expand $e^{-i\vec{k}\cdot\vec{r}} \approx 1 - i\vec{k} \cdot \vec{r}$. The first term will give the selection rule $i = f$ (since the ϕ_i's are orthonormal) and hence $\omega = \omega_{ba}$ exactly. The second term will have matrix elements of order $r\,\lambda$ which is $\ll 1$.

(i) Recoilless emission

Emission with $i = f$ is called recoilless emission because the atom has the same momentum distribution after the emission as before. The momentum of the photon is provided (or taken up in the case of absorption) by the trap itself. This is analogous to the Mössbauer effect in which the momentum is taken up by the crystal as a whole. There the confinement matrix element with $f = i$ is called Debye–Waller factor.

(ii) Sideband cooling

Strongly confined particles may be radiatively cooled to very low temperatures by a technique called sideband cooling, which we will see in detail in Chapter 11, "Cooling and Trapping." Briefly, a tightly confined particle is excited at frequency $\omega_c = \omega_{ba} - \omega_t$, i.e. on a motion-induced sideband. Since the subsequent spontaneous decay is most probably at ω_{ba}, one quantum of trap motion will be lost by the combined excitation/decay cycle. This works best when the spontaneous decay width Γ_s is $\ll \omega_t$, in which case it is possible to cool most of the particles to the lowest quantum state of the trap.

(iii) Dicke narrowing

It is not necessary to confine particles harmonically in order to achieve significant narrowing. In 1953 Dicke* pointed out that collisions could reduce the usual Doppler width substantially if two conditions were met—(i) the mean free path between collisions is much smaller than the wavelength, and (ii) the collisions do not destroy the coherence between the radiating states.

The essence of Dicke's argument was that gas collisions could be viewed as a succession of traps each with a different frequency (he considered traps with steep walls rather than harmonic springs, but this is immaterial). All particles would have a recoilless line at $\omega = \omega_{ba}$, and the average over the traps with different frequencies would average the other lines into a broad spectrum with approximately the original Doppler width. He also gave the

*R. H. Dicke, "The effect of collisions upon the Doppler width of spectral lines," *Phys. Rev.* **89**, 472–473 (1953).

results of a calculation in which the atom was allowed to diffuse randomly

$$I(\delta) = \frac{I_\circ(\Gamma/2\pi)}{\delta^2 + (\Gamma/2)^2}$$

which is a Lorentzian (!) with FWHM

$$\Gamma = k^2 D/2$$

where D is the self-diffusion coefficient. This linewidth is roughly $2.8L/\lambda$ times the usual Doppler width (L is the mean free path). Therefore, the linewidth is greatly reduced below the Doppler width when the condition $L \ll \lambda$ is satisfied.

3. Weak confinement — The classical regime

Now consider the case in which the particle is weakly confined so that the amplitude of oscillation is many wavelengths. In this case the maximum phase shift is large and the spectrum will contain many sidebands. We refer to this as the classical regime because the quantization of frequencies in the spectrum may be neglected while attention is concentrated on the overall lineshape. The viewpoint is completely justified for weak traps in which the trapping frequency is less than the spontaneous linewidth—then the sidebands are too close to be resolved and the spectrum will be continuous.

In this classical regime, the lineshape may be determined simply by examining the instantaneous frequency [Eq. (9.6)] and determining the fraction of the time it has each particular frequency. Consider the time interval 0 to π/ω_t during which the frequency has each value only once, at the time

$$t(\delta) = \omega_t^{-1} \cos^{-1}(\delta/\omega_m)$$

where $\delta = \omega - \omega_\circ$ and $\omega_m = kx_\circ\omega_t$ is the maximum deviation of $\omega(t)$ from ω.

The probability density for emission between δ and $\delta + d\delta$ is

$$P(\delta)d\delta = |t(\delta + d\delta) - t(\delta)| \ \frac{\pi}{\omega_t}$$

$$= \frac{\omega_t}{\pi} \left| \frac{dt}{d\delta} \right| d\delta$$

Using the derivative of $\cos^{-1} x$ as $-1/\sqrt{1 - x^2}$ we get

$$P(\delta) = \frac{(\omega_t/\pi)\omega_t^{-1}\omega_m^{-1}}{\sqrt{1 - (\delta/\omega_m)^2}} = \frac{1}{\pi\omega_m} \frac{1}{\sqrt{1 - (\delta/\omega_m)^2}}$$

with the condition that $|\delta| \leq \omega_m$. The possibility of $\delta > \omega_m$ is forbidden classically but slightly allowed quantum mechanically.

E. Gaussian beam optics

The study of Gaussian beams in optical systems is motivated by the fact that most lasers used in spectroscopy experiments oscillate with a Gaussian distribution of the electric field

$$E_s = \mathcal{E}_\circ \exp\left[-\frac{r^2}{w^2}\right]$$

As a consequence, the intensity is also Gaussian with its radial distribution given by

$$I(r) = I_\circ \exp\left[-\frac{2r^2}{w^2}\right]$$

The parameter w—called the **Gaussian beam radius**—is the radius at which the intensity has decreased to $1/e^2$ or 0.135 of its peak value. At $2w$, the intensity has reduced to 0.0003 of its peak value which is completely negligible. Thus nearly 100% of the power is contained within a radius of $2w$.

If we want the power contained within a radius r, that is easily obtained by integrating the intensity distribution from 0 to r

$$P(r) = P(\infty)\left[1 - \exp\left(-\frac{2r^2}{w^2}\right)\right]$$

where $P(\infty)$ is the total power in the beam. This implies that the total power is related to the maximum intensity as

$$I_\circ = P(\infty)\left[\frac{2}{\pi w^2}\right]$$

Propagation of Gaussian beams through an optical system is simplified by the fact that the Fourier transform of a Gaussian function is also Gaussian. Thus the transverse intensity distribution remains Gaussian at every point in the system, independent of the presence of optical elements such as lenses and mirrors. The only things that change as the beam propagates are the radius of the beam and the radius of curvature of the wavefront.

Imagine that we somehow create a light beam with a Gaussian distribution and a plane wavefront at the position $x = 0$—this point is called the **beam waist** because the beam is narrowest here, and the radius $w_\circ$ here is called the beam waist radius. The beam size and wavefront curvature vary with x, as shown in Fig. 9.4. The beam size will increase, slowly at first and then faster, eventually becoming proportional to x. The radius of curvature of the wavefront, which was infinite at $x = 0$, will become finite and initially decrease with x. At some point it will reach a minimum value then increase with increasing x, eventually becoming proportional to x.

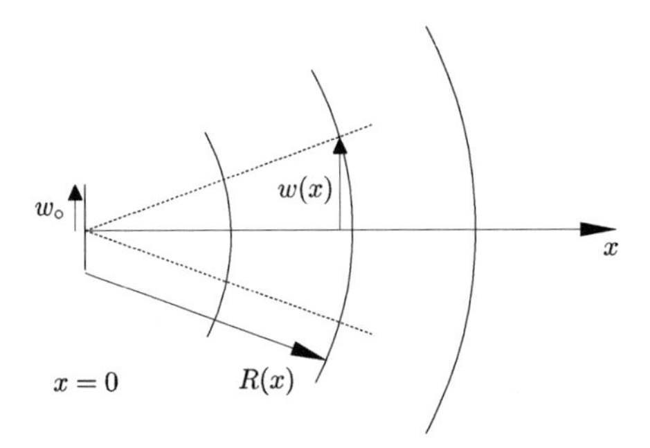

Figure 9.4: Divergence of a Gaussian beam as it propagates. The beam radius is smallest at the waist ($x = 0$) where the wavefront is plane.

The equations describing the Gaussian beam radius $w(x)$ and wavefront radius of curvature $R(x)$ are

$$w^2(x) = w_\circ^2 \left[1 + \left(\frac{\lambda x}{\pi w_\circ^2}\right)^2\right]$$

$$R(x) = x \left[1 + \left(\frac{\pi w_\circ^2}{\lambda x}\right)^2\right]$$

As expected from the diffraction of a wave, the beam diverges as if starting from a point, with the divergence depending on the waist radius and the wavelength. Since both the size and curvature depend on $w_\circ$ and λ in the same way, we are led to define a single parameter x_R—called the **Rayleigh range**—as

$$x_R = \frac{\pi w_\circ^2}{\lambda}$$

Thus x_R is the point at which the beam radius increases by a factor of $\sqrt{2}$ over its value at the beam waist. Similarly, the radius of curvature R has its minimum value at $x = x_R$.

F. Problems

1. Convolution of lineshapes

When two separate physical processes both contribute to the lineshape, the resultant lineshape is the convolution of the (normalized) distributions. Say $D_1(\omega - \omega_1)$ and $D_2(\omega - \omega_2)$ are the normalized lineshapes of the first and second type of process. The resultant lineshape is then their convolution

$$D_R(\omega - \omega_\circ) = \int_{-\infty}^{\infty} d\omega' \, D_1(\omega' - \omega_1) \, D_2(\omega' - \omega_2)$$

(a) Consider that D_1 and D_2 are both Lorentzian functions with FWHM's Γ_1 and Γ_2 respectively. Show that D_R is also a Lorentzian and find its FWHM.

(b) Do the same if D_1 and D_2 are Gaussian and Γ_1 and Γ_2 are the rms deviations.

(c) (i) Find the FWHM of a Gaussian with rms width Γ.

(ii) Find the rms width of a Lorentzian with FWHM Γ.

Solution

We make use of the convolution theorem, namely, that convolution in frequency domain is equivalent to multiplication in time domain.

(a) Two Lorentzians. Their normalized form is

$$D_i(\omega) = \frac{1}{\pi} \frac{\Gamma_i/2}{\omega^2 + (\Gamma_i/2)^2}$$

Fourier transform of $D_i(\omega)$ is

$$d_i(t) = e^{-\Gamma_i t/2}$$

Therefore

$$D_R(\omega) = D_i(\omega) \otimes D_2(\omega) \quad \stackrel{\text{FT}}{\Longleftrightarrow} \quad d_R(t) = d_1(t) \times d_2(t) = e^{-(\Gamma_1+\Gamma_2)|t|/2}$$

which shows that D_R is also Lorentzian with FWHM of $\Gamma = \Gamma_1 + \Gamma_2$.

(b) Two Gaussians. Their normalized form is

$$D_i(\omega) = \frac{1}{\sqrt{2\pi\Gamma_i^2}} \, e^{-\omega^2/2\Gamma_i^2}$$

Fourier transform of $D_i(\omega)$ is

$$d_i(t) = e^{-\Gamma_i^2 t^2/2}$$

Therefore

$$D_R(\omega) = D_1(\omega) \otimes D_2(\omega) \quad \stackrel{FT}{\Longleftrightarrow} \quad d_R(t) = d_1(t) \times d_2(t) = e^{-(\Gamma_1^2+\Gamma_2^2)t^2/2}$$

which shows that D_R is also a Gaussian function with rms width of $\Gamma = \sqrt{\Gamma_1^2 + \Gamma_2^2}$.

(c) (i) **FWHM of a Gaussian of rms width** Γ

A general Gaussian of amplitude A and width Γ has the form

$$D(\omega) = Ae^{-\omega^2/2\Gamma^2}$$

The maximum of this function is A and occurs at $\omega = 0$. Therefore the half maximum occurs at $\omega = \omega_{1/2}$ such that

$$Ae^{-\omega_{1/2}^2/2\Gamma^2} = \frac{A}{2}$$

which gives

$$\omega_{1/2}^2 = 2\Gamma^2 \ln 2 \qquad \Longrightarrow \qquad \omega_{1/2} = \pm\Gamma\sqrt{2\ln 2}$$

Therefore

$$\text{FWHM of Gaussian} = 2\Gamma\sqrt{2\ln 2}$$

(ii) **RMS width of a Lorentzian with FWHM** Γ

A general Lorentzian of amplitude A and FWHM Γ has the form

$$D(\omega) = A\frac{(\Gamma/2)^2}{\omega^2 + (\Gamma/2)^2}$$

The rms deviation is defined as

$$\text{rms} \equiv \sqrt{\langle (\omega - \langle\omega\rangle)^2 \rangle} = \sqrt{\langle\omega^2\rangle - \langle\omega\rangle^2}$$

For the Lorentzian, $\langle\omega\rangle = 0$, and

$$\langle\omega^2\rangle = A\left(\frac{\Gamma}{2}\right)^2 \int_{-\infty}^{\infty} d\omega' \frac{\omega'^2}{\omega'^2 + (\Gamma/2)^2} \qquad \longrightarrow \infty$$

Therefore the rms width of Lorentzian is undefined.

Physically this means that the Lorentzian function remains finite even when one goes far off resonance (its center frequency).

Chapter 10

Spectroscopy

LASERS have impacted our lives in a countless number of ways. Today they are found everywhere—in computer hard disk drives, CD players, grocery store scanners, and in the surgeon's kit. In research laboratories, almost everyone uses lasers for one reason or another. However, the greatest impact of lasers in physics has been in high-resolution spectroscopy of atoms and molecules. To see this, consider how spectroscopy was done before the advent of lasers. You would use a high-energy light source to excite all the transitions in the system, and then study the resulting emission "spectrum" as the atoms relaxed back to their ground states. This is like studying the modes of vibration of a box by hitting it with a sledgehammer and then separating the resulting sound into its different frequency components. A more gentle way of doing this would be to excite the system with a tuning fork of a given frequency. Then by changing the frequency of the tuning fork, one could build up the spectrum of the system. This is how you do laser spectroscopy with a tunable laser—you study the absorption of light by the atoms as you tune the laser frequency, and build up a resonance curve as you go across an atomic absorption.

In order to be able to do such high resolution laser spectroscopy, two things have to be satisfied. First, the atomic resonance should not be artificially broadened, e.g. due to Doppler broadening. Even with atoms at room temperature, the Doppler width can be 100 times the natural width, and can prevent closely spaced levels from being resolved. The second requirement is that the tunable laser should have a narrow "linewidth." The linewidth of the laser, or its frequency uncertainty, is like the width of the pen used to draw a curve on a sheet of paper. Obviously, you cannot draw a very fine curve if you have a broad pen. In this chapter, we will see how to narrow the linewidth of a diode laser in order to make it useful for spectroscopy. In addition, we will study techniques that make the spectrum Doppler-free. We will also study the phenomenon of nonlinear magneto-optic rotation, which has important applications in precision measurements.

A. Alkali atoms

The advent of low-cost diode lasers has revolutionized research involving atom-photon interactions. This is because the strongest transitions of most alkali atoms—the mainstay of such experiments—are accessible using diode lasers. In addition, all these alkali atoms (except Li) are sufficiently non-reactive with glass and have high enough vapor pressure at room temperature to allow vapor cells to be used for spectroscopy. In this section, we review the relevant properties of alkali atoms.

The alkali atoms belong to the first group of the periodic table: they are lithium, sodium, potassium, rubidium, cesium, and francium (radioactive). All of them are hydrogen-like in the sense that they have one valence electron, and inner ones just contribute to a core. Thus they have a ground state of $^2\mathrm{S}_{1/2}$. The first excited state corresponds to the valence electron being excited to a P orbital. Due to spin-orbit interaction this state splits into two states—$^2\mathrm{P}_{1/2}$ and $^2\mathrm{P}_{3/2}$—with the energy difference called the fine structure splitting. The $^2\mathrm{S}_{1/2} \to {}^2\mathrm{P}_{1/2}$ transition is called the D_1 line, while the $^2\mathrm{S}_{1/2} \to {}^2\mathrm{P}_{3/2}$ transition is called the D_2 line. Due to hyperfine interaction with the nucleus, the ground $^2\mathrm{S}_{1/2}$ state and the excited $^2\mathrm{P}_{1/2}$ state further split into two levels each, while the excited $^2\mathrm{P}_{3/2}$ state splits into three or four levels depending on the nuclear spin.

The relevant properties of the alkali atoms are listed in the table below. It is seen that the D lines* of all atoms except Na are accessible with diode lasers.

		$\mathrm{S}_{1/2}$	$\mathrm{P}_{1/2}$			$\mathrm{P}_{3/2}$		
	I	F	F	λ (nm)	$\Gamma/2\pi$ (MHz)	F	λ (nm)	$\Gamma/2\pi$ (MHz)
^{6}Li	1	$\frac{1}{2},\frac{3}{2}$	$\frac{1}{2},\frac{3}{2}$	671	5.9	$\frac{1}{2},\frac{3}{2},\frac{5}{2}$	671	5.9
^{7}Li	3/2	$1,2$	$1,2$	671	5.9	$0,1,2,3$	671	5.9
^{23}Na	3/2	$1,2$	$1,2$	589	9.8	$0,1,2,3$	589	9.8
^{39}K	3/2	$1,2$	$1,2$	770	6.0	$0,1,2,3$	767	6.0
^{41}K	3/2	$1,2$	$1,2$	770	6.0	$0,1,2,3$	767	6.0
^{85}Rb	5/2	$2,3$	$2,3$	795	5.8	$1,2,3,4$	780	6.1
^{87}Rb	3/2	$1,2$	$1,2$	795	5.8	$0,1,2,3$	780	6.1
^{133}Cs	7/2	$3,4$	$3,4$	894	4.6	$2,3,4,5$	852	5.2

*The latest technique to measure transition frequencies is by using a femtosecond frequency comb, which is detailed in Appendix D, "Frequency Comb."

To see the energy levels graphically, we show the low lying levels of the two isotopes of Li in Fig. 10.1. Li is unique in many ways, one of which is that the hyperfine levels in the $P_{3/2}$ state are inverted, with the smallest F value being the highest in energy. This is opposite of what is seen in other alkali atoms.

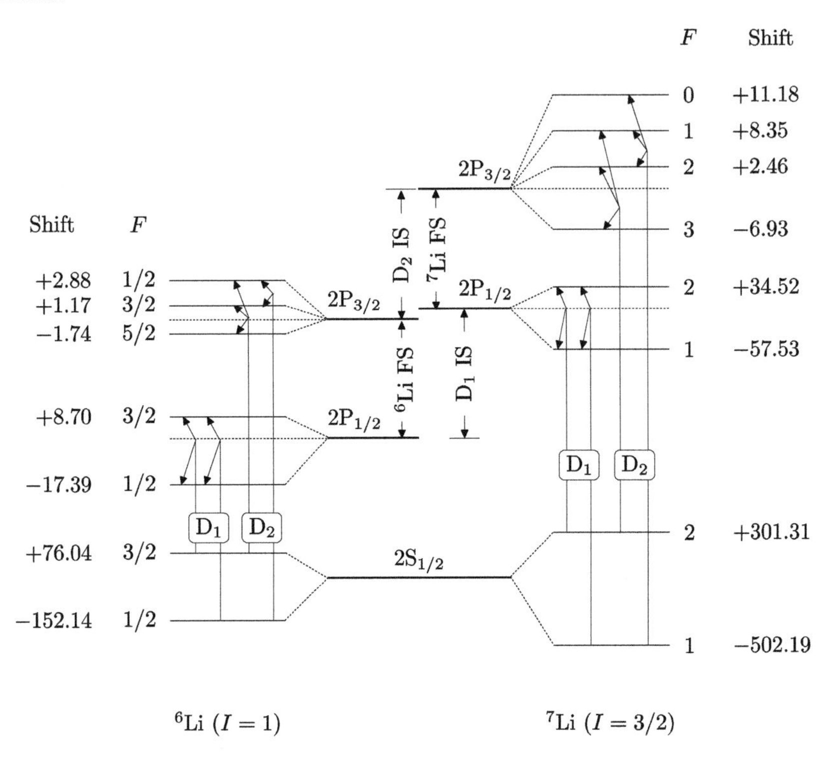

Figure 10.1: Energy levels of Li. Shown alongside each hyperfine level (labeled by its F value) is the shift in MHz from the unperturbed state. IS is isotope shift and FS is fine structure interval.

B. Experimental tools

In this section, we will see the different experimental tools used in a modern atomic and optical physics lab.

1. Diode laser

A commercial diode laser of the kind that is used in CD players has a linewidth of the order of a few GHz. But to be useful for atomic spectroscopy where transitions have linewidths of a few MHz, the laser linewidth should be reduced below 1 MHz. This is achieved by using optical feedback from a diffraction grating—called the Littrow configuration. This also serves the purpose of making the laser tunable by changing the angle of the grating. The grating is mounted on a piezo-electric transducer (PZT) so that the angle can be changed electronically.

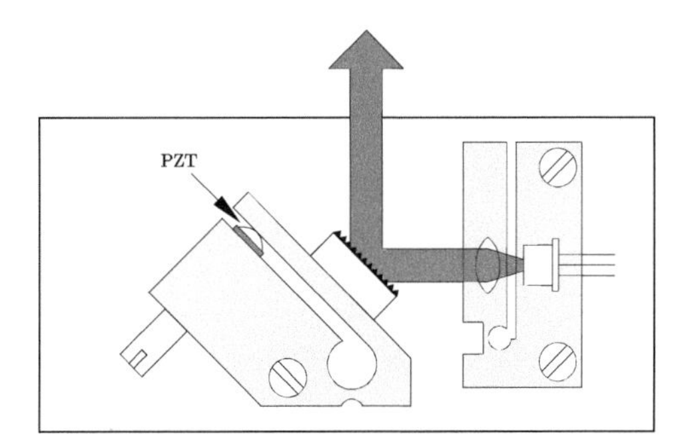

Figure 10.2: Diode laser stabilization in Littrow configuration. Optical feedback from a grating is used to reduce the linewidth of the laser. The grating is mounted on a piezoelectric transducer (PZT) to enable electronic tunning of the wavelength. This configuration is called an external cavity diode laser (ECDL).

The configuration, shown schematically in Fig. 10.2, is arranged so that the $-1^{\rm th}$ order diffraction from the grating is fed back to the laser, while the specular reflection from the grating is the output. From the grating equation we have

$$2d \sin\theta = m\lambda$$

where d is the spacing between successive lines of the grating, and θ is the angle of the $m^{\rm th}$ order diffraction. Since the specular reflection is the output beam, it is convenient to have θ close to 45°. This means that the grating used for accessing the D lines of Li (670 nm) has 2400 lines/mm, while the grating used for the D lines of K, Rb, and Cs (780–900 nm) has 1800 lines/mm. The power available after optical feedback is usually about 70% of the open loop power. Thus the linewidth is reduced by a factor of 1000 but the loss in power is only 30%, showing that this is not wavelength selection (as for a grating used with a white light source) but actual reduction in wavelength uncertainty of the laser. In effect, the grating along with the back facet of the diode forms a second lasing cavity—which is why this configuration is called an **external cavity diode laser** (ECDL)—and the longer cavity results in a smaller linewidth.

2. Lock-in amplifier

For many experiments, the laser needs to be locked to a resonance peak. The standard technique for doing this is to first do FM modulation of the laser frequency and then do lock-in detection at the modulation frequency. The laser frequency is modulated by varying the injection current into the diode laser. The signal is demodulated using a lock-in amplifier—consisting of a mixer, an amplifier, and a low-pass filter. It can be shown that output of the lock-in amplifier is the first derivative of the resonance curve. Thus if the signal is a peak, then the output of the lock-in amplifier (called the **error signal**) is dispersive. For example, if the signal is a Lorentzian peak with unit linewidth

$$\mathscr{L}(x) = \frac{1}{x^2 + 1/4}$$

then its derivative is

$$\frac{d\mathscr{L}}{dx} = -\frac{2x}{(x^2 + 1/4)^2}$$

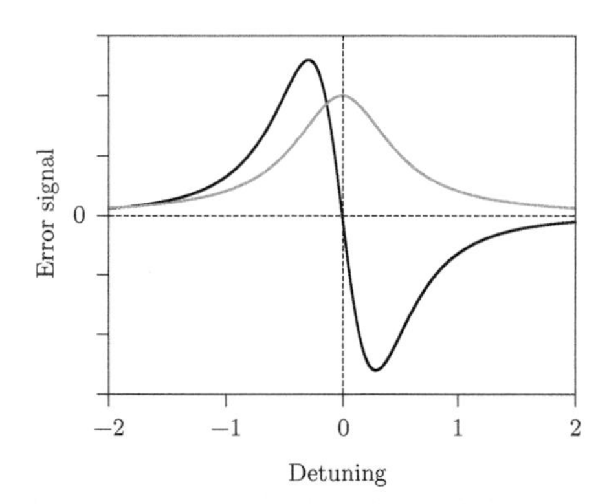

Figure 10.3: Error signal for locking the laser to a peak. The dispersive error signal is the output of a lock-in amplifier when the laser frequency is modulated. When fed back to the laser, it locks to the center of the peak shown in gray.

This is the form of the error signal, which is shown in Fig. 10.3. If the signal (with adequate gain) is now fed back to the laser, the frequency of the laser gets locked to the peak center.

3. Polarizing beam splitter cube — PBS

The polarizing beam splitter cube is an optical element with a coating inside that is designed to transmit one linear component of polarization and reflect the orthogonal component—usually p-polarization (plane polarized in the plane of incidence) is transmitted, while s-polarization (plane polarized normal to the plane of incidence) is reflected. If the coating is vertical, then the horizontal component is transmitted while the vertical component is reflected. This shows a PBS can be used to control the power going into an experiment—by having a $\lambda/2$ waveplate which rotates the polarization by some angle, the power can be controlled down to 0 since only $\cos\theta$ times the total power is transmitted. In addition, the PBS can be used to mix beams with orthogonal polarizations, or separate a beam into its orthogonal components.

4. Acousto-optic modulator — AOM

An acoustic-optic modulator is a device that uses traveling sound waves (usually at radio frequency) in a crystal to diffract a light beam passing through it. Its most common use in the lab is to shift the frequency of a laser beam passing through it, but it can also be used to control other properties of the light such as its intensity, phase, and spatial position. The mth order diffracted beam emerges at an angle θ that depends on the wavelength λ of the light relative to the wavelength Λ of sound in the medium, as follows

$$\sin\theta = \frac{m\lambda}{2\Lambda}$$

Since the light is diffracted by a moving wave, energy-momentum conservation requires that the frequency of the light be increased by that of the sound wave F

$$f_{\text{out}} \to f_{\text{in}} + mF$$

showing that the AOM can be used as a frequency shifter.

The diffraction efficiency depends on the amplitude of the sound wave, which is in turn determined by the intensity of the RF driver used to launch the wave in the crystal. Thus the properties of the light beam can be controlled by controlling the amplitude, phase, and frequency of the RF driver.

In addition, the deflection angle θ is determined by the frequency of the sound waves, so the light beam can be spatially modulated by varying the driver frequency.

For some applications, it is necessary to get variable frequency shifts without changing the beam direction. This is done by **double passing** through the AOM, which means passing through the AOM in both directions. As shown in Fig. 10.4, directional stability is guaranteed because the spatial shift in the first pass is reversed during the return. But this means that the input beam and return beam are exactly counter-propagating—this is solved by having a $\lambda/4$ waveplate after the AOM so that the return beam is orthogonally polarized and can be separated using a PBS. Double passing gives twice the frequency shift of the driver.

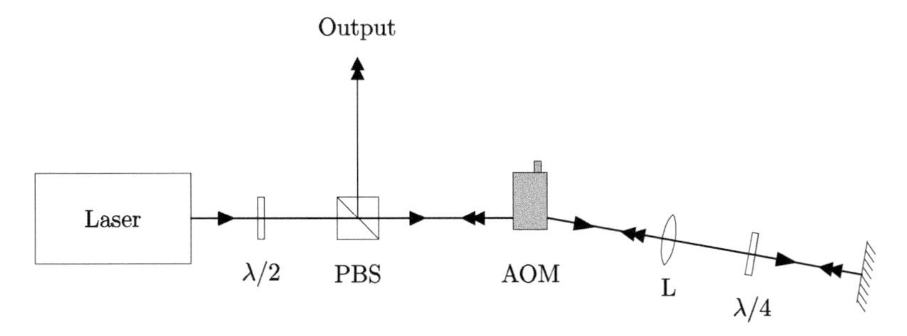

Figure 10.4: Double passing through an AOM. The $\lambda/4$ waveplate ensures that the return beam (which is exactly counter-propagating with the input and has no lateral shift) is orthogonally polarized and can be separated using a PBS.

The typical frequency shift from an AOM is about 100 MHz, and requires a few watts of RF drive power.

5. Faraday isolator

The stabilized diode laser used in spectroscopy can be destabilized if there is optical feedback from the elements in the experimental set up. This is prevented by having a Faraday isolator immediately after the laser. The Faraday isolator first has a polarizer which sets the polarization of the incoming light to be linear along some axis. This is followed by a crystal where a strong magnetic field along with the Faraday effect is used to rotate the plane of polarization by 45°. Any return beam undergoes the same rotation to make the total rotation equal to 90°, i.e. orthogonal in polarization to the incoming beam. This crossed polarization is then blocked by the input polarizer, thus preventing any light from feeding back to the laser. The typical isolation provided by a single isolator is 30 dB. If more isolation is needed, multiple isolators can be used.

C. Doppler-free techniques

1. Saturated absorption spectroscopy — SAS

When spectroscopy is done in a room temperature vapor cell, we have to consider the effect of Doppler broadening. From Chapter 9, "Lineshapes," we know that this is an example of inhomogeneous broadening, and the Maxwell–Boltzmann velocity distribution along with the first-order Doppler effect results in an absorption profile with a Gaussian lineshape. To give an idea of the deleterious effect of this broadening, consider that the Doppler width for ^{87}Rb atoms at room temperature on the D_2 line at 780 nm [shown in Fig. 10.6(a)] is about 600 MHz. But the natural linewidth of individual hyperfine transitions is only 6 MHz (with Lorentzian lineshape because this is an example of homogeneous broadening).

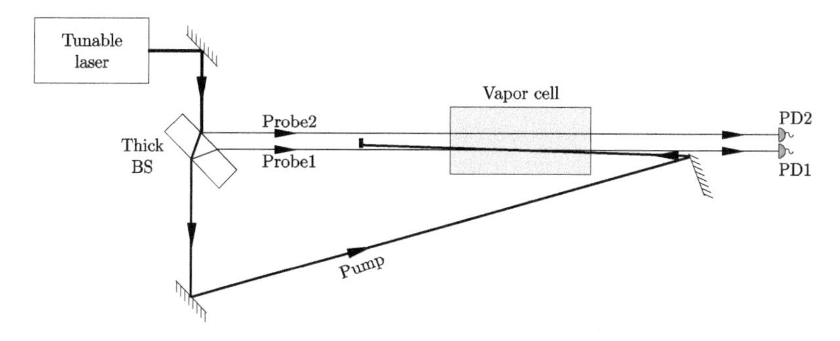

Figure 10.5: Experimental schematic for saturated absorption spectroscopy in a vapor cell. Two identical probe beams are generated using the thick beam splitter—one has the counter-propagating pump beam for the SAS, while the other is used for Doppler correction.

The standard method to resolve such a narrow transition embedded in a broad background is to use the technique of saturated absorption spectroscopy (SAS). The basic idea of an experiment to do SAS is shown schematically in Fig. 10.5. In addition to the probe beam, there is a much stronger pump beam that counter-propagates with respect to the probe, and plays the role of saturating the transition. Since the probability of transition is nonlinear in the light intensity, the probe shows decreased absorption when the pump is also resonant with the same transition. By counter-propagating the two beams, we ensure that the two beams are simultaneously resonant with zero-velocity atoms. The consequence of this on the probe spectrum is shown in Fig. 10.6. Probe absorption in the absence of the pump shows a broad Gaussian curve, as shown in Fig. 10.6(a). But a narrow dip appears at line center when the pump is turned on, as seen in Fig. 10.6(b). If the second probe beam is used to subtract the Doppler profile, then we get a single Lorentzian peak on a flat background in probe transmission. This is shown in Fig. 10.6(c).

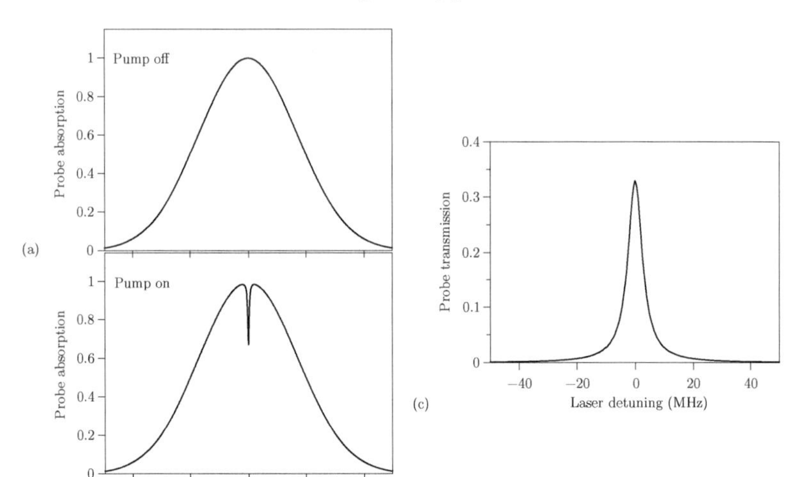

Figure 10.6: Saturated absorption spectrum. (a) Doppler broadened probe absorption spectrum (calculated for ^{87}Rb atoms at 300 K on the D_2 line at 780 nm) with counter-propagating pump beam off. (b) Dip in the center when the pump beam is turned on. (c) Single Lorentzian peak in probe transmission after the Doppler profile is subtracted.

2. Crossover resonances

Crossover resonances occur because of interaction of the pump-probe beams with non-zero velocity atoms, but with velocities such that they are within the Maxwell–Boltzmann distribution. Let us first consider the case of two excited states $|e_1\rangle$ and $|e_2\rangle$ coupled to the same ground state $|g\rangle$, as shown in Fig. 10.7(a). Consider that the laser frequency is exactly between the two excited states, and a velocity group such that the Doppler shift is equal to half the difference between the two excited states. For atoms with this velocity moving towards the probe beam and away from the pump beam, the Doppler shift is such that the probe is resonant with the $|g\rangle \to |e_2\rangle$ transition while the pump beam is resonant with the $|g\rangle \to |e_1\rangle$ transition. Saturation by the pump beam thus causes a peak to appear in the probe transmission spectrum. This peak is called an **excited state crossover resonance**, and is a spurious peak that appears when the laser frequency is exactly between two real peaks. The peak also has a contribution from atoms with the same velocity but moving in the opposite direction—the probe beam is now resonant with $|g\rangle \to |e_1\rangle$ transition while the pump beam is resonant with the $|g\rangle \to |e_2\rangle$ transition. This shows that two velocity groups contribute to each crossover resonance compared to one (zero-velocity) group for the real peaks. Thus the crossover resonances (which appear between every pair of real peaks) are generally more prominent than the real peaks.

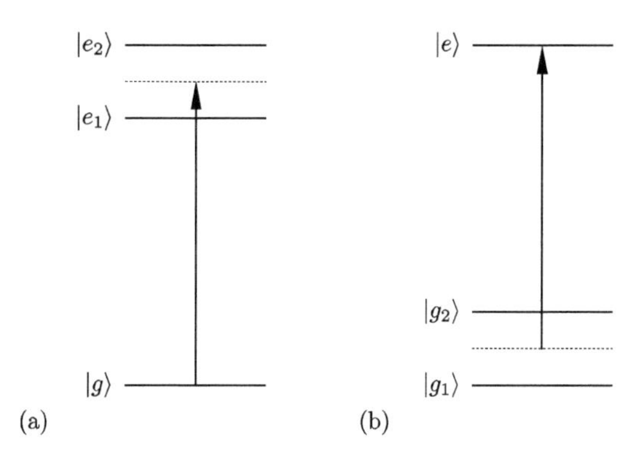

Figure 10.7: Crossover Resonances: (a) Excited state crossover resonance occurs when the laser frequency is midway between two excited states. (b) Ground state crossover resonance when the laser frequency is midway between two ground states coupling to the same excited state.

We next consider the phenomenon of ground state crossover resonances. These appear when two ground states $|g_1\rangle$ and $|g_2\rangle$ are coupled to the same excited state $|e\rangle$, as shown in Fig. 10.7(b). We again consider that the laser frequency is exactly between the two transition frequencies, and a velocity group whose Doppler shift is equal to half the separation between the two ground states. Thus for atoms moving with this velocity towards the probe beam and away from the pump beam, the probe is resonant with the $|g_1\rangle \rightarrow |e\rangle$ transition while the pump is resonant with the $|g_2\rangle \rightarrow |e\rangle$ transition. The strong pump now causes optical pumping which increases population in the $|g_1\rangle$ state. Thus the probe shows enhanced absorption, which is opposite to the enhanced transparency seen in the excited state crossover resonance. Hence these resonances appear as negative peaks in the probe transmission spectrum. As with the excited state crossover resonance, a second velocity group moving in the opposite direction also contributes to the peak but with the roles of the pump and probe beams being interchanged. This again makes the resonance more prominent.

As an example of these two kinds of crossover resonances, we consider the SAS spectrum for the D_1 line in ^{7}Li. Both the ground state and the excited state have two hyperfine levels each, with $F = 1$ and $F = 2$. The excited state hyperfine interval is 92 MHz, and has an excited state crossover resonance because the velocity required to get a Doppler shift of 46 MHz is only 31 m/s. The ground state hyperfine interval is 803 MHz, which implies that a velocity of 269 m/s will give the required Doppler shift. This is well within the velocity profile of Li vapor because it is a light atom, and the spectrum shows a ground state crossover resonance. Thus the spectrum shown in Fig. 10.8 hence has three sets of peaks. Each set comprises of three peaks, with the excited state crossover resonance in between labeled as $F_g \rightarrow F_e = (1, 2)$. The middle set is the ground state crossover resonance labeled as $F_g = (1, 2) \rightarrow F_e$, and is inverted as expected.

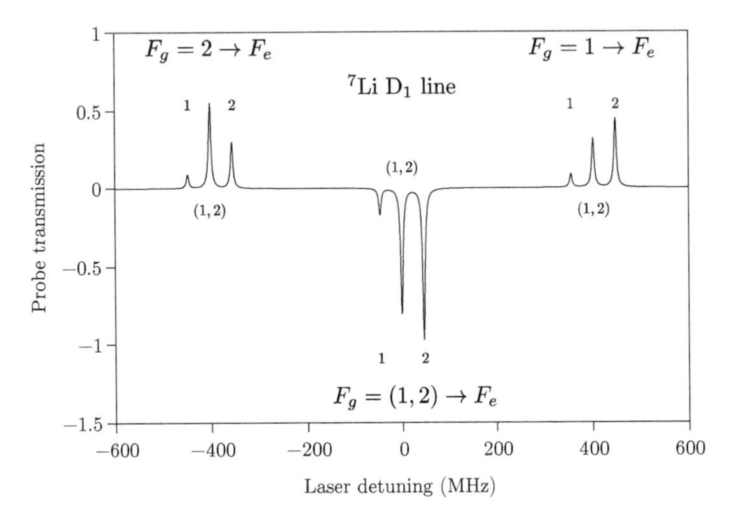

Figure 10.8: Doppler-corrected SAS spectrum for the D_1 line in ^{7}Li. The spectrum is calculated based on experimental data used for getting the relative heights of the peaks. The spectrum shows 3 sets of peaks. The first and third sets show the excited state crossover resonance in between the two real peaks. The middle set is a ground state crossover resonance, which is inverted and more prominent as explained in the text.

The spectrum shown in the figure is based on experimental data, and the experimental spectrum is used to get the relative heights of the peaks. Though Li is the simplest alkali atom, and many *ab initio* theoretical calculations have been done on it, experimentally it is the most challenging. This is because of its high reactivity with all kinds of glasses which precludes the use of vapor cells. Therefore the SAS experiments were done in a special Li spectrometer consisting of a pyrex cell, connected to a resistively heated Li source and a turbomolecular pump to maintain pressure below 10^{-7} torr.

3. Eliminating crossover resonances using copropagating SAS

Crossover resonances are generally not a problem in spectroscopy. But they can become an issue if the energy levels are spaced by a few linewidths— they then swamp the real peaks and prevent them from being resolved. One way to eliminate them is to use pump and probe beams that are copropagating instead of counter-propagating as in normal SAS. But the price one has to pay for this is that two independent lasers are needed. The experimental schematic is shown in Fig. 10.9. The spectrum is obtained by scanning only the pump laser while keeping the probe laser locked. This has the additional advantage of making the spectrum Doppler-free. In effect, the locked probe talks to one velocity group (primarily the zero-velocity group), and the signal remains flat until the pump also comes into resonance with the same velocity group. Thus there are no crossover resonances in between the real peaks.

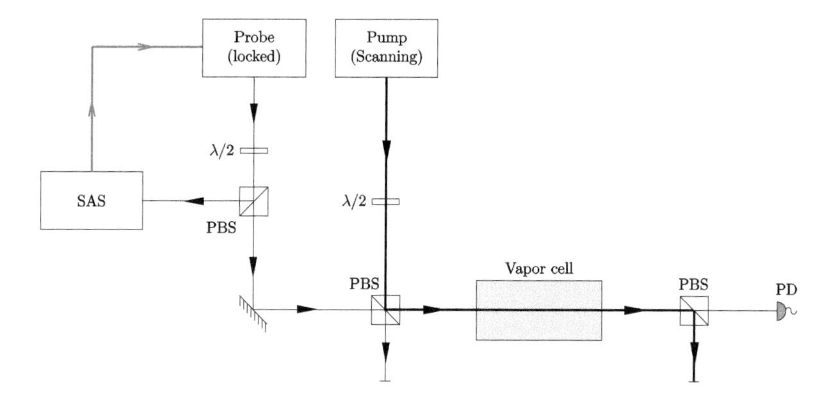

Figure 10.9: Experimental schematic for copropagating SAS. The setup uses two lasers, with the probe locked (using SAS) and the pump scanning.

To see the advantages of this scheme, we compare the spectra obtained with the two techniques for the $F_g = 2 \rightarrow F_e$ transitions in the ^{87}Rb D_2 line. We first consider the spectrum taken with normal SAS shown in Fig. 10.10(a). As expected, there are six peaks—three real peaks corresponding to the $F_e = 1, 2, 3$ hyperfine levels of the excited state; and three spurious crossover resonances that are more prominent. In addition to the usual saturation effects, probe transparency is caused by optical pumping into the $F_g = 1$ ground hyperfine level for open transitions, i.e. those involving the $F_e = 1, 2$ levels of the excited state. The linewidth of the peaks is about 12 MHz, compared to the natural linewidth of 6 MHz. This increase is typical in SAS and arises due to a misalignment angle between the pump and probe beams (which causes non-zero-velocity atoms to contribute), and power broadening by the pump beam.

Now let us consider the spectrum as shown in Fig. 10.10(b) taken with the copropagating configuration. The probe beam is locked to the $F_g = 2 \rightarrow F_e = 3$ transition, while the pump beam is scanned across the set of $F_g = 2 \rightarrow F_e = 1, 2, 3$ transitions. Since the probe is locked, its transmitted signal corresponds primarily to absorption by zero-velocity atoms making a transition to the $F_e = 3$ level. The signal remains flat (or Doppler-free)

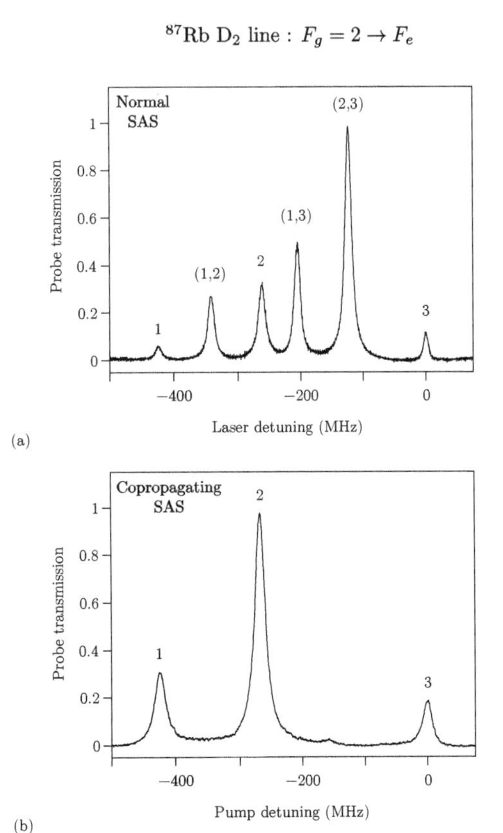

Figure 10.10: Comparison of experimental spectra in ^{87}Rb for the D$_2$ line. (a) Spectrum taken with normal SAS, showing crossover resonances and requiring Doppler correction. (b) Spectrum taken with the copropagating SAS technique, showing no crossover resonances and being inherently Doppler-free. There is one spurious peak (due to interaction with non-zero-velocity atoms) at -157 MHz which is very small; the others lie outside the spectrum as explained in the text.

until the pump also comes into resonance with transition for the same zero-velocity atoms. Thus there are three peaks at the location of the hyperfine levels, with no crossover resonances in between. The hyperfine peaks are located at -423.6 MHz, -266.7 MHz, and 0, all measured with respect to the frequency of the located probe laser. The primary cause for probe transparency is the phenomenon of electromagnetically induced transparency (EIT) in this V-type system, as discussed in Chapter 8, "Coherence." In addition, there are effects of saturation and optical pumping as in normal SAS, but these are less important than EIT.

To see the effect of non-zero-velocity atoms on the spectrum, we note that there will be two additional velocity classes that absorb from the locked probe beam—both moving in the same direction as the probe, but with velocities such that one drives transition to the $F_g = 2$ level (266.7 MHz lower), and the second to the $F_g = 1$ level (423.6 MHz lower). Each of these will cause three additional transparency peaks from the mechanisms discussed above. The first velocity class (moving at 208 m/s) will cause peaks at -156.9 MHz, 0, 266.7 MHz, i.e. a set of peaks shifted up by 266.7 MHz. The second velocity class (moving at 330 m/s) will cause peaks at 0, 156.9 MHz, 423.6 MHz, i.e. a set of peaks shifted up by 423.6 MHz. Thus there will be seven peaks in all—three real and four spurious. However, only the peak at -156.9 MHz will appear within the spectrum (caused by the probe beam driving the $F_g = 2 \rightarrow F_e = 2$ transition and the pump driving the $F_g = 2 \rightarrow F_e = 1$ transition), the other three spurious peaks will be outside the spectrum to the right hand side. This is indeed what is observed in the experimental spectrum shown in Fig. 10.10(b).

4. Two-photon Doppler-free

A two-photon transition can be made inherently Doppler-free by having the two photons come from two beams that are counter-propagating with respect to each other. Since only zero-velocity atoms interact with both beams, the resulting spectrum is not Doppler broadened. As an example, we consider the $S_{1/2} \rightarrow D_{5/2}$ transition at 778 nm in ^{87}Rb. The experimental schematic to study this is shown in Fig. 10.11(a). The cell has to be heated to increase the number density of atoms and get a good signal. A sample spectrum of the $F_g = 2 \rightarrow F_e$ hyperfine transition is shown in Fig. 10.11(b). The four peaks corresponding to the four hyperfine levels in the excited state are well resolved. Because the two beams have to be perfectly counter-propagating to get a good spectrum, the diode laser is extremely sensitive to feedback, which makes the spectrum unstable. Therefore, the experiment requires the use of a Faraday isolator in front of the laser.

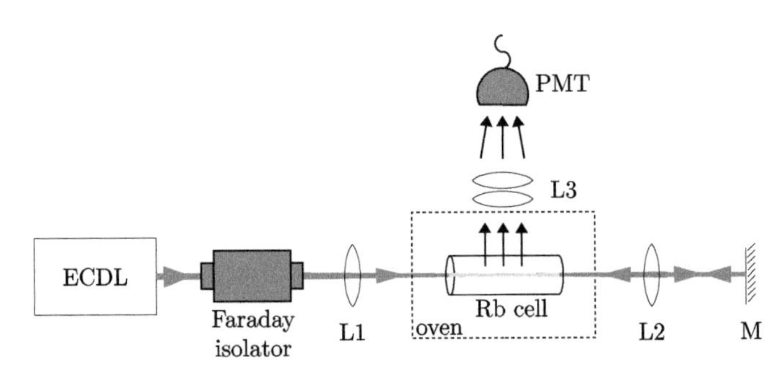

(a)

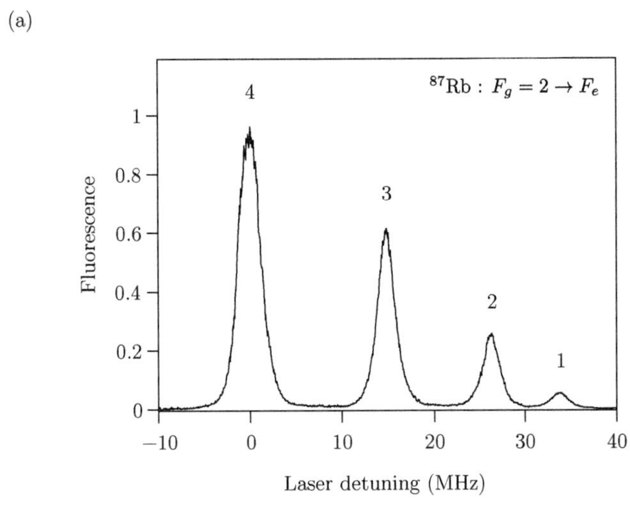

(b)

Figure 10.11: Two-photon spectroscopy. (a) Experimental setup showing the Faraday isolator to prevent feedback into the laser. (b) Experimental spectrum for the $F_g = 2 \to F_e$ hyperfine component of the $S_{1/2} \to D_{5/2}$ transition in ^{87}Rb.

D. Nonlinear magneto-optic rotation — NMOR

The well-known phenomenon of the rotation of the plane of polarization of near resonant light passing through atomic vapor in the presence of a longitudinal B field is called magneto-optic rotation (MOR) or the **Faraday effect**. It arises due to the birefringence in the medium induced by the B field. This phenomenon of MOR shows nonlinear effects when the light is sufficiently strong, as can be produced by a laser for example. In the simplest manifestation of nonlinear magneto-optic rotation (NMOR), the strong light field aligns the atoms by inducing (through optical pumping) $\Delta m = 2$ coherences among the ground state magnetic sublevels. The aligned atoms undergo Larmor precession about the longitudinal B field, and cause additional rotation due to the precessed birefringence axis. The effect is nonlinear because the degree of optical alignment depends on the light intensity. The study of NMOR is important because it has important applications such as sensitive magnetometry, search for a permanent electric dipole moment, precision measurements, and magnetic resonance imaging (MRI).

The mechanism for the origin of NMOR shows that the width of the resonance is going to be limited by the decoherence time of the atomic alignment. If the experiments are done in a vapor cell at room temperature, then the mean free path between collisions is several orders of magnitude larger than the cell size. Thus the coherence is destroyed by spin-exchange collisions not between atoms but with the cell walls. It has been known for a long time that the use of paraffin coating on the cell walls reduces the ground-state depolarization rate and enhances the optical pumping signals. Therefore these experiments are typically done using paraffin-coated vapor cells.

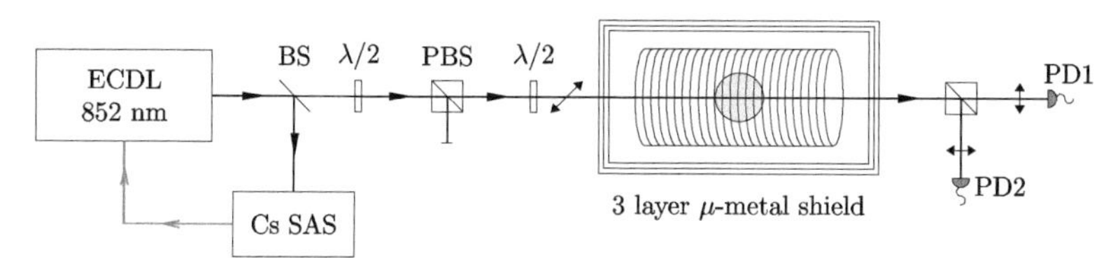

Figure 10.12: Experimental schematic for observing NMOR. The vapor cell is spherical in shape, has paraffin coating on the walls, is placed in a three-layer magnetic shield to reduce stray fields, and is in the center of a solenoid used to apply a longitudinal magnetic field.

A typical experimental schematic for studying NMOR (on the D_2 line of Cs) is shown in Fig. 10.12. Because the experiments use very small magnetic fields it is important to shield Earth's field and other stray fields. This is achieved by placing the cell inside a three-layer magnetic shield made of μ-metal, which reduces the ambient fields to less than 0.1 mG. The vapor cell with paraffin coating on the walls is spherical in shape so that the recoil velocity after each wall collision is in a random direction. It is kept in the middle of a solenoidal coil that allows the application of a uniform longitudinal magnetic field. The probe beam is chosen to have linear polarization at 45° with respect to the horizontal, and separated into its horizontal and vertical components using a PBS after the cell. The two intensities I_x and I_y are measured using photodiodes (PD1 and PD2), and the rotation angle calculated as

$$\Phi = \frac{1}{2}\sin^{-1}\frac{I_x - I_y}{I_x + I_y} \approx \frac{I_x - I_y}{2(I_x + I_y)}$$

where the approximate result is valid when the rotation angle is small (of the order of a few mrad in these experiments).

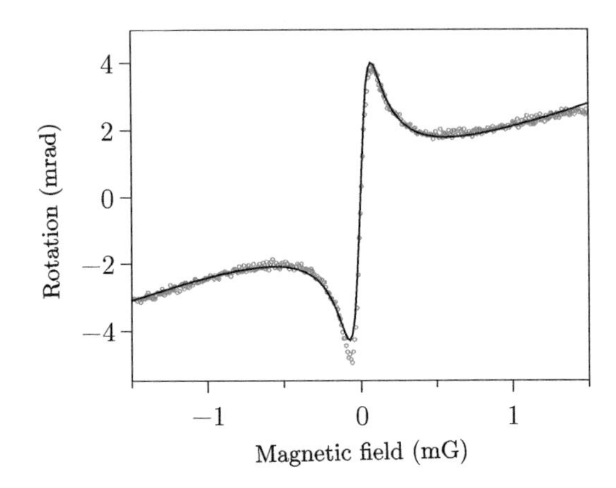

Figure 10.13: NMOR resonance in the D_2 line of ^{133}Cs as a function of applied magnetic field. Open circles are the experimental data, while the solid line is the curve-fit using Eq. (10.2) in the text.

The observed NMOR resonance on the $F_g = 4$ ground hyperfine level of ^{133}Cs is shown in Fig. 10.13. To understand the lineshape, consider that the origin of the NMOR effect is the creation of a ground state $\Delta m = 2$ coherence, and the precession of this coherence in the magnetic field. From Eq. (4.9) of Chapter 4, we know that the Zeeman shift of an $|F, m\rangle$ sublevel in a magnetic field $B_\circ\,\hat{z}$ is $g_F \mu_B m B_\circ$. Hence the $\Delta m = 2$ alignment/coherence will precess at $g_F \mu_B\, \Delta m\, B_\circ/\hbar$, or a frequency of $2\omega_L$ when we define the Larmor precession frequency as

$$\omega_L = g_F \mu_B B_\circ/\hbar$$

If the alignment is assumed to decohere exponentially at a rate $\Gamma_r/2$, then the instantaneous rotation angle is

$$\phi_{\text{nmor}} = C e^{-\Gamma_r t/2} \sin\left(2\omega_L t\right)$$

where C is a proportionality constant that depends on the degree of alignment. What we observe is the time-averaged rotation, which is the integral of the instantaneous rotation

$$\Phi_{\text{nmor}} \equiv \langle \phi_{\text{nmor}} \rangle = C \int_0^\infty e^{-\Gamma_r t/2} \sin(2\omega_L t)\, dt$$

This can be derived as

$$\Phi_{\text{nmor}} = C \frac{2\omega_L}{(2\omega_L)^2 + \Gamma_r^2/4} \tag{10.1}$$

Thus, the lineshape of the NMOR resonance is dispersive and centered at $\omega_L = 0$ (or equivalently at $B_\circ = 0$). If we include a linear term to account for normal MOR effect, then the combined lineshape for the rotation is

$$\Phi_{\text{total}} = A\omega_L + C \frac{2\omega_L}{(2\omega_L)^2 + \Gamma_r^2/4} \tag{10.2}$$

The experimental data, shown with open circles in Fig. 10.13, bear out this expectation. The solid line is a fit to the above equation, and describes the data well.

1. Chopped NMOR

Since NMOR has important applications in precision measurements, any method to improve the signal-to-noise ratio (SNR) is advantageous. One way to improve SNR is to chop the laser beam on and off, and then do lock-in detection at the chopping frequency. The basic idea in this technique is that of a repeated Ramsey SOF measurement of the Larmor precession frequency. During the on time of the beam, the atoms are optically pumped into the $\Delta m = 2$ coherence of the ground state. During the off time, they freely precess around the B field at the Larmor frequency. If the on-off modulation frequency matches the Larmor precession frequency, then the atoms are realigned exactly when the light field comes back on thus resonantly enhancing the optical pumping process. The process continues again for the next on-off cycle, so this is like a repeated Ramsey method. The resonance actually appears when the modulation frequency is equal to $2\times$ the Larmor frequency—the factor of 2 appears because the atomic alignment has two-fold symmetry.

To get the lineshape with the chopped NMOR technique, we look at the effect of chopping on the NMOR signal given in Eq. (10.1). To first order, the effect of the chopping is to modulate the atomic alignment at this frequency. Thus, if the chopping frequency is ω_m then the constant C becomes $C \cos \omega_m t$. The lock-in amplifier demodulates this signal at ω_m to get the in-phase and out-of-phase quadratures. It is straightforward to see that the time-averaged rotation angle for the in-phase component is

$$\Phi_{\text{ip}} = C\left[\frac{D(2\omega_L)}{2} + \frac{D(2\omega_L - \omega_m)}{4} + \frac{D(2\omega_L + \omega_m)}{4}\right]$$

where

$$D(x) = \frac{x}{x^2 + \Gamma_r^2/4}$$

is the same dispersive function that appeared in Eq. (10.1). Thus the in-phase rotation for chopped NMOR has two additional peaks compared to normal NMOR, with the same dispersive lineshape but centered at $\omega_L = \pm\omega_m/2$ and having half the amplitude.

A similar analysis shows that the out-of-phase rotation angle is given by

$$\Phi_{\text{op}} = C\left[-\frac{\mathscr{L}(2\omega_L - \omega_m)}{4} + \frac{\mathscr{L}(2\omega_L + \omega_m)}{4}\right]$$

where

$$\mathscr{L}(x) = \frac{\Gamma_r}{x^2 + \Gamma_r^2/4}$$

is a Lorentzian function. Thus we see that the out-of-phase component has two Lorentzian peaks (with opposite signs) centered at $\omega_L = \pm\omega_m/2$. The linewidth of each peak is Γ_r determined by the same decoherence rate as far as normal NMOR. Hence the advantage of the chopped NMOR technique is that the out-of-phase component gives Lorentzian peaks and at non-zero values of the B field. This results in better SNR because noise at higher frequencies is generally smaller. In addition, the Lorentzian lineshape makes it easier to determine the peak center.

The experimental spectra confirm the above analysis. Shown in Fig. 10.14 are the above lineshapes, calculated to match experimental data. As seen, the in-phase component has the same linear variation as normal NMOR (due to the MOR effect); there is no linear part for the out-of-phase component, which is another advantage of the chopped NMOR technique.

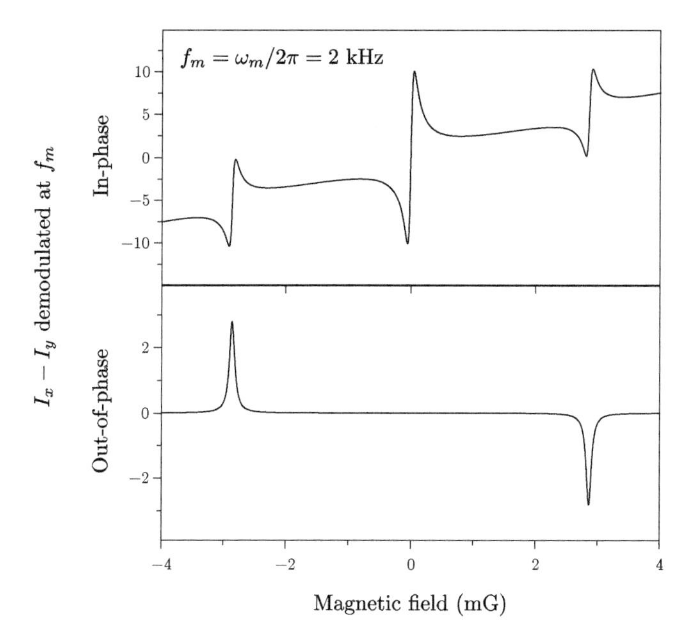

Figure 10.14: Chopped NMOR resonance demodulated at f_m showing the in-phase and out-of-phase components. The lineshapes and locations are derived in the text.

E. Problems

1. Diode laser linewidth

The linewidth of a diode laser after feedback stabilization is of the order of 1 MHz. Design an experimental scheme to measure this linewidth.

Solution

The linewidth of the laser can be measured by interfering two beams that have a phase difference much greater than that corresponding to the coherence time, and taking a Fourier transition of the beat (interference) signal. This is because the phase of the two such beams will be uncorrelated, and the beat sigal will reflect the phase uncertainty in the laser. The beat signal of two perfectly correlated beams is a pure sine wave, which becomes increasingly fuzzy as the correlation decreases.

If the linewidth is 1 MHz, then the coherence time is

$$\tau_c = \frac{1}{2\pi \times 1 \text{ MHz}} = 1.6 \times 10^{-7} \text{ s}$$

The corresponding phase difference is

$$\Delta\phi = \omega\tau_c = \frac{2\pi}{\lambda} c\tau_c$$

which implies a path length difference in the interferometer of

$$\Delta\ell = \frac{\lambda}{2\pi}\Delta\phi = c\tau_c = 48 \text{ m}$$

Such a large path difference cannot be produced in the lab in a normal interferometer except if one of the arms has an optical fiber of sufficient length.

Therefore, the experimental scheme consists of the following steps.

(i) The laser beam is split into two parts.

(ii) The first beam has an AOM in its path so that it has a known frequency offset of 20 MHz. The second beam is sent through an optical fiber that is a few kms long to get the required phase difference.

(iii) The two beams are beat on a fast photodiode with response time fast enough to measure a frequency of 20 MHz (of the order of 1 ns).

(iv) A Fourier transform of the photodiode time-domain signal is taken. The Fourier transform will be centered at 20 MHz and have a width equal to the linewidth of the laser.

2. Vapor cells

(a) If the lifetime of the $P_{3/2}$ state of Na is 16.3 ns, what is the natural linewidth (in Hz) of the D_2 line?

(b) If the wavelength of this line is 589 nm, what is the corresponding saturation intensity (in mW/cm^2)?

(c) If you make a vapor cell of Na for probe-absorption spectroscopy that is maintained at 100°C, what is the one-dimensional rms velocity? What is the Gaussian width (FWHM) of the probe spectrum corresponding to this velocity distribution?

(d) The typical vapor pressure inside the above cell is 0.1 μtorr.

(i) What is the mean free path between collisions if the atomic size is 1 Å?

(ii) For a cell length of 5 cm, what is the percentage absorption for a probe beam?

Solution

(a) The linewidth is related to the lifetime as

$$\Gamma = \frac{1}{\tau} \qquad \Longrightarrow \qquad \frac{\Gamma}{2\pi} = \frac{1}{2\pi\tau} = 9.98 \text{ MHz}$$

(b) The saturation intensity is defined as

$$I_s = \frac{\hbar\omega^3}{12\pi c^2}\Gamma = \frac{h\pi c}{3\lambda^3}\Gamma = 6.22 \text{ mW/cm}^2$$

(c) For a probe beam (propagating along z) passing through a vapor cell containing a gas of atoms of mass M at temperature T, the relevant velocity distribution is a one-dimensional Maxwell–Boltzmann distribution given by

$$f(v_z)\,dv_z = \sqrt{\frac{M}{2\pi k_B T}} \exp\left(-\frac{Mv_z^2}{2k_B T}\right) dv_z$$

The rms velocity is defined as

$$v_{\text{rms}} \equiv \sqrt{\langle v^2\rangle - \langle v\rangle^2}$$

For this distribution, $\langle v\rangle = 0$ and

$$\begin{aligned}\langle v^2\rangle &= \int_{-\infty}^{\infty} dv_z\, v_z^2 f(v_z) \\ &= \sqrt{\frac{M}{2\pi k_B T}} \int_{-\infty}^{\infty} dv_z\, v_z^2 \exp\left(-\frac{Mv_z^2}{2k_B T}\right)\end{aligned}$$

Substituting

$$x^2 = \frac{Mv_z^2}{2k_BT}$$

gives

$$2xdx = \left(\frac{M}{2k_BT}\right) 2v_z dv_z \qquad \Longrightarrow \qquad dv_z = \frac{2k_BTx}{Mv_z} dx$$

Therefore

$$\begin{aligned}
\langle v^2 \rangle &= \sqrt{\frac{M}{2\pi k_BT}} \left(\frac{2k_BT}{M}\right) \int_{-\infty}^{\infty} dx\, x\, v_z e^{-x^2} \\
&= \sqrt{\frac{M}{2\pi k_BT}} \left(\frac{2k_BT}{M}\right) \int_{-\infty}^{\infty} dx\, x^2 \sqrt{\frac{2k_BT}{M}} e^{-x^2} \\
&= \frac{2k_BT}{M} \frac{1}{\sqrt{\pi}} \int_{-\infty}^{\infty} dx\, x^2\, e^{-x^2} \\
&= \frac{k_BT}{M} \qquad \text{using} \qquad \int_{-\infty}^{\infty} x^2\, e^{-x^2}\, dx = \frac{\sqrt{\pi}}{2}
\end{aligned}$$

The atomic mass of Na is 23 amu, so at a temperature of $T = 100°\text{C} = 373$ K, the rms velocity is

$$v_{\text{rms}} = \sqrt{\frac{k_BT}{M}} = 367.2 \text{ m/s}$$

Since the Doppler shift for an atom with velocity v is v/λ, the resultant distribution will be a Gaussian of the form (with center frequency $\omega_\circ$)

$$\mathscr{D}(\omega) = A \exp\left[-\frac{(\omega - \omega_\circ)^2}{2\omega_D^2}\right]$$

where ω_D is the rms width related to the rms velocity as

$$\omega_D = 2\pi \frac{v_{\text{rms}}}{\lambda_\circ}$$

The maximum of this function is A and occurs at $\omega = \omega_\circ$. Therefore the half maximum occurs at $\omega_{1/2} = (\omega - \omega_\circ)$ such that

$$A \exp\left[-\frac{\omega_{1/2}^2}{2\omega_D^2}\right] = \frac{A}{2}$$

which gives

$$\omega_{1/2}^2 = 2\omega_D^2 \ln 2 \qquad \Longrightarrow \qquad \omega_{1/2} = \pm\omega_D \sqrt{2 \ln 2}$$

Therefore the linewidth of the probe profile is

$$\text{FWHM of Gaussian} = 2\frac{\omega_D}{2\pi}\sqrt{2\ln 2} = 2\frac{v_{\text{rms}}}{\lambda_\circ}\sqrt{2\ln 2} = 1.47 \text{ GHz}$$

(d) (i) The mean free path is

$$\Lambda = \frac{k_B T}{\sqrt{2}\,\pi d^2 P}$$

where d is the size of the atom, and P is the pressure.

Using values of size 1 Å and pressure 0.1 μtorr, we get

$$\Lambda = 8698 \text{ m}$$

showing that it is orders-of-magnitude larger than the typical cell size.

(ii) The intensity after the cell is related to the incident intensity as

$$I_c = I_\circ e^{-\mathrm{OD}}$$

where OD is the optical density defined as

$$\mathrm{OD} = N\sigma\ell$$

Here N is atomic density inside the cell, ℓ is the length of the cell, and σ is the scattering cross-section. The scattering cross-section at intensity I and detuning δ is related to the on-resonance scattering cross-section $\sigma_\circ$ as

$$\sigma = \frac{\sigma_\circ}{1 + (2\delta/\Gamma)^2 + I/I_s}$$

Therefore at low intensity and on resonance, $\sigma = \sigma_\circ$.

From the definition of the cross-section, it is

$$\sigma = \frac{\text{Scattering rate}}{\text{Incident photon flux}} = \frac{R_{\text{sc}}}{I/\hbar\omega}$$

Since the saturation intensity is defined as

$$R_{\text{sc}} = \frac{\Gamma}{2} \qquad \text{when} \qquad I = I_s$$

we get

$$\sigma_\circ = \frac{\hbar\omega\Gamma}{2I_s} = \frac{3\lambda^2}{2\pi}$$

For a pressure of 0.1 μtorr

$$N = \frac{P}{k_B T} = 9.6 \times 10^7 \text{ atoms/cc}$$

Therefore for a cell length of 5 cm, the optical density is

$$\mathrm{OD} = N\sigma_\circ\ell = 0.795$$

and the percentage absorption is

$$100\,\frac{I_c}{I_\circ} = 100\,e^{-\mathrm{OD}} = 45.2\,\%$$

Chapter 11

Cooling and Trapping

So far we have seen how to address the internal degrees of freedom of an atom. Now we will see how to control the external degrees of freedom of an atom—its position and momentum—both with and without light. When light is involved, the study is called "laser cooling and trapping." This is one of the most active areas of research in AMO physics, especially after the observation of Bose–Einstein condensation (BEC) in 1995 in a dilute vapor of Rb atoms.

In this chapter, we will discuss the two kinds of forces of near-resonant light on atoms—the **spontaneous** or scattering force, which is used for laser cooling; and the **stimulated** or dipole force, which is a conservative force and hence can be derived from a potential, with the potential being proportional to the light intensity. An atom in an optical lattice experiences a periodic potential, akin to the periodic potential experienced by electrons in a crystalline lattice. Hence cold atoms in an optical lattice realize the kind of Hamiltonians normally associated with condensed matter systems; in effect they **emulate** a quantum many body system. Such quantum emulation promises not only the resolution of many long-standing puzzles in condensed matter physics (e.g. high T_c superconductivity), but also provides new opportunities to realize quantum states that are not seen in conventional systems such as solids. In addition, we will see how the dipole force can be used in optical tweezers—a tool that has important applications in biology and medicine. Finally, we will discuss ion traps which have wide-ranging applications in mass measurements, atomic clocks, and quantum computation.

A. Spontaneous force

Laser cooling of neutral atoms, first proposed by Ted Hänsch and Art Schawlow in 1975, uses the spontaneous force to cool an atomic cloud. The force arises because each photon carries a momentum of $\hbar k$ which, by momentum conservation, gets transferred to the atom each time it absorbs or emits a photon. The momentum transfer is in the direction of the laser beam for absorption, and in a random direction for emission. Thus, after n absorption-emission cycles, the average momentum transferred is $n\hbar k$ in the direction of the laser beam, because the average for the emission cycles is zero. The random nature of the spontaneous emission process is important for laser cooling because the entropy of the atomic system has to be decreased to achieve cooling, and this decrease in entropy has to be compensated by an increase in entropy somewhere, the somewhere being the photon field in this case. In addition, the irreversibility of spontaneous emission is important in defining a thermodynamic direction to the cooling process.

1. Doppler cooling

The simplest form of laser cooling is called Doppler cooling. To understand it, let us first consider it in one dimension (1D). Imagine a simple case where the atom has just two states $|g\rangle$ and $|e\rangle$ with a transition frequency $\omega_\circ$, as shown in Fig. 11.1. Now consider that the atom is bombarded with identical laser beams from the left and the right sides, both of which are detuned below resonance. If the atom is stationary, Fig. 11.1(a) shows that the scattering rate (and hence the force) of the two laser beams is equal. However, if an atom is moving to the right with a small velocity v, then the laser beam on the right is Doppler shifted closer to resonance compared to the laser beam on the left. As a consequence, the atom scatters more photons from the right beam and feels a force to the left. The opposite happens for an atom moving to the left—it Doppler shifts the left beam closer to resonance, scatters more photons from that beam, and thus feels a force to the right. Therefore the force always opposes the motion, and behaves like a frictional force that results in cooling. Another way to think about it is that the detuning below resonance means that the photon has less energy than is required for the atomic transition, it makes up for this shortfall using the kinetic energy of the atom, thus reducing the kinetic energy and cooling the atom in the process. Such a configuration of laser beams has been colorfully called "optical molasses" to highlight the fact that the atom feels a viscous drag on its motion. And the 1D model considered above can be readily extended to 3D by having three sets of counter-propagating beams in the three orthogonal directions.

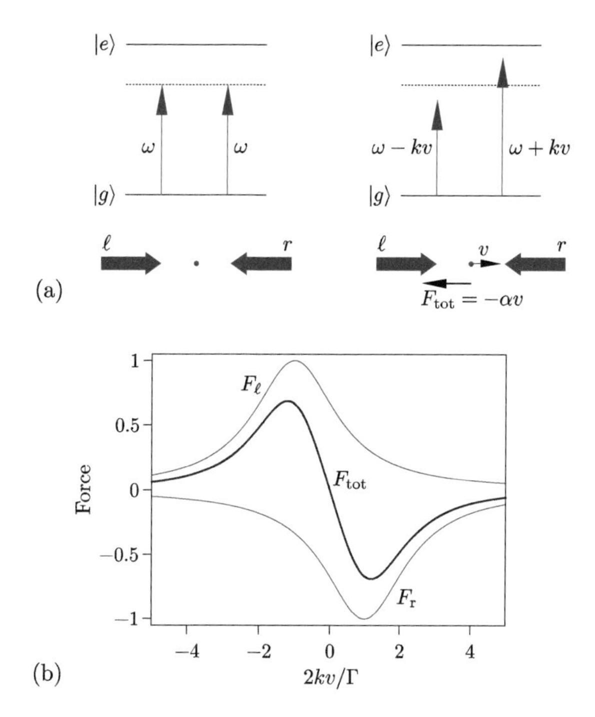

Figure 11.1: Laser cooling of atoms in 1D using optical molasses. The two laser beams are equally detuned below resonance. (a) The two forces are equal when the atom is stationary; but become unequal when the atom is moving with a velocity v, because one beam is Doppler-shifted closer to resonance. (b) Forces due to the left beam, right beam, and their sum, as a function of velocity, for the configuration shown in (a). The detuning is chosen to be $\Delta = -2kv/\Gamma$.

The above analysis can be quantified by considering the force due to each laser beam, which is just the photon momentum times the scattering rate. The scattering rate has already been derived in Eq. (6.32) of Chapter 6. Thus the force due to the left beam for an atom moving with a velocity v is

$$F_\ell(v) = +\hbar k \frac{\Gamma}{2} \frac{I/I_s}{1 + I/I_s + [2(\delta - kv)/\Gamma]^2} \tag{11.1}$$

where the detuning has been changed to $(\delta - kv)$ to transform from the lab to the atom's frame. Similarly the force from the right beam is

$$F_r(v) = -\hbar k \frac{\Gamma}{2} \frac{I/I_s}{1 + I/I_s + [2(\delta + kv)/\Gamma]^2}$$

The total force on the atom is the sum of these two forces, which, in the limit of low intensity and small atomic velocity, can be approximated as

$$F = F_\ell + F_r \approx 4\hbar k \frac{I}{I_s} \frac{kv(2\delta/\Gamma)}{[1 + (2\delta/\Gamma)^2]^2}$$

Thus the total force is linear in δ and v, and is a frictional force ($\propto -v$) when the detuning is negative.

The two forces as a function of velocity along with their sum are shown in Fig. 11.1(b). Analysis of the above equation shows that the lowest temperature—called the **Doppler cooling limit**—is reached when the detuning $\delta = -\Gamma/2$, and the temperature reached is

$$k_B T_{\text{min}} = \hbar\Gamma/2$$

This is the detuning chosen for the curves shown in the figure. The total force is clearly linear with a negative slope for small v. In order to give a numerical estimate of this temperature, for ^{87}Rb atoms cooled on the D_2 line, $\Gamma/2\pi = 6$ MHz and $T_{\text{min}} = 140$ μK.

Like other alkali atoms, ^{87}Rb also has two hyperfine levels in the ground state. Cooling is done using the **closed** $F_g = 2 \to F_e = 3$ transition, closed because atoms in the excited state can only decay to the $F_g = 2$ level. Still there is a finite probability of off-resonant excitation to the $F_e = 2$ level ($\sim$ 40 linewidths away so the probability of excitation is very small), from where atoms can decay to the $F_g = 1$ level and be lost from the cooling process. Thus after a while all the atoms get optically pumped to the $F_g = 1$ level and are lost. To recover them, a second laser beam resonant with the $F_g = 1 \to F_e = 2$ transition, called the **repumping beam**, is mixed with the cooling beam. Because the transition uses the $F_e = 2$ level of the excited state, atoms have a finite probability of decaying to the $F_g = 2$ level and coming back to the cooling cycle.

Though the Doppler cooling limit is a low temperature, much lower temperatures can be reached with the technique of polarization gradient cooling discussed in the next section.

2. Polarization gradient cooling

Polarization gradient cooling is also called sub-Doppler cooling because it allows one to reach temperatures below the Doppler cooling limit, or Sisyphus cooling for reasons that will become clear soon. The cooling mechanism requires the presence of magnetic sublevel structure in the transition used for cooling, coupled with varying light shifts for the different sublevels in the presence of different polarizations of light. Therefore it is most easily understood for a $F_g = 1/2 \to F_e = 3/2$ transition in one dimension. The laser configuration consists of two counter-propagating beams with orthogonal linear polarizations—the so-called lin ⊥ lin configuration. If we take the direction of propagation to be along the z axis, the two polarizations to be along the x and y directions with a phase difference of $-\pi/2$ at $z=0$, and an amplitude of $\mathcal{E}_\circ$ for each beam, then the total electric field is

$$\begin{aligned}\vec{E}(\vec{r},t) &= \mathcal{E}_\circ \left[e^{i(kz-\omega t)}\hat{x} - ie^{i(-kz-\omega t)}\hat{y}\right] \qquad + \text{ c.c.} \\ &= \sqrt{2}\mathcal{E}_\circ \left[\cos(kz)\frac{\hat{x}-i\hat{y}}{\sqrt{2}} + i\sin(kz)\frac{\hat{x}+i\hat{y}}{\sqrt{2}}\right] e^{-i\omega t} \qquad + \text{ c.c.}\end{aligned}$$

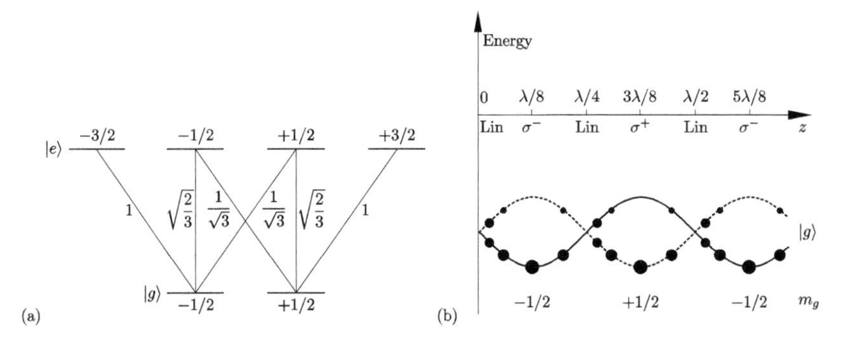

Figure 11.2: Energy level scheme for understanding polarization gradient cooling. (a) Sublevel structure and Clebsch-Gordan coefficients for a $F_g = 1/2 \rightarrow F_e = 3/2$ transition. (b) Light shifts and steady-state populations (filled circles) for the sublevels of the $F_g = 1/2$ ground state, in the presence of red detuning and lin ⊥ lin configuration which causes the polarization to vary with a period of $\lambda/2$. The most populated sublevel is always the one with the largest negative shift.

Thus the polarization cycles from linear to circular. This is shown in Fig. 11.2—starting with linear at $z=0$, it becomes σ^+ at $\lambda/8$, then orthogonal linear at $\lambda/4$, then σ^- at $3\lambda/8$, and back to linear at $\lambda/2$ from where it starts repeating. This polarization gradient causes different light shifts for the two sublevels of the $F_g = 1/2$ ground state considered for the cooling because the magnitude of the shift depends on the strength of the transition, which in turn depends on the polarization of light and the relevant Clebsch-Gordan coefficient. The C-G coefficients for the different transitions are shown in Fig. 11.2(a). For this case, a red detuned beam with σ^+ polarization down shifts the $m_g = +1/2$ sublevel three times more than the $m_g = -1/2$ sublevel, while σ^- polarization down shifts the $m_g = -1/2$ sublevel more than the $m_g = +1/2$ sublevel by the same amount. In between where the light is linearly polarized, both sublevels have the same light shift and the energies cross each other. The modulated light shifts of the two sublevels, and their steady-state populations are shown in Fig. 11.2(b). The population is maximum for the sublevel with the largest negative shift.

The other thing to note in understanding this cooling mechanism is the process of optical pumping, which drives the population toward its steady-state value. If the optical pumping rate is fast compared with the time it takes for the atom to travel a distance equal to $\lambda/2$, then the population is nearly the equilibrium population, so that the atom finds itself pumped to the bottom of the potential hill as soon as it reaches the top—this is reminiscent of the Greek mythological character Sisyphus who is always pushing a stone uphill because he finds himself back at the bottom as soon as he reaches the top. This is shown in Fig. 11.3 and the reason why polarization gradient cooling is also called Sisyphus cooling.

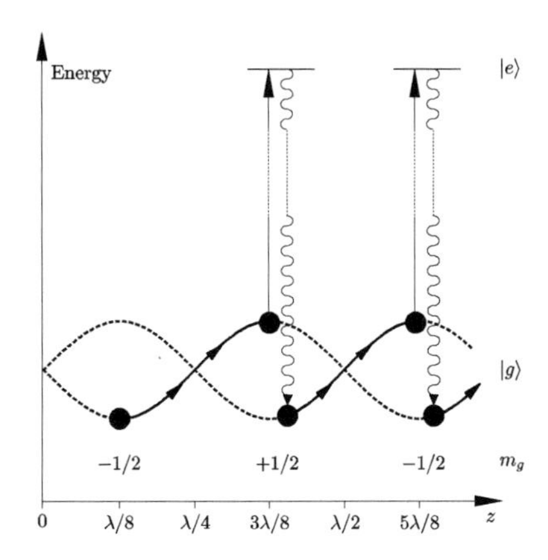

Figure 11.3: Polarization gradient cooling for the energy level structure shown in Fig. 11.2. The atom is always climbing a potential hill because it gets optically pumped to the bottom as soon as it reaches the top.

We are now in a position to understand the cooling mechanism, by referring to Fig. 11.3. Consider an atom moving with a velocity v to the right starting at $z = \lambda/8$ starting in the $m_g = -1/2$ sublevel. It climbs the potential hill toward $z = 3\lambda/8$ by converting a part of its kinetic energy into potential energy. At $z = 3\lambda/8$, it gets optically pumped to the $m_g = +1/2$ sublevel. This process takes away energy from the atom because the energy of the absorbed photon is lower than that of the spontaneously emitted anti-Stokes Raman photon. From $z = 3\lambda/8$ to $5\lambda/8$ it starts climbing the hill again, but this time in the other sublevel. This process continues till the atom's energy becomes too small to climb the next hill. Thus the minimum atomic kinetic energy is of the order of the modulation depth $U_\circ$ in the light shift, so the limit of polarization gradient cooling is $k_B T_{\text{min}} \sim U_\circ$, which is one to two orders of magnitude smaller than the Doppler cooling limit. Since the light shift varies as $\Gamma/|\delta|$, lower temperatures are reached when the (red) detuning is larger. The typical detuning used in polarization gradient cooling is two to three times Γ. In addition, the above analysis shows that it requires careful nulling of B fields so that the sublevels are degenerate in the absence of light.

Finally, we note that the simplifying assumption of a $F_g = 1/2 \to F_e = 3/2$ transition is not so limiting—it works for other values of F_g as long as there are multiple magnetic sublevels.

3. Magneto-optic trap — MOT

So far we have seen how to use laser beams to localize the atoms in momentum space, i.e. cool them by reducing their velocity spread. But in order to localize them in real space (or trap them), we need to provide a restoring force that pushes the atoms toward a particular point in space. This is most easily done by adding a quadrupole magnetic field to the 3D molasses configuration. The magnetic field, produced using a pair of anti-Helmholtz coils, has a linear variation along the three axes, with the field gradient along the z axis equal to -2 times that along the x and y axes so that it satisfies $\nabla \cdot \vec{B} = 0$. The laser beams are chosen to have opposite circular polarizations along each of the three orthogonal directions. This scheme, shown in Fig. 11.4(a), is called a magneto-optic trap (MOT), and is the workhorse of all laser cooling experiments.

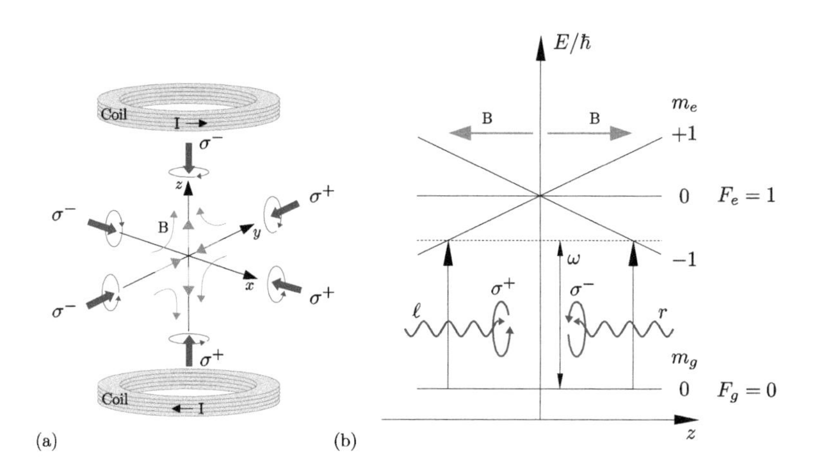

Figure 11.4: The magneto-optic trap (MOT). (a) The 3D MOT requires the addition of a quadrupole B field (produced using a pair of anti-Helmholtz coils) to a molasses configuration with opposite circular polarizations along the three axes. (b) Principle of operation of the MOT using the z axis as an example. The Zeeman shifts of the m sublevels in the linear B field, and selection rules for σ polarization, mean that the ℓ beam is closer to resonance on the left hand side, while the r beam is closer to resonance on the right hand side.

The principle of the MOT can be understood by considering what happens along the z axis, as shown in the Fig. 11.4(b). For simplicity we consider an $F_g = 0 \rightarrow F_e = 1$ transition. The linearly varying B field means that the magnetic sublevels have linearly varying Zeeman shifts, and intersect at the origin where the field is 0. The laser beam on the left is chosen to have σ^+ polarization and drives transitions with the selection rule $\Delta m = +1$, while the laser beam on the right has σ^- polarization and drives transitions with the selection rule $\Delta m = -1$. If the laser frequency is detuned below resonance, then for an atom on the right hand side, the r beam is closer to resonance compared to the ℓ beam. The imbalance in scattering rates results in a net force that pushes the atom toward the origin. The opposite happens for an atom on the left hand side—the ℓ beam is closer to resonance compared to the r beam and again pushes the atom toward the origin. Thus we have created a restoring force pointing toward the origin. Note that the laser cooling (both Doppler and polarization gradient) described earlier still works because the laser beams are red detuned. Furthermore, the assumption of an $F_g = 0 \rightarrow F_e = 1$ transition is not so restrictive, and the scheme works for most values of angular momenta used in experiments.

Experimental details

Specific numbers are for the ^{87}Rb D$_2$ line (5S$_{1/2}$ $\rightarrow$ 5P$_{3/2}$ transition) at 780 nm.

(i) The experiments are done inside an ultra-high vacuum (UHV) chamber maintained at a pressure below 10^{-8} torr using an ion pump. In fact, the lower the pressure in the chamber the longer the lifetime in the MOT, because collisions between trapped atoms and background atoms are reduced.

(ii) The three pairs of cooling laser beams along the three axes should be sufficiently big so that the overlap region at the trap center is about 2 cm in size.

(iii) The intensity in each beam should be a few times the saturation intensity, which for ^{87}Rb atoms is 1.6 mW/cm^2.

(iv) The detuning of the beams should be about -2Γ so that polarization gradient cooling can work. The mechanism for polarization gradient cooling in $\sigma^+\sigma^-$ configuration is different from the lin $\perp$ lin case discussed earlier, and is covered in detail in a paper by Dalibard and Cohen-Tannoudji.* But the important point from an experimental point of view is that it works.

*J. Dalibard and C. Cohen-Tannoudji, "Laser cooling below the Doppler limit by polarization gradients: simple theoretical models," *J. Opt. Soc. Am.* B **6**, 2023–2045 (1989).

(v) To get an estimate of the required gradient of the quadrupole B field, we say that it should be such that the Zeeman shift at a distance of 1 cm (the radius of the laser beams) is equal to the detuning. For ^{87}Rb atoms cooled on the $|F_g = 2, m_g = +2\rangle \rightarrow |F_e = 3, m_e = +3\rangle$ hyperfine transition, the Zeeman shift is 1.4 MHz/G. Therefore for a detuning of −12 MHz (-2Γ), the required field gradient is 8.6 G/cm. Note that the operation of the MOT is not very sensitive to the field gradient, and this calculation is just to get a rough estimate.

(vi) The fluorescence from the trapped atoms is imaged using a CCD camera. The number of atoms is estimated using a calibrated photodiode, after taking into account the solid angle subtended by the collection optics at the location of the trapped atoms and the scattering rate.

(vii) The repumping beam has to be mixed with the cooling beam to prevent atoms from being lost due to optical pumping.

The **capture velocity** v_c of the MOT is defined as the velocity at which the Doppler shift takes the atom out of resonance by one linewidth, which implies

$$v_c = (|\delta| + \Gamma)\frac{\lambda}{2\pi}$$

It is called capture velocity because the atomic velocity has to be of this order so that it can be cooled and captured. For Rb atoms trapped in a MOT with a detuning of $-2\,\Gamma$, v_c is 14 m/s. Since this velocity is small compared to the average velocity in hot vapor, the number of atoms with velocity less than this value is quite small. This necessitates the use of two techniques to load the MOT.

(i) Through the low-velocity tail of hot vapor produced using what is called a getter source. Such a MOT is called a **vapor cell** MOT. Its main advantage is easy loading, but it comes at the cost of high pressure in the MOT region, and consequent low lifetime. If a long lifetime is required, then the atoms either have to be pushed to a higher vacuum region using laser beams or physically transported by moving the B field.

(ii) By using a Zeeman slower to reduce the velocity of atoms coming out of an oven, which will be discussed in the next part. The advantage of a MOT loaded from a Zeeman-slowed beam is that the MOT is in a high vacuum region.

In Fig. 11.5 we show an image of ^{87}Rb atoms trapped in a vapor cell MOT. The vacuum chamber had a pressure of around 10^{-9} torr. The atoms were cooled on the $F_g = 2 \rightarrow F_e = 3$ transition with a detuning of −12

MHz. The three pairs of cooling beams were formed by retro-reflecting incoming circularly polarized beams through quarter wave plates (so that the return beam had opposite circular polarization). Each incoming beam had a power of around 4 mW and a Gaussian profile with $1/e^2$ diameter of 18 mm. This implies that the maximum intensity at beam center was 3.1 mW/cm^2 compared to the saturation intensity of 1.6 mW/cm^2. Each beam was mixed with a repumping beam resonant with the $F_g = 1 \rightarrow F_e = 2$ transition. The repumping beam had the same size but with a maximum intensity of 2 mW/cm^2. The field gradient was 10 G/cm. The resulting cloud shown in the image has about 5×10^7 atoms at a temperature of ~ 100 μK. The cloud size is roughly 3 mm.

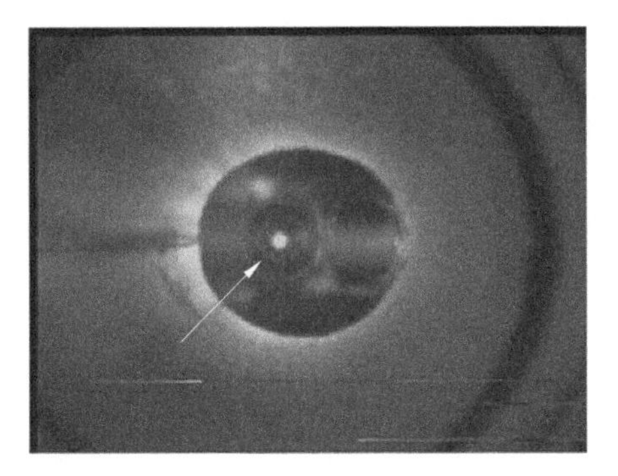

Figure 11.5: CCD image of ^{87}Rb held in a MOT. The cloud has approximately 5×10^7 atoms at a temperature of 100 μK, and a size of 3 mm.

4. Zeeman slower

The probability density function (pdf) of the velocity distribution of atoms coming out of an oven at temperature T is

$$f(v) = 2\left(\frac{m}{2k_BT}\right)^2 v^3 \exp\left(-\frac{mv^2}{2k_BT}\right)$$

where m is the mass of the atom. The cumulative probability defined as

$$P(v) = \int_0^v f(v')\,dv'$$

gives the total fraction of atoms with velocity up to v.

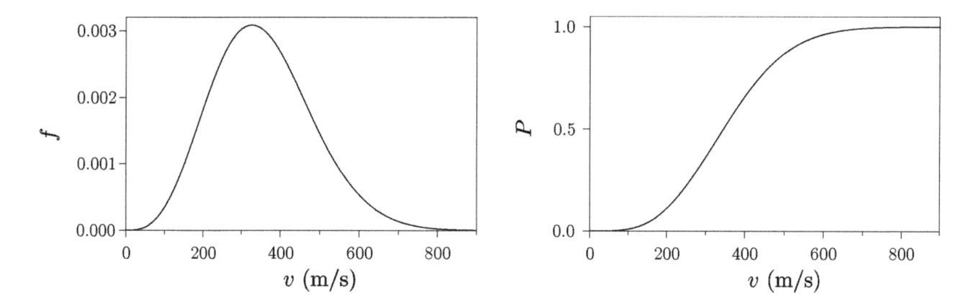

Figure 11.6: Probability density function f and integrated probability P as a function of velocity for ^{87}Rb coming out of an oven at 100°C.

Plots of $f(v)$ and $P(v)$ for ^{87}Rb atoms coming out of an oven at 100°C are shown in Fig. 11.6. The curve for the cumulative probability shows that a negligible fraction of atoms have velocity less than 14 m/s, which is the capture velocity of a Rb MOT with -2Γ detuning. An easy way to have a sizable fraction of atoms with velocity below this value is to slow them down using the spontaneous force from a counter-propagating laser beam. But this has the problem that the change in Doppler shift as the velocity decreases means that the atoms soon go out of resonance. For example, if we consider an atom to be out of resonance if the Doppler shift exceeds one linewidth, then for ^{87}Rb atoms slowed on the D_2 line, a reduction in velocity by 4.7 m/s is enough to take them out of resonance. An elegant way to compensate for this varying Doppler shift is to use the Zeeman shift in a magnetic field, with the magnetic field profile chosen to match the velocity at every point as the atoms slow down—such an arrangement is called a **Zeeman slower**.

From Eq. (4.9) of Chapter 4, we see that the Zeeman shift for a $|F_g, m_g\rangle \to |F_e, m_e\rangle$ transition in a field of magnitude $B_\circ$ is

$$W_B = (g_{F_e} m_e - g_{F_g} m_g)\,\mu_B B_\circ$$

Thus the condition for the Doppler shift being compensated by the Zeeman shift (taking the direction of slowing to be along the z axis) becomes

$$\omega_\circ + \frac{W_B}{\hbar}\frac{B(z)}{B_\circ} = \omega + \frac{2\pi v(z)}{\lambda_\circ}$$

where ω is the frequency of the slowing laser. This gives

$$B(z) = \frac{\hbar B_\circ}{W_B}\left[\delta + \frac{2\pi v(z)}{\lambda_\circ}\right] \tag{11.2}$$

The atoms get decelerated because of the spontaneous force. Therefore the acceleration can be written in a manner similar to Eq. (11.1) as

$$a = \frac{F_{\text{spont}}}{m} = \frac{\hbar k}{m}\frac{\Gamma}{2}\frac{I/I_s}{1 + I/I_s + (2\delta/\Gamma)^2}$$

If we assume that the deceleration is constant throughout the slowing process, then the velocity at a distance z for an atom with initial velocity v_i is

$$v(z) = \sqrt{v_i^2 - 2az}$$

If the atoms are brought to rest after a distance $L_\circ$ then $a = v_i^2/2L_\circ$, so

$$v(z) = v_i\sqrt{1 - \frac{z}{L_\circ}}$$

Substituting in Eq. (11.2) we get

$$B(z) = \frac{\hbar B_\circ}{W_B}\left[\delta + \frac{2\pi v_i}{\lambda_\circ}\sqrt{1 - \frac{z}{L_\circ}}\right]$$

which shows that the required profile is parabolic.

Depending on the laser detuning, the following three kinds of slower designs are possible.

(i) **Decreasing field slower** — This has a field maximum at the beginning and zero at the end, with a laser detuning of zero. The main disadvantage of this design is that the slowing beam is on resonance with the trapped atoms, and can therefore disturb the MOT unacceptably.

(ii) **Increasing field slower** — This has a zero at the beginning and increases to a maximum at the end, with the laser detuning chosen so that it compensates for the Doppler shift of the initial velocity. The main disadvantage of this design is that the large field near the end causes an unacceptably large fringing field at the location of the MOT.

(iii) **Spin-flip slower** — This is a compromise between the above two designs, and has non-zero fields at the beginning and the end (with opposite signs) and goes to zero in between. Apart from overcoming the disadvantages of the above two designs, the spin-flip design also consumes less electrical power in the slowing coils. This is because the field magnitude can be made half of that in the above designs while keeping their difference (which determines v_i) the same.

Actually, if the atoms are brought to complete rest, then they will not make it to the MOT. Therefore, they are given a small final velocity v_f by truncating the slower at a length L that is smaller than $L_\circ$. In terms of the initial and final velocities, the magnetic fields at the end points of the slower from Eq. (11.2) are

$$B_i = B(0) = \frac{\hbar B_\circ}{W_B}\left[\delta + \frac{2\pi}{\lambda_\circ} v_i\right]$$

$$B_f = B(L) = \frac{\hbar B_\circ}{W_B}\left[\delta + \frac{2\pi}{\lambda_\circ} v_f\right]$$

The B field profile in terms of these two field values is

$$B(z) = B_f + (B_i - B_f)\sqrt{1 - \frac{z}{L}}$$

It is important to note that all atoms with velocity up to v_i are slowed by the slower. If an atom starts with a smaller initial velocity, then it begins slowing down from the point where the B field is right. In effect each atom goes through every value of velocity below its starting value, and the number of slowed atoms corresponds to the cumulative probability as the slowing progresses.

The profile for slowing ^{87}Rb atoms from 400 m/s to 10 m/s is shown in Fig. 11.7. From the cumulative probability in Fig. 11.6 this corresponds to capturing 67% of atoms coming out of the oven. The laser is detuned by −200 MHz from the $|F_g = 2, m_g = 2\rangle \rightarrow |F_e = 3, m_e = 3\rangle$ transition of the D_2 line, and has σ^+ polarization. The solid line in the figure is the desired profile. It is realized by wrapping welding cables around the outside of the vacuum chamber in a tapered manner—the calculated field from one such realization is shown using open circles, and closely matches the required profile. The measured field after wrapping the cables is almost identical to the calculation.

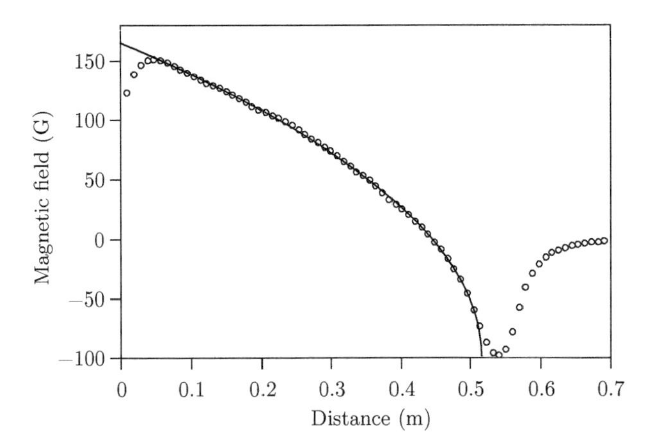

Figure 11.7: Spin-flip Zeeman slower for slowing ^{87}Rb atoms from 400 m/s to 10 m/s with a laser detuned by −200 MHz from the $F_g = 2 \rightarrow F_e = 3$ transition of the D_2 line. Solid line is the required profile, while open circles represent the calculated field from a physical realization with welding cables wrapped on the outside of the vacuum chamber.

5. Atomic fountain

Many precision experiments (e.g. atomic clocks) are done with cold atoms launched vertically in a fountain from a MOT. The most common method of launching atoms while keeping them cold is to use the idea of **moving molasses**. We have already seen that a pair of counter-propagating laser beams with identical detunings of $-\Gamma/2$ will result in atoms with $\langle v \rangle = 0$ cooled to the Doppler limit. If we want atoms to be launched vertically with a velocity of $v_\circ$, then the two detunings should be $-\Gamma/2$ in a frame moving up at $v_\circ$. Thus the detunings in the laboratory frame are

$$\begin{aligned} \delta_{\rm up} &= -\Gamma/2 + v_\circ/\lambda \\ \delta_{\rm down} &= -\Gamma/2 - v_\circ/\lambda \end{aligned}$$

showing that the two laser beams should have unequal detunings for launching.

B. Stimulated force

Like the spontaneous force, this force also arises because of momentum transfer from the laser to the atom, but the difference is that the emission process is stimulated. The atom is driven coherently between the ground and excited states by the laser. The incoherent nature of spontaneous emission from $|e\rangle$ is ignored, which becomes an increasingly better approximation as the detuning is increased. We thus see that it is a conservative force and can be derived from a potential. The force is also called the dipole force because it arises due to the induced dipole moment in the atom by the light. Since the induced dipole moment is proportional to the amplitude of the incident electric field $\mathcal{E}_\circ$, the energy (or potential) is proportional to $\mathcal{E}_\circ^2$ or the light intensity. Thus it can result in a force if the intensity is inhomogeneous—at the focus of a Gaussian beam for example.

To get a quantitative measure of the force, we go back to the energy shift of an atomic state in the presence of an oscillating E field in the dressed atom picture given in Eq. (3.28) of Chapter 3. We take the detuning to be to be negative (red detuned) so that the atom is in a strong field seeking state, and therefore feels a force toward the intensity maximum of the light. For this case the light shift is

$$U = \frac{\hbar}{2}\left(\omega_R' - |\delta|\right) = \frac{\hbar}{2}\left(\sqrt{\Gamma^2 \frac{I}{2I_s} + \delta^2} - |\delta|\right)$$

where we have related ω_R^2 to I and I_s using Eq. (6.31) of Chapter 6. The above is also called the "dressed state potential" because the dipole force can be derived by differentiating it.

In the limit of large detuning $|\delta|/\Gamma \gg 1$, the potential can be approximated as

$$U \approx \frac{\hbar\Gamma^2}{8\delta}\frac{I}{I_s}$$

which shows explicitly that the force requires an intensity gradient.

1. Dipole trap

A simple way to create an intensity gradient is by focusing a Gaussian beam. Atoms (in the strong field seeking state) are then trapped both in the transverse direction (because of the Gaussian profile of the beam), and in the longitudinal direction (because the focusing creates an intensity maximum at the beam waist)—such a dipole trap is called a **far off resonance trap (FORT)**. But the trap stiffness is very different in the two directions—in the transverse direction the size is determined by the waist radius $w_\circ$ which is of order microns for a tightly focused beam; while in the longitudinal direction the size is determined by the Rayleigh range $\pi w_\circ^2/\lambda$ which is 10 to 100 times larger.

This difference should be clear from Fig. 11.8(a), where we show the potential experienced by ^{87}Rb atoms at the focus of an Nd-YAG laser beam at 1064 nm. Using a typical value of 3 W for the beam power and that it is focused to a waist radius of 5 µm, we get the maximum intensity at beam center as 7.64×10^6 W/cm^2 and the corresponding Rayleigh range as 74 µm. The size of the potential in the two directions in the figure shows that the trap is very weak in the longitudinal direction. This is solved by using a crossed dipole trap so that atoms are trapped tightly in all directions at the intersection of the two beams, as shown in Fig. 11.8(b). The depth of the potential from Fig. 11.8(a) is 9.7 mK, which shows that atoms have to be precooled using laser cooling to be successfully trapped in a dipole trap.

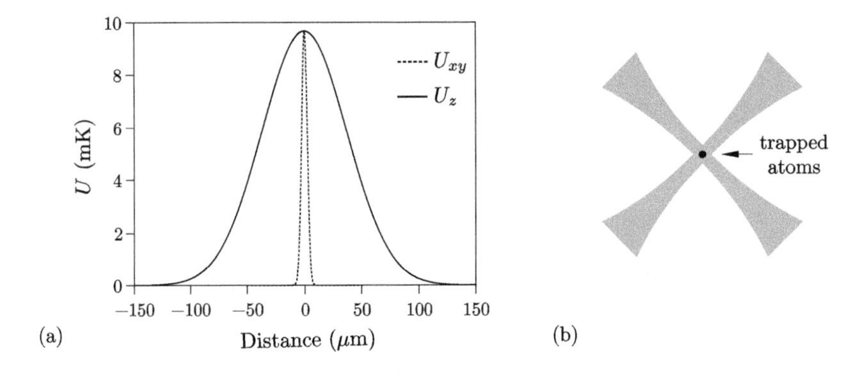

Figure 11.8: (a) Dipole potential in the transverse and longitudinal directions for trapping ^{87}Rb atoms with an Nd-YAG laser beam at 1064 nm, using 3 W power focused to a spot size of $w_\circ = 5$ µm. The trap is much weaker in the longitudinal direction. (b) Crossed dipole trap used to provide tight confinement in all directions.

C. Magnetic trapping and evaporative cooling

Magnetic trapping of neutral atoms uses the fact that atoms have a magnetic moment $\vec{\mu}$ and that it interacts with an external B field through a $-\vec{\mu} \cdot \vec{B}$ term in the Hamiltonian. Atoms with spin oriented along the B field (weak field seekers) can be trapped if the field is inhomogeneous and has a minimum (in magnitude) at some point. Atoms with anti-parallel spin (strong field seekers) are not trapped. If the atom has $F = 1/2$, then there are only two spin states: the $m = +1/2$ state is trapped while the $m = -1/2$ state is untrapped.

The simplest way to create a field minimum is to have a quadrupole B field produced using a pair of anti-Helmholtz coils—called a **quadrupole trap**. Spin-polarized atoms are then trapped near the center. But the field minimum is actually equal to zero. In fact this zero field point is the major drawback of the quadrupole trap because atoms can be lost from here due to the fact that there is no quantization axis and the trapped and untrapped states are degenerate, raising the possibility of non-adiabatic spin flip as the atoms pass this point—such losses are called Majorana spin-flip losses. The losses become more and more significant as the atoms get colder because they spend more time near this point. One would think that this problem can be solved by adding a small constant field to the configuration, but a moment's introspection shows that it only shifts the zero point to a new location.

The other configuration used for magnetic trapping, which has a non-zero field minimum and thus eliminates the problem of spin-flip losses, is the **Ioffe–Pritchard trap**. The trap is shown schematically in Fig. 11.9. It consists of a pair of circular coils (carrying current in the same direction), and four current carrying rods lying symmetrically on a circle with alternating current directions as shown in Fig. 11.9(a). Along the z axis the field due to the four rods is zero, therefore only the coils contribute. Then the field near the center can be expanded in a series as

$$|\vec{B}| = B_z(z) = \frac{\mu I R^2}{(R^2 + A^2)^{3/2}} \left[1 + 3z^2 \frac{(4A^2 - R^2)}{2(A^2 + R^2)^2} + \dots \right]$$

where I is the current through the coils, R is their radius, and $2A$ is their separation.

As a specific example, we consider the case where $I = 100$ A, $R = 1.5$ cm, and $A = 2.25$ cm. In these mixed units where the current is measured in amperes and lengths in centimeters, the constant μ is $4\pi/10$. The calculated field along the z direction for this configuration is shown in Fig. 11.9(b)—it shows that the field has a minimum value of 14.3 G. The field profile in the radial direction (not shown) has a similar shape with the same minimum. But the price one has to pay for this non-zero minimum is that the field has a parabolic shape and not linear as in a quadrupole trap. This means that in order to get the same tightness of confinement in the two cases, the magnitude of the field for the Ioffe-Pritchard trap has to be much larger.

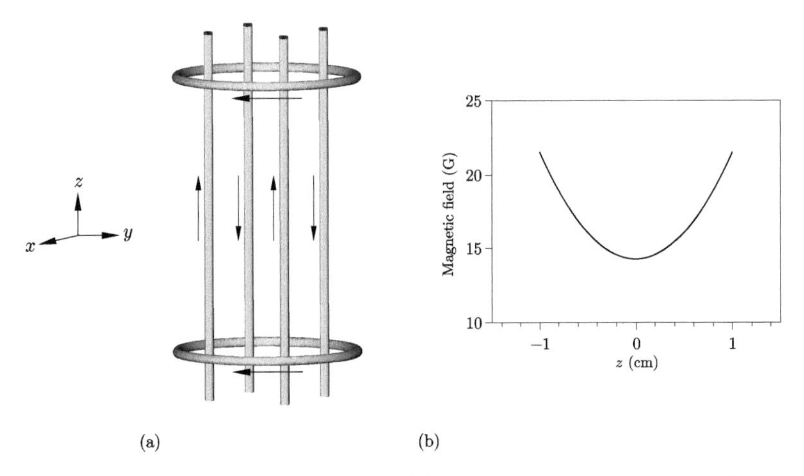

Figure 11.9: Ioffe-Pritchard trap. (a) Configuration consisting of two circular coils carrying current in the same direction, and four current carrying rods lying symmetrically on a circle with alternate current directions. (b) Field along the z direction for the case of current 100 A, coil radius 1.5 cm, coil separation 4.5 cm, and rods on a circle of radius 1 cm. The field has a non-zero minimum of 14.3 G.

As with dipole traps, atoms have to be pre-cooled before being loaded into a magnetic trap because the trap depths are less than 1 K. But once loaded, they can be cooled further using the technique of **evaporative cooling**. This is a technique that works by removing the hottest atoms from the trap and allowing the remaining atoms to thermalize to a lower temperature. Rethermalization of the remaining atoms requires that the atoms be trapped tightly so that the collision rate is high compared to the evaporation rate.

We will see more about magnetic trapping and evaporative cooling in the next section on Bose–Einstein condensation.

D. Bose–Einstein condensation

This section is a bit historical since the intention is to narrate the evolution of Bose–Einstein condensation (BEC).

The story of BEC begins in 1924 when the young Indian physicist S. N. Bose gave a new derivation of the Planck radiation law. He was able to derive the law by reducing the problem to one of counting or statistics: how to assign *indistinguishable* particles (photons) to cells of energy $h\nu$ while keeping the total energy constant. Einstein realized the importance of the derivation for developing a quantum theory of statistical mechanics. He argued that if the photon gas obeyed the statistics of Bose, so should material particles in an ideal gas. Carrying this analogy further, he showed that the quantum gas would undergo a phase transition at a sufficiently low temperature when a large fraction of the atoms would condense into the lowest energy state. This is a phase transition in the sense of a sudden change in the state of the system, just like steam (gaseous state) changes abruptly to water (liquid state) when cooled below 100°C. But it is a strange state because it does not depend on the interactions of the particles in the system, only on the fact that they obey a kind of quantum statistics.

In modern physics, the phenomenon is understood to arise from the fact that particles obeying Bose–Einstein statistics (called bosons) "prefer" to be in the same state. This is unlike particles that obey Fermi–Dirac statistics (fermions), and therefore the Pauli exclusion principle, which states that no two of them can be in the same state. With this property of bosons in mind, imagine a gas of bosons at some finite temperature. The particles distribute the total energy amongst themselves and occupy different energy states. As the temperature is lowered, the desire of the particles to be in the same state starts to dominate, until a point is reached when a large fraction of the particles occupies the lowest energy state. If any particle from this state gains some energy and leaves the group, the other particles quickly pull it back to maintain their number. This is a Bose–Einstein condensate, with the condensed particles behaving like a single quantum entity.

The point at which "the desire for the particles to be in the same state starts to dominate" can be made more precise by considering the quantum or wave nature of the particles in greater detail. The basic idea is shown in Fig. 11.10. From the de Broglie relation, each particle has a wavelength $\lambda_{\rm dB}$ given by h/mv, where m is the mass and v is the velocity. At high temperatures, the average thermal velocity of the particles is high and they behave like billiard balls. As the temperature is lowered, the mean velocity decreases and the de Broglie wavelength increases, and the particles start to behave like extended wavepackets. BEC occurs when $\lambda_{\rm dB}$ becomes comparable to the average interparticle separation so that the wavefunctions of the particles overlap. The average interparticle separation for a gas with number density n is $n^{-1/3}$, and from kinetic theory the mean de Broglie wavelength of gas particles at a temperature T is $h/(2\pi mkT)^{1/2}$. For the wavefunctions to overlap, this product should be of order 1. A more rigorous analysis shows that BEC occurs when the dimensionless phase-space density $n\lambda_{\rm dB}^3$ exceeds 2.612.

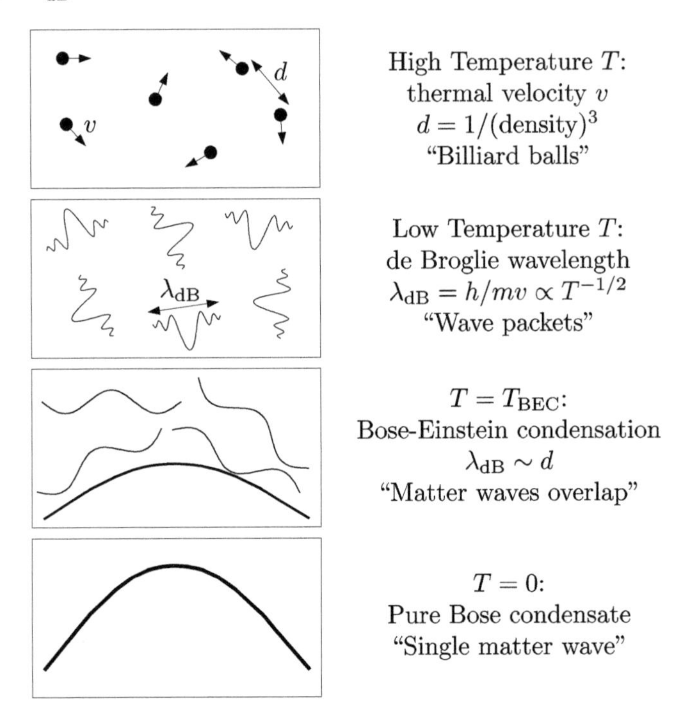

Figure 11.10: Bose–Einstein condensation. As the temperature is lowered, the particles go from behaving as billiard balls, to behaving as extended wavepackets, to having overlapping wavefunctions when BEC occurs, to having a single wavefunction at $T = 0$.

In the early days, it was believed that BEC was only a theoretical prediction and was not applicable to real gases. However, the observation of superfluidity in liquid He in 1938 made people realize that this was a manifestation of BEC, even though it occurred not in an ideal gas but in a liquid with fairly strong interactions. BEC in a non-interacting gas was now considered a real possibility. The first serious experimental quest started in the early 1980s using spin-polarized atomic hydrogen. There were two features of hydrogen that were attractive: it was a model system in which calculations could be made from first principles, and it remained a gas down to absolute zero temperature without forming a liquid or solid. Spin-polarized H could also be trapped in a quadrupole magnetic trap. Using a dilution refrigerator, the gas was cooled to about 1 K and then loaded into the magnetic trap.

One of the major developments to come out of this effort was the proposal in 1986 by Harald Hess, then a post-doc with Dan Kleppner at MIT, to use evaporative cooling to lower the temperature and reach BEC. The idea in evaporative cooling, as mentioned earlier, is to selectively remove the hottest atoms from the trap, and then allow the remaining atoms to thermalize. Since the remaining atoms have lower energy, they thermalize to a lower temperature. This is similar to cooling coffee in a cup: the hottest particles near the top are blown to take away the heat, while the remaining particles get colder. The MIT group of Kleppner and Greytak demonstrated evaporative cooling of spin-polarized H by lowering the height of the magnetic trap below the kinetic energy of the hottest atoms. By 1992, they had come within a tantalizing factor of 3 of observing BEC but were stopped short due to technical problems.

Meanwhile, a parallel effort in observing BEC using alkali atoms was getting underway. The main impetus for this was to see if the tremendous developments that occurred in the late 1980s in using lasers to cool atomic clouds could be used to achieve BEC. Alkali atoms could be maintained in a gaseous state if the density was low, typically less than 10^{14} atoms/cc. But this meant that BEC would occur only at temperatures below 1 μK. As we have seen, laser cooling, particularly polarization gradient cooling, had indeed achieved temperatures in the range of a few μK, with a corresponding increase in phase-space density of about 15 orders of magnitude. However, the lowest temperature attained experimentally was limited by heating due to the presence of scattered photons in the cloud. One advance to this problem came from the MIT group of Dave Pritchard. His then post-doc, Wolfgang Ketterle, proposed using a special magneto-optic trap in which the repumping laser was blocked near the center. This meant that the coldest atoms got "shelved" in the other hyperfine state and became dark—they called it the **dark spot MOT**. Since the shelved atoms do not see the light, they do not get heated out of the trap. This helped improve the density by another order of magnitude, but BEC was still a factor of million away.

Pritchard's group at MIT also demonstrated magnetic trapping of sodium at around the same time. Pritchard and his student Kris Helmerson proposed a new technique for evaporative cooling in such a trap: RF-induced evaporation. Instead of lowering the magnetic field to cause the hottest atoms to escape, as was done in the spin-polarized hydrogen experiments, they proposed using an RF field tuned to flip the spin of the hottest atoms, and hence drive them into the untrapped state. The beauty of this technique is that the RF frequency determines which atoms get flipped, while the trapping fields remain unchanged. Pritchard's group was however unable to demonstrate evaporative cooling in their magnetic trap because the density was too low.

Laser cooling and evaporative cooling each had their limitations because they required different regimes. Laser cooling works best at low densities, while evaporative cooling works at high densities when collisions enable rapid rethermalization. Therefore, in the early 1990s, a few groups started using a hybrid approach to achieve BEC, i.e. first cool atoms to the microkelvin range using laser cooling, and then load them into a magnetic trap for evaporative cooling. By the year 1994, two groups were leading the race to obtain BEC: the Colorado group of Cornell and Wieman, and the MIT group of Ketterle. Both groups had demonstrated RF-induced evaporative cooling in a magnetic trap, but found that there was a new limitation, namely the zero field point at the bottom of the quadrupole trap from which atoms leaked out.

Ketterle's solution to plug the hole was to use a tightly focused Ar-ion laser beam at the trap center. The dipole force from the laser beam kept the atoms out of this region, and, since the laser frequency was very far from the resonance frequency of the sodium atoms used in the experiment, it did not cause any absorption or heating. The technique proved to be an immediate success and gave Ketterle's team an increase of about 3 orders of magnitude in phase-space density. But again technical problems limited the final observation of BEC.

Cornell had a different solution to the leaky trap problem: the time-orbiting potential (TOP) trap. His idea can be understood in the following way. If you add a constant external field to the quadrupole trap configuration, the hole does not disappear, it just moves to a new location depending on the strength and direction of the external field. Atoms will eventually find this new hole and leak out of it. However, Cornell's idea was that if you move the location of the hole faster than the average time taken for atoms to find it, the atoms will be constantly chasing the hole and never find it! A smooth way to achieve this is to add a rotating field that moves the hole in a circle. The time-averaged potential is then a smooth pseudo-potential well with a non-zero minimum.*

Plugging the leaky trap proved to be the final hurdle in achieving BEC. In July 1995, Cornell and Wieman announced that they had observed BEC in a gas of ^{87}Rb atoms. The transition temperature was a chilling 170 nK, making it the coldest point in the universe! The researchers had imaged the cloud by first releasing the trap and allowing the cloud to expand, and then illuminating it with a pulse of resonant light. The light absorbed by the cloud cast a shadow on a CCD camera. The "darkness" of the shadow gave an estimate of the number of atoms at any point in the image. Since the atoms expanded ballistically from nearly zero size, the final spatial location in the image was proportional to the initial velocity, i.e. the image was a map

*This pseudo-potential is similar to the one in a Paul trap for charged particles, which we will see later.

of the velocity distribution in the cloud. Looking at cuts through the images to get the velocity distribution, they found that the appearance of the condensate was marked by the atoms having a thermal velocity distribution above the transition temperature to having a bimodal distribution with a large peak at the center below the transition temperature. Calculated distributions of what they obtained is shown in Fig. 11.11, the experimental data matched this prediction.

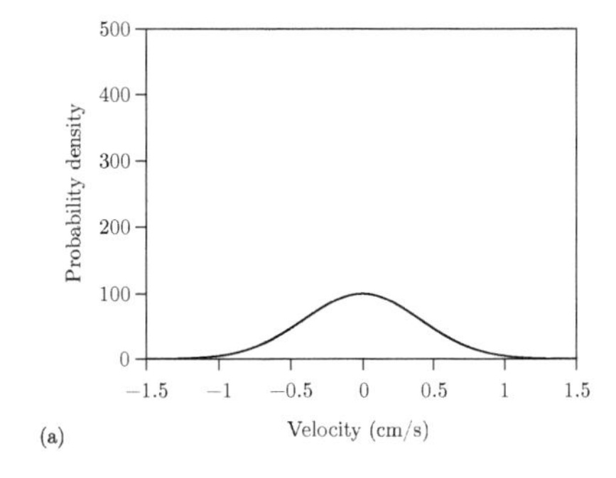

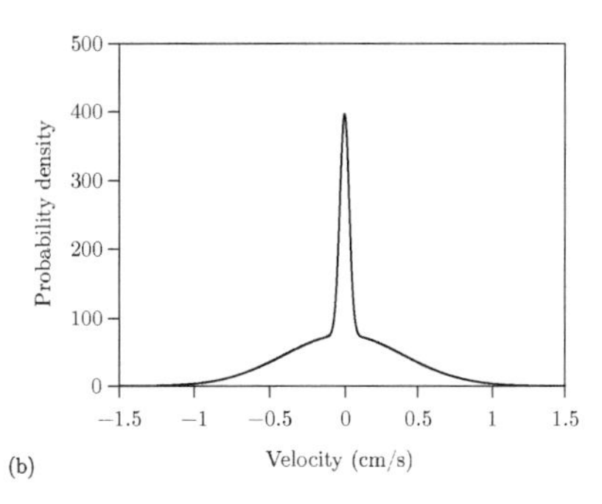

Figure 11.11: Velocity distribution of ^{87}Rb atoms near the Bose–Einstein condensation transition. (a) Above the transition temperature (170 nK) atoms have a thermal distribution. (b) Below the transition temperature, the distribution is bimodal with 25% of atoms in the narrow peak near the center corresponding to the condensed phase.

There was no doubt that the central peak corresponded to atoms in the ground state of the trap because it was asymmetric, exactly as predicted by quantum mechanics due to the fact that the trap strength was not the same in the two directions. By contrast, the thermal distribution was always symmetric, showing that the atoms in the central peak were non-thermal. Furthermore, as the temperature was lowered below the transition temperature, the density of atoms in the central peak increased abruptly, indicating a phase transition. Thus there was no room for skepticism that they had achieved BEC.

Soon after this, Ketterle's group observed BEC in a cloud of ^{23}Na atoms. As against the few thousand condensate atoms in the Colorado experiment, they had more than a million atoms in the condensate. This enabled them to do many quantitative experiments on the fundamental properties of the condensate. For example, they were able to show that when two condensates were combined, they formed an interference pattern, indicating that the atoms were all phase coherent. They were also able to extract a few atoms from the condensate at a time to form a primitive version of a pulsed **atom laser**, i.e. a beam of atoms that are in the same quantum state. They could excite collective modes in the condensate and watch the atoms slosh back and forth. These results matched the theoretical predictions very well.

BEC in atomic gases has since been achieved in several laboratories around the world, including from the lab of my first PhD student, Umakant Rapol, at IISER Pune in India (images from their experiment shown in Fig. 11.12).

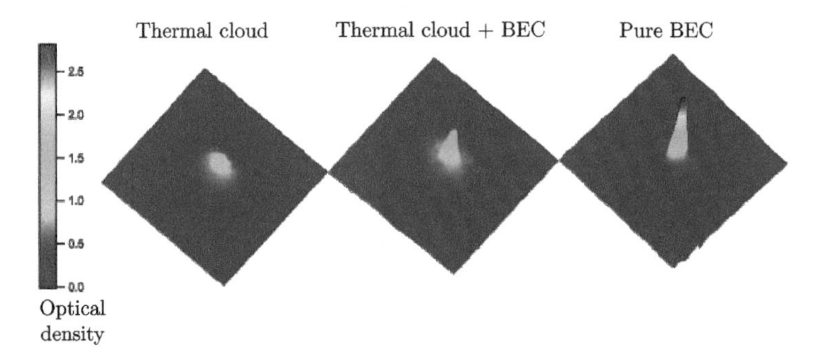

Figure 11.12: Velocity distribution of ^{87}Rb atoms showing a phase transition from thermal cloud to BEC as the RF scalpel frequency is lowered. Images taken by absorption imaging of near-resonant light falling on a CCD camera. Image size 0.8 mm $\times$ 0.8 mm. Time of flight of ballistic expansion after release from the trap is 21 ms.

Apart from Rb and Na, it has been observed in all the other alkali atoms. The H group at MIT achieved it in 1998. Metastable He has also been cooled to the BEC limit. A Rb BEC has also been obtained by evaporative cooling in a crossed dipole trap, thus eliminating the need for strong magnetic fields and allowing atoms to be condensed independent of their spin state. Such an all-optical trap has been used to condense the two-electron atom Yb. The variety of systems and techniques to get BEC promises many applications for condensates. The primary application, of course, is as a fertile testing ground for our understanding of many-body physics, bringing together the fields of atomic physics and condensed-matter physics. In precision measurements, the availability of a giant coherent atom should give enormous increase in sensitivity. BECs could also impact the emerging field of nanotechnology since the ability to manipulate atoms greatly increases with their coherence.

In recognition of their pioneering work on BECs, Cornell, Wieman, and Ketterle were awarded the 2001 Nobel Prize in physics.

E. Optical tweezers

The optical tweezers is an invention by Art Ashkin and colleagues working at AT&T Bell Labs. They were studying the manipulation of neutral atoms in a dipole trap, when they realized that the same force could be used to trap micron-sized particles. The trapping force is produced using a tightly focused laser beam, usually realized by sending an IR laser beam through a high numerical aperture (NA) microscope objective. The laser wavelength λ is about 1 µm, while the size d of the trapped particle ranges from 1 to 10 µm, therefore a ray-optics picture (valid for $d \gtrsim \lambda$) can be used to understand the trapping force.

The basic idea is shown in the Fig. 11.13. The momentum change associated with refraction of rays produces a restoring force toward the intensity maximum located at the beam waist, thus trapping the particle both in the transverse and longitudinal directions. Instead of red detuning as required for trapping atoms, particles trapped in an optical tweezers trap have to be transparent to the light with a refractive index greater than the surrounding medium. As before, the trap stiffness is very different in the two directions, but the weak nature in the longitudinal direction is not a problem for most applications. In addition, the trapped particle does not need to be separately cooled—the damping force required for the trap to work is provided by having the particle immersed in a viscous medium. For example, many experiments are done with polystyrene beads immersed in water, where the water provides the viscous damping.

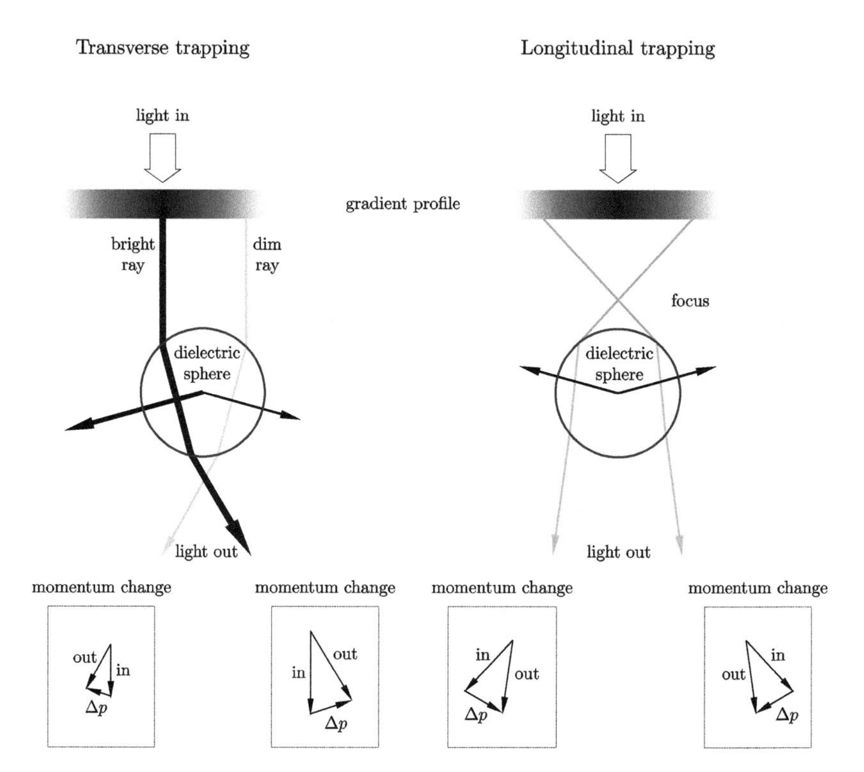

Figure 11.13: Ray-optics picture for trapping of a dielectric sphere in an optical tweezers trap at the intensity maximum. Left hand side shows trapping in the transverse direction due to the recoil momentum from the refracted rays being toward the beam center for a particle displaced from it, because the two rays have different intensities. Right hand side shows trapping in the longitudinal direction because the recoil momentum from the refracted rays is toward the focus for a particle displaced from it.

A typical experimental set up is shown in Fig. 11.14(a). The high NA objective is usually a 100×, 1.4 NA, oil-immersion objective. The laser beam from an Nd-YAG laser at 1064 nm is fed into an **inverted** microscope, where the direction of gravity helps in bringing the particles closer to the trap. The trapped particle is visualized using a video camera, while quantitative measurements of its position are measured using a quadrant photodiode (QPD) with a typical sensitivity of 25 mV/µm. If a polystyrene bead immersed in water is used as the trapped particle, then it can be attached to other things such as DNA strands which can then be manipulated. Alternately, single cells or organelles within cells can be directly manipulated. As an example, a single red blood cell (RBC), which has a size of 5 to 8 µm, is shown before and after trapping in Fig. 11.14(b). Optical tweezers can then be used as a sensitive probe to study the changes in the RBC due to infection, by the malarial parasite for instance. This is just one of the myriad applications of optical tweezers in biology and medicine, which take advantage of the fact that the technique is non-invasive and ultra-sensitive.

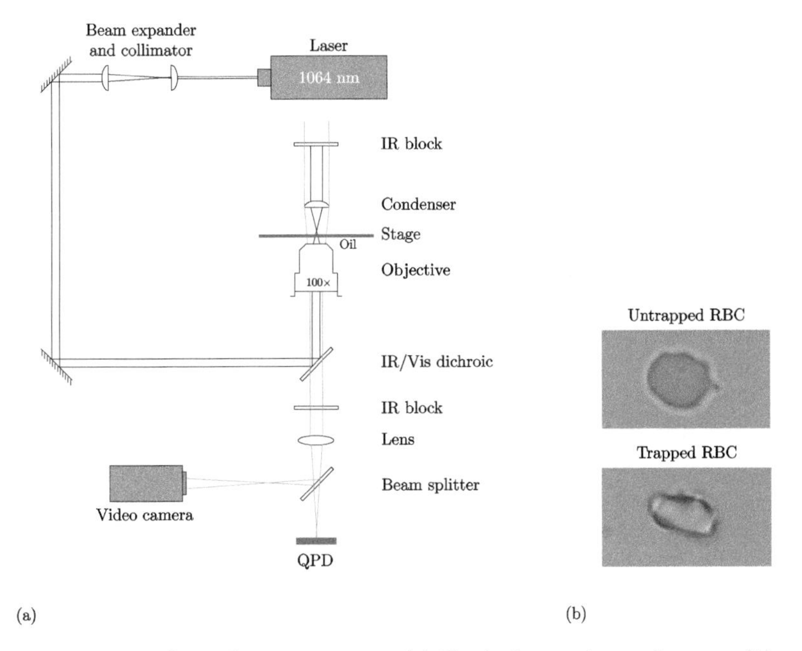

Figure 11.14: Optical tweezers trap. (a) Typical experimental setup. (b) Single red blood cell (RBC) upon being trapped reorients itself because of its flat shape.

Calibrating the trap

Many applications of optical tweezers require a precise knowledge of its strength. Assuming that the trap is harmonic, its strength is characterized by a spring constant k. This can be measured in the following ways.

(i) Escape force method

In this method one measures the minimum force required to free a particle from the trap using the viscous drag of the medium. The force on the particle is applied by moving the medium with respect to the trap, either by moving the sample stage while keeping the trapping laser stationary, or by moving the laser with a stationary stage. The velocity of the particle, measured using the video camera in the setup, is increased continuously until the particle escapes. For a spherical particle the viscous force is given by

$$F_{\text{viscous}} = 6\pi\eta a v$$

where η is the viscosity of the medium, a is the diameter of the particle, and v is the velocity. Thus by knowing the escape velocity one can estimate the trapping force. This is a somewhat approximate method and gives the trap strength with about 10% accuracy, because the particle escapes from the edge—the optical properties near the edge are often not the same as that at the center.

(ii) Drag force method

This method relies on applying a known force to the trapped particle and measuring its displacement from the trap center. If the applied force is F and displacement of the particle is d then the trap stiffness is just $k = F/d$. The force is applied by moving the medium with a constant velocity and using $F = 6\pi\eta av$, while its position is recorded using the QPD shown in the setup. To ensure that one remains in the linear regime of force versus displacement, the particle should not reach the edge of the trap. In addition, the viscocity is different near the walls of the chamber; hence it is better to stay away from them.

(iii) Equipartition method

An alternate method, which does not require knowledge of the viscosity or shape/size of the particle, is to measure the thermal fluctuations in the position of the trapped particle. For a particle bound in a harmonic potential and at a temperature T, the equipartition theorem tells us that

$$\frac{1}{2}k\left\langle x\right\rangle^2=\frac{1}{2}k_BT$$

where x is the particle's position as measured by the QPD. Thus a statistical analysis of the QPD data gives us k. Although the method does not use η or a, it requires the QPD to be well calibrated in terms of relating its data points to position values, and the electronics to have sufficient bandwidth so that fluctuations at high frequencies can be measured. Furthermore, other sources of noise (such as from the electronics) add to the fluctuation in the data and hence influence the estimate of k, so they must be minimized. Since the method relies on the potential being harmonic, one has to ensure that the trapped particle remains near the center.

(iv) Power spectrum method

This is another statistical method using the QPD data, but has the advantage that the QPD does not have to be accurately calibrated. It relies on measuring the power spectrum of position fluctuations in the trap. The equation of motion of a particle within a harmonic potential and inside a viscous medium (neglecting the inertial term because the particle is in an overdamped medium) is given by

$$\gamma\,\dot{x}(t)+k\,x(t)=F(t)$$

where $F(t)$ is the Langevin force, and γ is Stoke's viscous drag coefficient which is $3\pi a\eta$ for a spherical particle of diameter a.

From the above equation one sees that the Fourier transform $X(f)$ of $x(t)$, and the corresponding power spectrum $P(f)$, for a particle at temperature T obey

$$P(f)=\left\langle X(f)X^*(f)\right\rangle=\frac{k_BT}{\pi^2\gamma}\left(\frac{1}{f_c^2+f^2}\right)$$

This shows that the power spectrum has a Lorentzian lineshape, with a corner frequency f_c. The corner frequency is related to the trap stiffness as

$$k=2\pi\gamma f_c$$

Thus one can calculate the trap stiffness by measuring f_c and knowing γ. f_c can be determined by recording the position fluctuations with the QPD, finding the power spectrum of the data, and fitting the above Lorentzian lineshape to the spectrum. The advantage of this method is that one does not have to calibrate the QPD since any arbitrary scaling of the data will give the same corner frequency. In addition, if the tweezer is slightly misaligned, then the power spectrum will not have a Lorentzian lineshape—so this technique can be used to ensure that the trap is aligned properly.

(v) Step response method

In this method one measures the trap stiffness by looking at the response of the trapped particle to a sudden (step) displacement of the trap. If the size of the step is $x_\circ$, then the response of the particle in a harmonic potential is an exponential build-up from 0 to $x_\circ$ given by

$$x(t) = x_\circ \left[1 - \exp\left(-\frac{kt}{\gamma} \right) \right]$$

Here also one does not require the QPD to be calibrated since any scaling of x will not affect the determination of k. As before one has to ensure that the particle remains in the harmonic regime near the trap center.

In all the above methods except for the **equipartition method**, a knowledge of the viscosity of the medium is needed. Thus there should be an independent method to determine η. For a polystyrene bead in water at room temperature, $\eta = 1002\ \mu\text{Ns/m}^2$. The typical trap strength for trapping such a particle with a $100\times$ objective is $k \approx 8\ \mu\text{N/m}$ with 25 mW of laser power incident on the sample plane.

F. Ion trapping

In this section we will discuss the two kinds of ion traps—Penning and Paul. The Penning trap was invented by Hans Dehmelt and used to make a precision measurement of the electron g factor using a single trapped electron. The Paul trap was invented by Wolfgang Paul and used for many spectroscopic measurements on trapped ions. It is generally easier to trap ions compared to neutral atoms because their charged nature means that they interact strongly with external fields.

1. Penning trap

The Penning trap uses **static** electric and magnetic fields to trap charged particles in all three dimensions. A strong uniform magnetic field along the z axis confines the particle radially, while a weak quadrupole electric field provides a linear restoring force in the axial direction. The equipotential surfaces for a quadrupole electric field are hyperbolae of rotation. The trap is formed with copper electrodes that are machined to have this shape, as seen from the cross-sectional view in Fig. 11.15. The top and bottom electrodes are called endcaps, while the central one is called the ring electrode. The holes in the endcaps allow ions to be injected into the trap. The trap is kept in a vacuum chamber, which is in turn kept in a liquid He bath so that the cryogenically pumped vacuum gives a lifetime of several days for the trapped ions.

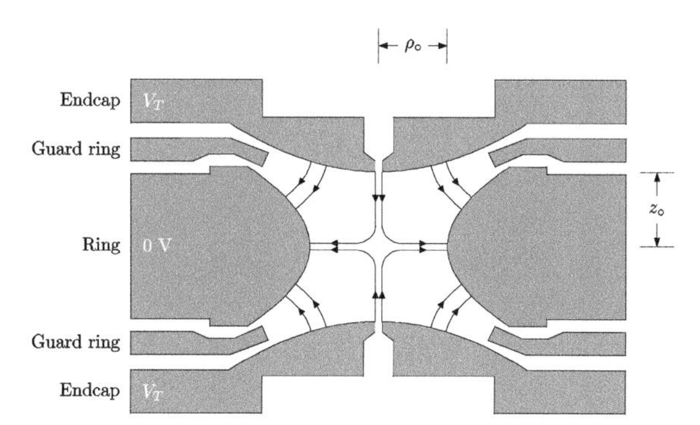

Figure 11.15: Cross-sectional view of a Penning trap. The electric field lines for the quadrupole field are as shown.

If we assume that the electrodes are infinite in size so that the electric field is perfect, then the electrostatic potential (in cylindrical coordinates) inside the trap is

$$\phi(z, \rho) = \frac{z^2 - \frac{1}{2}\rho^2}{2d^2} V_T$$

where V_T is the applied voltage between the ring and endcaps, z is the axial position, ρ is the radial position, and d is the characteristic size of the trap given by

$$d^2 \equiv \frac{z_\circ^2}{2} + \frac{\rho_\circ^2}{4}$$

with $z_\circ$ and $\rho_\circ$ as defined in the figure.

For simplicity, we consider that a single ion of mass m and charge e is trapped in the trap. Then its equation of motion is

$$m\frac{d^2\vec{r}}{dt^2} = e\vec{E}(\vec{r}) + \frac{e}{c}\vec{v} \times \vec{B}(\vec{r})$$

In an ideal trap, the motion decomposes into three normal modes. In the axial direction the linear electric field gives rise to harmonic motion (called the **axial mode**) at a frequency defined by

$$\omega_z^2 = \frac{eV_T}{md^2}$$

The two modes in the radial plane are the trap cyclotron mode—at a frequency ω_c' corresponding to the normal cyclotron oscillation around the magnetic field lines, but slightly modified due to the electric field in the trap—and the magnetron mode—a slow drift at ω_m due to the $\vec{E} \times \vec{B}$ field away from trap center. The eigenfrequencies can be derived from the radial equation of motion in the trap

$$\frac{d^2\vec{\rho}}{dt^2} - \omega_c\frac{d\vec{\rho}}{dt} \times \hat{z} - \frac{1}{2}\omega_z^2\vec{\rho} = 0$$

where $\omega_c = eB/mc$ is the frequency of free-space cyclotron motion around the B field.

We guess solutions of the form $\mathrm{Re}\{\rho_\circ e^{i\omega t}\}$ and plug it into the above equation to obtain the characteristic equation

$$\omega^2 - \omega_c\omega + \frac{1}{2}\omega_z^2 = 0$$

This yields the two solutions

$$\begin{aligned} \omega_c' &= \frac{1}{2}\left(\omega_c + \sqrt{\omega_c^2 - 2\omega_z^2}\right) \\ \omega_m &= \frac{1}{2}\left(\omega_c - \sqrt{\omega_c^2 - 2\omega_z^2}\right) \end{aligned} \tag{11.3}$$

The three trap modes behave as harmonic oscillators but differ greatly in the partition of energy between kinetic and potential. The axial motion has equal average kinetic and potential energy as for a mass bound harmonically on a spring. The cyclotron motion is mainly the circular motion in a magnetic field at a high speed, so the energy is predominantly kinetic. On the other hand, the magnetron motion is a slow drift and the energy is almost entirely potential. In fact, the potential energy (and the total energy) in the magnetron mode decreases as its radius increases, even as its kinetic energy increases. Therefore an ion at the center of the trap is in unstable equilibrium on top of this potential hill and does not leave the trap only because it has no way of losing energy and momentum.

One important application of Penning traps is in precision mass measurement. From the cyclotron frequency of eB/mc, we see that we can get the ratio of two atomic masses by measuring the ratio of their cyclotron frequencies in the same B field. This is relatively easy in an ideal trap as seen from the expression in (11.3) for the radial modes. But even in the presence of real life non-idealities, such as a small misalignment of the magnetic and electric field axes (tilt), or machining imperfections in the electrodes leading to an eccentricity in the hyperboloids, we can obtain the cyclotron frequency from the following invariance theorem

$$\omega_c^2 = \omega_c'^2 + \omega_z^2 + \omega_m^2$$

For a typical ion of mass 28 amu, the trap cyclotron frequency is about 4.5 MHz in a magnetic field of 8.5 T. The trap voltage is adjusted to make the axial frequency 160 kHz, which gives a magnetron frequency of 2.8 kHz. This results in the following hierarchy of frequencies $\omega_c' \gg \omega_z \gg \omega_m$. Then, from the above relation, one only needs to measure ω_c' to the precision desired for the mass measurement; ω_z and ω_m need to be measured to correspondingly lower precision.

2. Paul trap

The Paul trap is an example of dynamic stabilization. It uses the same quadrupole E field configuration as for the Penning trap shown in Fig. 11.15, but with an oscillating potential applied between the endcaps and the ring electrode. The field at the center of the trap in the presence of a DC field is a saddle point, with a force toward the center (trapping force) in one direction but away from the center (destabilizing force) in the orthogonal direction.* But the particle can be stably trapped if the saddle point is rotated fast enough that the particle does not have enough time to roll off, which can be achieved by having an oscillatory potential. The idea is shown in Fig. 11.16. The rotation of the saddle point creates a pseudopotential which is a harmonic well in all directions.

*This can also be seen as a consequence of Earnshaw's theorem which states that it is impossible to create a point of stable equilibrium in a source free region.

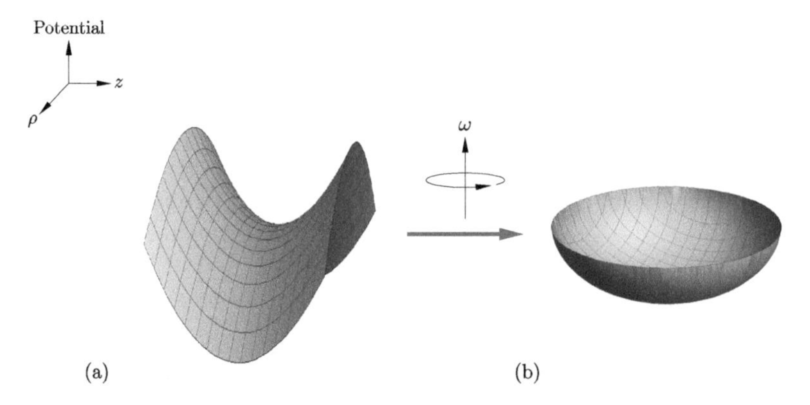

Figure 11.16: Dynamic stabilization in a Paul trap. (a) The quadrupole potential has a saddle point in the presence of a DC field. (b) The presence of an AC field is equivalent to rotating the saddle, which creates a pseudopotential well in both directions.

Mathematically, this situation is achieved by replacing in the Penning trap

$$V_T \qquad \rightarrow \qquad V_\circ - V_1 \cos\omega t$$

where $V_\circ$ is the amplitude of the DC field, and V_1 is the amplitude of the AC field oscillating at ω. The resulting equations of motion for the ρ and z components are

$$\frac{d^2\rho}{dt^2} + \left(\frac{e}{m\rho_\circ^2}\right)(V_\circ - V_1\cos\omega t)\rho = 0$$
$$\frac{d^2 z}{dt^2} + \left(\frac{2e}{m\rho_\circ^2}\right)(V_\circ - V_1\cos\omega t)z = 0$$

If we now make the following substitutions

$$a_z = -2a_\rho = -\frac{8eV_\circ}{m\rho_\circ^2\omega^2}$$
$$q_z = -2q_\rho = -\frac{4eV_1}{m\rho_\circ^2\omega^2}$$
$$\zeta = \frac{\omega t}{2}$$

then the equations of motion reduce to the following **Mathieu equations**

$$\frac{d^2\rho}{d\zeta^2} + (a_\rho - 2q_\rho\cos 2\zeta)\rho = 0$$
$$\frac{d^2 z}{d\zeta^2} + (a_z - 2q_z\cos 2\zeta)z = 0$$

As expected, the Mathieu equations have stable solutions for only certain values of a and q, which in turn depend on m and ω, and shows that the saddle must be rotated sufficiently fast for a given mass to make the motion stable. The first stability region in aq space, shown in Fig. 11.17, is the most important since the trap is usually operated with $a = 0$ (i.e. $V_\circ = 0$). Then the stable operation requires q to be less than 0.85. This is achieved by selecting a suitable combination of V_1 and ω for a given mass m.

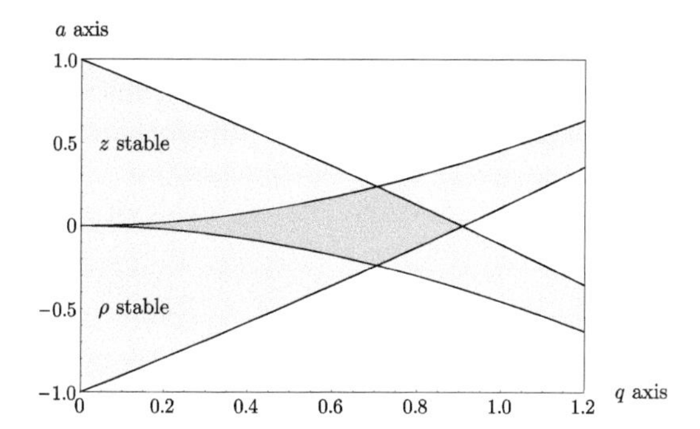

Figure 11.17: Mathieu stability diagram. First region of stability of the Mathieu equation in aq space. The intersecting shaded region is where both z and ρ motions are stable, and where the trap is operated.

One application of Paul traps is as a mass analyzer—to get a mass spectrum of fragments from the dissociation of a sample. In this case the frequency of operation is constantly reduced to make the trap unstable for all masses below a certain value. As soon as a given fragment becomes unstable, it hits the endcaps where it produces a current that can be measured.

3. Mode coupling in a Penning trap

When a Penning trap is used for mass measurement, usually only the axial motion is detected by measuring the image current induced in the endcaps by the oscillating charge. The radial modes are detected by coupling to the axial mode. This mode coupling is an application of the classical coupled oscillator system dealt with in Chapter 2, "Preliminaries."

The detection makes the axial motion that of a damped oscillator. The equilibrium temperature is determined by the Johnson noise of the detector, which is 4.2 K if the trap is kept in a liquid He bath. The radial modes are nominally undamped (which is important for precision measurements) since the trap is azimuthally symmetric, and radiation damping at these frequencies takes hundreds of thousands of years. The azimuthal symmetry is only broken by the split guard ring electrodes. By applying an RF voltage at a frequency ω_p across the two halves of the upper guard ring, a time varying, diagonal quadrupole potential varying as $zx \cos \omega_p t$ is produced. To an ion in, for instance, a large cyclotron orbit, such a field gives kicks in the axial direction and couples the two modes. These kicks are in phase if $\omega_p = \omega_c' - \omega_z$. Under such resonant coupling, the classical action swaps back and forth between the two modes at a frequency determined by the strength of the coupling—the Rabi frequency for an analogous two-state atomic system coupled by a laser field.

We can carry this analogy to the avoided crossing of dressed states in a two-state quantum system that we saw in Fig. 5.5 of Chapter 5, "Resonance." Recall that the energy levels are shifted from their uncoupled values, and the splitting on resonance is equal to $\hbar\omega_R$. When the cyclotron and axial mode are coupled by the RF field, the new normal modes in the trap represent a superposition of the two modes. The role of the populations in the dressed states is now played by the amplitudes in the normal modes. The dressed state analysis tells us that the amplitudes have equal axial and radial components on resonance, but become predominantly one or the other far from resonance. Thus, sufficiently close to resonance, both normal modes are excited and can be detected through the axial component of their motions. As the coupling frequency is tuned through the resonance, the two modes repel each other. From the splitting on resonance, we get the Rabi frequency. Once we know the Rabi frequency for a given coupling field strength, we know the exact time-amplitude product to apply different pulses, e.g. a π-pulse to get a complete swap of the mode amplitudes.

The coupling also results in cooling of the radial modes. An initially hot cyclotron mode is cooled by having the coupling drive on until it cools by coupling to the 4.2 K axial detector bath. The cooling limit is obtained from a thermodynamic argument as follows. The entropy change associated with the emission of one RF coupling photon is

$$\Delta S = \frac{\hbar\omega_z}{T_z} - \frac{\hbar\omega_c'}{T_c}$$

This process continues until, at equilibrium, there is a reversible process with no net change in entropy, so that

$$\frac{\hbar\omega_z}{T_z} = \frac{\hbar\omega_c'}{T_c} \qquad \Longrightarrow \qquad T_c = \frac{\omega_c'}{\omega_z} T_z$$

The same argument holds for the magnetron mode except that, because it is at the top of a potential hill, its energy is positive and its "temperature" negative. An ion in a large magnetron orbit is "cooled" to the center of the trap with a coupling field at $\omega_z + \omega_m$.

4. Sideband cooling

Laser cooling in ion traps, first proposed by Dehmelt and Wineland, is called **sideband cooling**, and is actually a quantized version of Doppler cooling that we saw in the beginning of this chapter. For the trapped particle, the motional degrees of freedom are quantized into discrete energy levels, which are evenly spaced by ω_t if the trap potential is harmonic with oscillation frequency ω_t. The idea in sideband cooling, as shown in Fig. 11.18(a), is to excite the trapped particle at a frequency $\omega_c = \omega_\circ - \omega_t$, i.e. on a motion-induced sideband. Since the subsequent spontaneous decay is most probably at a frequency $\omega_\circ$, one quantum of trap motion will be lost in each combined excitation/decay cycle. This works best when the spontaneous decay width Γ_s is $\ll \omega_t$, in which case it is possible to cool the particle to the lowest quantum state of the trap.

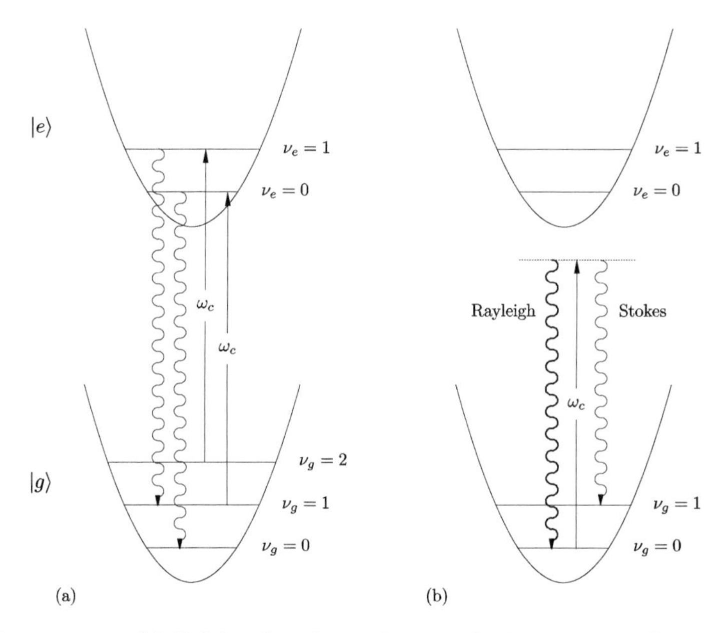

Figure 11.18: (a) Sideband cooling scheme. The harmonic oscillator energy levels are shown for the trapped particle in the ground $|g\rangle$ and excited $|e\rangle$ states. The cooling frequency is $\omega_c = \omega_\circ - \omega_t$ so that an atom in vibrational level ν_g of state $|g\rangle$ is excited to level $\nu_e - 1$ of state $|e\rangle$, from where it decays predominantly to level $\nu_g - 1$ of state $|g\rangle$ resulting in cooling. (b) Heating occurs because the laser at ω_c causes spontaneous Raman scattering whose Stokes component corresponds to transfer from level $\nu_g = 0$ to level $\nu_g = 1$. The stronger Rayleigh scattering neither heats nor cools.

The ultimate cooling limit is set by Raman scattering in which atoms initially in the $\nu_g = 0$ level absorb a laser photon at $\omega_\circ - \omega_t$ and emit a

Stokes line at $\omega_\circ - 2\omega_t$, winding up in the $\nu_g = 1$ level, as shown in Fig. 11.18(b). Both the $\nu_e = 0$ and $\nu_e = 1$ levels of the excited state can serve as a near-resonant intermediate state. The cooling limit is reached when the cooling rate R_c times the number n_1 in the $\nu_g = 1$ level is equal to the Raman heating rate R_H times the number $n_\circ$ in the $\nu_g = 0$ level. Roughly speaking, R_c is the on-resonant excitation rate ω_R^2/Γ_s [from Eq. (6.16) of Chapter 6] times the branching matrix element β to go from $\nu_g = 1$ to $\nu_g = 0$, while R_H is (taking only intermediate vibrational level $\nu_e = 0$ into account) the off-resonant excitation rate $\Gamma_s\omega_R^2/(2\delta)^2$ times the branching ratio β to obtain the probability of winding up in the $\nu_g = 1$ level.

Equilibrium requires

$$n_\circ R_H = n_1 R_c \qquad \Longrightarrow \qquad \frac{n_\circ \Gamma_s \omega_R^2 \beta}{(2\omega_t)^2} = \frac{n_1 \omega_R^2 \beta}{\Gamma_s}$$

where we have used $\delta \equiv \omega_c - \omega_\circ = -\omega_t$. Thus

$$\frac{n_1}{n_\circ} = \frac{\Gamma_s \omega_R^2 \Gamma_s}{4\omega_t^2 \omega_R^2} = \left(\frac{\Gamma_s}{2\omega_t}\right)^2$$

independent of the intensity (which we assumed was $\ll \Gamma_s$ to avoid power broadening).

This gives an estimated temperature of (from $n_1 = n_\circ e^{-\hbar\omega_t/k_B T}$)

$$k_B T = \frac{\hbar\omega_t}{\ln(n_\circ/n_1)} = \frac{\hbar\omega_t}{2\ln(2\omega_t/\Gamma_s)}$$

The minimum value of this temperature is reached when $\omega_t = e\Gamma_s/2$, and has a value

$$k_B T_{\min} = \frac{e\hbar\Gamma_s}{4}$$

This limit is similar to the limit of $\hbar\Gamma_s/2$ for Doppler cooling of neutral atoms that we saw earlier.

A single laser cooled ion is a spectroscopist's dream.

(i) It is usually is inside an ultra-high vacuum chamber, so that perturbations due to collisions with background atoms is negligible.

(ii) It is cold, therefore the second-order Doppler shift is negligible.

(iii) The cold ion is tightly confined (in the Lamb–Dicke regime), so that its emission is recoilless which is important for applications such as next-generation optical clocks.

As an example of laser cooling of ions, we consider the case of Ca^+. Neutral Ca is a two electron atom, but in the singly ionized state it is a one electron atom with energy level structure similar to that of the alkali atoms. The relevant low lying energy levels of Ca^+ are shown in Fig. 11.19. Laser cooling is done on the E1 allowed $S_{1/2} \to P_{1/2}$ transition at 397 nm. The linewidth of the $P_{1/2}$ state is $\Gamma/2\pi = 20$ MHz, which gives a cooling limit of 100 μK. From the $P_{1/2}$ state the ion has a finite probability of decaying to the metastable $D_{3/2}$ state. Selection rules show that an ion in this state can only decay to the ground state by an E2 transition. As a consequence, the lifetime of this state is about 1 s, and an ion decaying to this state is lost from the cooling cycle. Therefore, as for laser cooling of alkali atoms, a **repumping** laser driving the $D_{3/2} \to P_{1/2}$ transition (at 866 nm) is required to bring the ion back to the cooling cycle.

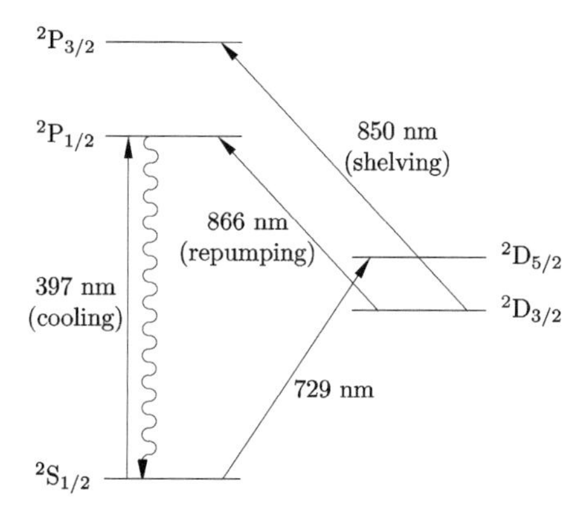

Figure 11.19: Low lying energy levels of Ca^+, showing the laser wavelengths required for cooling, repumping, and shelving.

A separate laser driving the $D_{3/2} \to P_{3/2}$ transition at 850 nm is called the **shelving** laser (which will be important for quantum computation that we will see in detail in the next section). From the $P_{3/2}$ state, the ion has a finite probability of decaying to the $D_{5/2}$ state, where it remains "shelved" until it decays to the ground state. As with the other D state, decay to the ground state is through E2 radiation, so the lifetime is of order 1 s. The ion on the cooling cycle is called **bright** because it fluoresces from the laser radiation, while the ion in the $D_{5/2}$ state is called **dark** because it does not interact with the laser. This dark to bright transition shows an interesting property of quantum systems, namely a **quantum jump**. The signature of this jump is that the transition is sudden, so that the on-off periods have no transition time between them. The statistical average of all the off periods over a long observation time gives the lifetime of the $D_{5/2}$ state.

When the $S_{1/2} \to P_{1/2}$ is used for cooling, the lowest energy is of the order of the linewidth of 20 MHz. Assuming that trap frequency is a few MHz, this means that vibrational levels up to $\nu_g \approx 10$ are occupied (on average).

To cool to the lowest vibrational level of the trap and hence be in the Lamb–Dicke regime, sideband cooling on the $S_{1/2} \rightarrow D_{5/2}$ transition is required because the $D_{5/2}$ state has a linewidth of order of 1 Hz.

5. Quantum computation in a linear Paul trap

One important application of trapped ions is in quantum computation. The bits in a quantum computer, called **qubits**, can take on any superposition value between 0 and 1, while the bits of a classical computer can be only one of the two. This feature, along with other quantum mechanical aspects like entanglement, can be exploited to compute some things much faster than can be done on a classical computer. The trapped laser cooled ion is an almost ideal qubit because it is a single particle in a perturbation free environment, and its energy levels can be addressed very precisely using lasers. The 0 and 1 state are the ground state and a metastable state respectively. Ca^+ ions in a linear Paul trap are now the species of choice for quantum computing because the transitions, as we have already seen, are accessible with low-cost diode lasers.

The linear Paul trap is an elongated version of the 3D trap discussed earlier—it uses the same oscillatory potential to trap ions in the xy plane, but along the z direction the ions are nominally untrapped and there is only a weak static potential to prevent then from escaping. The simplest configuration, shown schematically in Fig. 11.20, consists of four cylindrical electrodes divided into three segments. The AC potential required for providing radial confinement is applied to all the three segments. For axial confinement, the outer two segments are maintained at a slightly higher potential compared to the middle segment. A string of ions, each one acting as a qubit, is then trapped near the central axis with interion repulsion keeping them apart but interacting. Conditional operations on the qubits are induced using collective modes of the string.

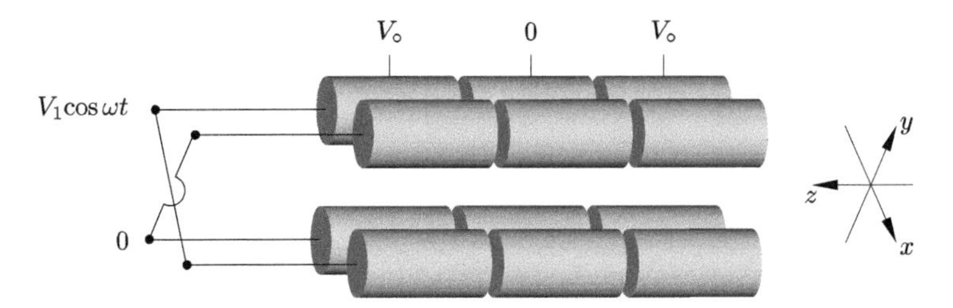

Figure 11.20: Linear Paul trap. Each electrode is a cylinder divided into three segments, which are isolated to have different potentials. Radial (xy) confinement is provided by the AC field (as in the 3D Paul trap) applied to all three segments. In the axial (z) direction, the ion is nominally untrapped and only weakly confined using a small DC potential on the outer two segments.

G. Problems

1. Number of atoms in a MOT

The number of atoms N trapped in a MOT can be estimated using a calibrated photodiode. What is the total optical power P falling on the photodiode if it subtends a solid angle of Ω at the cloud, I is the beam intensity of each of the six trapping beams, and δ is the detuning.

Solution

The scattering rate for a single atom from each trapping beam is

$$R_{\rm sc} = \frac{\Gamma}{2}\left[\frac{I/I_s}{1+(2\delta/\Gamma)^2+I/I_s}\right]$$

Thus the total scattering rate from six beams is $6R_{\rm sc}$.

Noting that the energy of one photon is hc/λ, the total power falling on the photodiode from N atoms is

$$P = \frac{hc}{\lambda}\frac{\Omega}{4\pi}6R_{\rm sc}\,N$$

2. Absorption imaging

Let the atomic density at a given point in a cloud of trapped atoms be N. If you send resonant light of intensity $I_\circ$, its intensity reduces to I_N after absorption. Find N from I_N.

Solution

The intensity after absorption is related to the incident intensity as

$$I_N = I_\circ e^{-\rm OD}$$

where OD is the optical density defined as

$$\mathrm{OD} = N\sigma\ell$$

Here N is the density, ℓ is the thickness of the cloud at that point, and σ is the (resonant) scattering cross-section.

From the definition of the cross-section, it is

$$\sigma = \frac{\text{Absorption rate}}{\text{Incident photon flux}} = \frac{R_{\rm abs}}{I/\hbar\omega}$$

Since the saturation intensity is defined such that the absorption rate is $\Gamma/2$ when $I = I_s$, we get the on-resonance scattering cross-section as

$$\sigma = \frac{\hbar\omega\Gamma}{2I_s} = \frac{3\lambda^2}{2\pi}$$

Therefore the atomic density is given by

$$N = \frac{1}{\sigma\ell}\ln\left(\frac{I_\circ}{I_N}\right) = \frac{2\pi}{3\lambda^2\ell}\ln\left(\frac{I_\circ}{I_N}\right)$$

Appendix A

Standards

EVER since humans started living in community settlements, day-to-day activities have required the adoption of a set of standards for weights and measures. For example, an everyday statement such as "*I went to the market this morning and bought 2* **kgs** *of vegetables; the market is 1* **km** *away and it took me 15* **mins** *to get there*" uses the three standards of **mass**, **length**, and **time**. Without a common set of measures for these quantities, we would not be able to convey the quantitative meaning of this statement. The scientific study of such measures is called *metrology*, and it is an important part of modern industrial societies. In most countries, standards for weights and measures are maintained by national institutions, in coordination with similar bodies in other countries. The definitions and maintenance of standards are improving constantly with progress in science and technology.

In olden times, each local community had its own set of measures defined by arbitrary man-made artifacts. As long as inter-community trade was not common, this proved to be sufficient. Usually kings and chiefs used their power to set the standards to be employed in their territories. For example, the measure of length "inch" was probably defined as the size of the thumb of a tribal chief and the length of his foot gave birth to a standard "foot." As trade and commerce increased it became necessary to introduce more global standards. In addition, with the rise of rationalism and modern science in 17th century Europe, the limitations of such arbitrary definitions soon became apparent. For instance, the results of controlled scientific experiments in different laboratories could not be compared unless all scientists agreed on a common set of standards. Here, we will see the evolution of **rational** standards for time, length, and mass, from those early days to their modern scientific definitions.

Modern standards have to satisfy several important requirements.

(i) They should be **invariant**, i.e. the definition should not change with time.

(ii) They should be **reproducible**, i.e. it should be possible to make accurate and faithful copies of the original.

(iii) They should be on a **human scale**, i.e. of a size useful for everyday purposes.

(iv) They should be **consistent** with physical laws, i.e. the minimum number of independent units should be defined and other units derived from these using known physical laws.

As we will see in this appendix, it is only in recent times that these requirements have been met in a consistent manner. An important way to achieve these goals is to define standards based on fundamental constants of nature. This helps us get away from artifacts and makes the definition universal.

1. Time standards

Time keeping is as old as the earth itself since all it requires is a periodic process. The rotation and revolution of the earth give rise to daily and yearly cycles, respectively, and this has been used by nature as a clock much before humans evolved. Many living organisms, including humans, show diurnal (24 hour) rhythms regulated by the sun, and seasonal patterns that repeat annually. It is therefore natural that the earliest man-made clocks also relied on the sun. One example is the sundial. It consisted of a pointer and a calibrated plate, on which the pointer cast a moving shadow. Of course this worked only when the sun is shining. The need to tell the time even when the sun was not out, such as on an overcast day or in the night, caused man to invent other clocks. In ancient Egypt, water clocks known as *clepsydras* were used. These were stone vessels with sloping sides that allowed water to drip at a constant rate from a small hole near the bottom. Markings on the inside surface indicated the passage of time as the water level reached them. While sundials were used during the day to divide the period from sunrise to sunset into 12 equal hours, water clocks were used from sunset to sunrise. However, this definition made the "hours" of the day and night different, and also varying with the season as the length of the day changed. Other more accurate clocks were used for measuring small intervals of time. Examples included candles marked in increments, oil lamps with marked reservoirs, hourglasses filled with sand, and small stone or metal mazes filled with incense that would burn at a certain rate.

Time measurements became significantly more accurate only with the use of the pendulum clock in the 17th century. Galileo had studied the pendulum as early as 1582, but the first pendulum clock was built by Huygens only in 1656. As we now know, the "natural" period of the pendulum clock depends only on the length of the pendulum and the local value of g, the acceleration due to gravity.* Huygens' clock had an unprecedented error of less than 1 minute per day. Later refinements allowed him to reduce it to less than 10 seconds a day. While very accurate compared to previous clocks, pendulum clocks still showed significant variations because even a few degrees change in the ambient temperature could change the length of the pendulum due to thermal expansion. Therefore clever schemes were developed in the 18th and 19th centuries to keep the time period constant during the course of the year by compensating for seasonal changes in length. Pendulum clocks were also not reproducible from one place to another because of variations in the value of g. Finally, this and all other mechanical clocks

*This is the familiar result from high school that $T = 2\pi\sqrt{\ell/g}$. The fact that the period of the pendulum is independent of the mass of the bob has a lot of important physics buried in it. It comes from the principle of equivalence, which states that the inertial mass (which is what you use when calculating momentum mv or acceleration F/m) is exactly equal to the gravitational mass (which determines weight). This principle was exploited by Einstein to formulate the general theory of relativity.

(e.g. based on oscillations of a balance wheel as in wrist watches) suffered from unpredictable changes in time-keeping accuracy with wear and tear of the mechanical parts.

Mechanical time keeping devices were useful in telling the time whenever we wanted. As a time standard, however, the rotation rate of the earth proved to be more regular than anything man made. Despite the many advances in technology of mechanical clocks, they were still less accurate than the earth "day." Moreover, the success of astronomical calculations in the 19th century led scientists to believe that any irregularities in the earth's rotation rate could be adequately accounted for by theory. Therefore, until 1960, the unit of time "second" was defined as the fraction 1/86400 of the "mean solar day" as determined by astronomical theories. The earth's rotation rate was the primary time standard, while mechanical clocks were used as secondary standards whose accuracy was determined by how well they kept time with respect to earth's rotation.

As clock accuracies improved in the first half of the 20th century, especially with the development of the quartz crystal oscillator, precise measurements showed that irregularities in the rotation of the earth could not be accounted for by theory. In order to define the unit of time more precisely, in 1960 a definition given by the International Astronomical Union based on the tropical year was adopted. However, scientists were still looking for a truly universal standard based on some physical constant. They were able to do this in the late 1960s based on the predictions of quantum mechanics. As you might know, quantum mechanics was developed in the early part of the 20th century to explain the discrete energy levels and spectral lines in atoms. Planck's famous relation between the energy and frequency of a photon, $E = h\nu$, which signaled the birth of quantum mechanics, implies that atoms have a unique internal frequency corresponding to any two energy levels. Since these energy levels are characteristic of an atom anywhere in the universe, a definition based on the atom's internal frequency would be truly universal. The ability to measure such frequencies accurately was developed only about 50 years ago. Soon after, in 1967 the following definition of the unit of time was adopted: *the second is the duration of 9192631770 periods of the radiation corresponding to the transition between the two hyperfine levels of the ground state of the cesium-133 atom.*

Modern atomic clocks are built according to the above definition. The clock consists of a vacuum chamber with a cesium atomic beam, and a radio-frequency oscillator which is tuned to drive the atoms between the two hyperfine levels. There is maximum transfer of energy from the laboratory oscillator to the atoms when the atoms are in **resonance**, i.e. the oscillator frequency matches the internal frequency of the atom. A feedback circuit locks these two frequencies and ensures that the laboratory clock does not drift with respect to the atom frequency. With this feedback, the best atomic clocks (used as primary standards) are so precise that they lose less than 1 second in a million years! The time standard is also universal in the sense that anybody who builds an atomic clock will get exactly the same frequency because all atoms are identical, and their behavior under controlled experimental conditions is the same anywhere in the universe.

Life in the modern high-technology world has become crucially dependent on precise time. Computers, manufacturing plants, electric power grids, satellite communication, all depend on ultra-precise timing. One example that will highlight this requirement is the Global Positioning System (GPS), which uses a grid of satellites to tell the precise location of a receiver anywhere on earth. Transport ships plying the vast oceans of the world now almost completely rely on the GPS system for navigation. The system works by triangulating with respect to the three nearest satellites. The distance to each satellite is determined by timing the arrival of pulses traveling at the speed of light. It takes a few millionths of a second for the signal to reach the receiver, which gives an idea of how precise the timing has to be to get a differential reading between the three signals. Such demands of modern technology are constantly driving our need for ever more precise clocks. The cesium fountain clock at National Institute of Standards and Technology (NIST) in the United States, and other national labs, is one example of how the latest scientific research in laser cooling has been used to improve the accuracy of the clock by a factor of 10. Another promising technique is to use a single laser-cooled ion in a trap, and define the second in terms of the energy levels of the ion. The ion trap represents an almost ideal perturbation-free environment where the ion can be held for months or years. Perhaps one day in the not-so-far future we will see ion-trap clocks in all homes!

2. Length standards

As mentioned before, for a long time length standards were based on arbitrary measures such as the length of the arm or foot.* The first step in the definition of a rational measure for length was the definition of the "meter" by the French in 1793, as a substitute for the yard. It was defined such that the distance between the equator and the north pole, as measured on the great circle through Paris, would be 10,000 kilometers. This gave a convenient length scale of one meter, which was very close to the old yard *but now invariant.* A prototype meter scale made of a platinum-iridium bar was kept in Paris. The alloy was chosen for its stability and exceptionally low thermal expansivity. Copies of this scale were sent to other nations and were periodically recalibrated by comparison with the prototype.

The limitations of the artifact meter scale in terms of invariance became apparent as more precise experiments started to be conducted in the 20th century. Again the results of quantum mechanics provided a solution. Each photon has not only a frequency ν but also a wavelength λ, with the two related by the speed of light c,

$$\nu\lambda = c$$

Therefore if an atom is excited, it decays to the ground state by emitting photons of well defined wavelengths corresponding to the energy difference between the two levels. These photons form a unique line spectrum, or wavelength signature, of the particular atom. The wavelength of photons can be measured precisely in optical interferometers by counting fringes as a function of path length difference in the two arms. Each fringe corresponds to a path length difference of λ. Therefore, in 1960, the artifact meter scale was replaced by a definition based on the wavelength of light—1 meter was defined as 1650763.73 wavelengths of the orange-red line in the radiation spectrum from electrically excited krypton-86 atoms. Anybody who wanted to make a standard meter scale could do it by comparing to the krypton line, thus making it universally reproducible.

The advent of lasers has made the measurement of wavelength in optical interferometers very precise. In addition, the frequency of the laser can be measured accurately with respect to atomic clocks. In order to eliminate the fact that the definition of the meter was tied to the wavelength of a particular line from the krypton atom, the meter was redefined in 1983 using the above frequency-wavelength relation: *1 meter is the length of the path traveled by light in vacuum during a time interval of 1/299792458 of a second.* It is important to note that this definition makes the speed of light

*Even today, roadside flower sellers in Madras will sell flowers strung together by the *mozham*, defined as the length of the seller's arm from the elbow to the tip of the extended middle finger, obviously varying from one person to the next.

exactly equal to 299792458 m/s and demonstrates our faith in the special theory of relativity, which postulates that the speed of light in vacuum is a constant. As our ability to measure the meter becomes more accurate, it is the definition of the meter that will change in order to keep the numerical value of the speed of light a constant. In this sense, we have actually dispensed with an independent standard for length and made it a derived standard based on the standard of time and a fundamental constant of nature c.

Using iodine-stabilized HeNe lasers, the wavelength of light and thus the definition of the meter is now reproducible to about 2.5 parts in 10^{11}. In other words, if we were to build two meter scales based on this definition, their difference would be about one million times smaller than the thickness of a human hair!

3. Mass standards

In olden societies, mass standards were based on artifacts such as the weight of shells or of kernels of grain. The first scientific definition of mass adopted in the 18th century was the "gram," defined as the mass of 1 cubic centimeter of pure water at 4°C. This definition made the density of water exactly 1 g/cc. The definition survived till almost the end of the 19th century. It was replaced in 1889 when the 1st General Conference on Weights and Measures sanctioned an international prototype kilogram to be made of a cylinder of platinum-iridium alloy and kept at the International Bureau of Weights and Measures in France. It was declared that henceforth the prototype would be the unit of mass. Among the base units, mass is the only one that is still based on an artifact and not on some fundamental property of nature. Environmental contamination and material loss from surface cleaning are causing the true mass of the kilogram to vary by about 5 parts in 10^8 per century relative to copies of the prototype in other nations. There are many physical constants that depend on mass, and the drift of the mass standard means that these constants have to be periodically revised to maintain consistency within the SI system.

There are several proposals to replace the mass standard with a universal one based on physical constants. There are two major approaches taken in this effort—one based on electrical measurements, and the other based on counting atoms. In the electrical measurement approach, an electrical force is balanced against the gravitational force on a kilogram mass. The electrical force is determined by the current and voltage used to produce the force. As we will see later, electrical standards are now based on fundamental constants, therefore the kilogram will also be related to fundamental constants.

The atom counting approach is a promising proposal that uses the tremendous advances made in silicon processing technology in recent decades. It is now possible to make large single crystals of silicon with very high purity and a defect rate that is less than one part in 10^{10}, i.e. atoms in the crystal are stacked perfectly and less than one atom in 10 billion is out of place! Such single crystals also cleave along certain symmetry planes with atomic precision, i.e. the crystal facet after a cut is atomically flat. Laser interferometry can be used to measure the distance between the outer facets of the crystal very precisely. This yields the precise volume of the crystal. Similarly, X-ray diffraction can be used to measure the spacing between successive atoms (lattice spacing) very precisely. The volume and the lattice spacing effectively tell us how many atoms there are in the crystal. From the definition of a mole we know that N_A silicon atoms will have a weight of $M_{\mathrm{Si}} \times 10^{-3}$ kilograms, where N_A is Avogadro's constant and M_{Si} is the atomic mass of silicon. Therefore, a knowledge of the physical constants N_A and M_{Si}, combined with length measurements for the size of the crystal, will give the mass of the sample in kilograms. The present limitation in using a silicon mass standard is the precise knowledge of N_A. However, there are several experiments currently being performed that might yield a more precise and useful value for this constant. When this happens, the kilogram would be redefined as being the mass of a specific number of silicon atoms, and we would have eliminated the only remaining artifact standard.

4. Electrical standards

We finally discuss the use of fundamental constants in defining electrical standards. Electrical units can all be related to the base units of mass, length, and time through physical laws. For example, Coulomb's law for the force between two charges q_1 and q_2 separated by a distance r is expressed as

$$F = K\frac{q_1 q_2}{r^2}$$

The proportionality constant K appearing in the above equation can be interpreted in two ways. From a physicist's point of view, it is just a matter of definition and can be set to 1 without any change in the underlying physics. In such a case, the units for measuring charges are so defined that q^2/r^2 has the dimensions of force. Thus charge becomes a derived unit which can be related to the dimensions of mass, length, and time through the above equation. From a practical point of view (which is followed in the SI system of units), it is useful to assign an independent unit for charge (coulomb in the SI system).* The constant K then serves to match the dimensions on both sides of the equation. Thus in the SI system $K = 1/4\pi\epsilon_\circ$ with $\epsilon_\circ$ having units of farad/meter. It should be emphasized that both points of view are valid; however the latter introduces concepts such as permittivity of vacuum which needlessly complicates our understanding of the physics.

Whatever the point of view, the units have to be consistent with Coulomb's law. Therefore, two unit charges placed unit distance apart in vacuum should experience a force of K units. This is how electrical standards have to be finally related to the mechanical standards in a consistent manner. While it may not be practical to measure forces on unit charges this way, other consequences of the electromagnetic laws make a practical comparison of electrical and mechanical units possible. For example, it is possible to measure electrical forces in current carrying conductors by balancing them against mechanical forces. This would be a realization of the ampere.

These traditional methods of defining electrical standards showed a lot of variation over time. However, recently it has been possible to base the definitions of electrical quantities using invariant fundamental constants. Two effects are used for their definition: the Josephson effect, and the quantum Hall effect. The Josephson effect relates the frequency of an AC current generated when a DC voltage is applied across a tunnel junction between two superconductors. The frequency of the current is given by

$$f = \frac{2e}{h}V$$

*In SI units, it is actually the ampere that is the base unit; it is defined as the constant current that has to flow along two infinite parallel conductors placed 1 meter apart so as to produce a force of 2×10^{-7} N/m. In this system, 1 coulomb is 1 ampere-second.

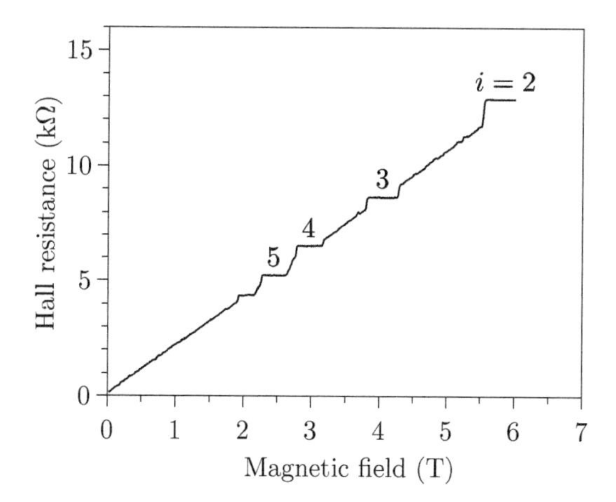

Figure A.1: Quantum Hall effect. The plot shows the steps in the Hall resistance of a semiconductor sample as a function of the magnetic field. The steps occur at values given by the fundamental constant h/e^2 divided by an integer i. The figure shows the steps for $i = 2, 3, 4,$ and 5. The effect is called the quantum Hall effect and was discovered by von Klitzing, for which he won the Physics Nobel Prize in 1985. Since 1990, the effect has been used for a new international standard for resistance called 1 klitzing, defined as the Hall resistance at the fourth step $= h/4e^2$. The data are from my colleague Prof. Aveek Bid, based on a sample he made at Weizmann Institute in Israel, highlighting the truly universal nature of this standard.

where e is the charge on the electron. In SI units, a DC voltage of 1 µV produces a current with a frequency of 483.6 MHz. The DC voltage can therefore be related to the time or frequency standard through the fundamental constant $h/2e$. The quantum Hall effect is a phenomenon discovered by von Klitzing in 1980. He showed that at low temperatures and high magnetic fields, the Hall resistance in certain semiconductor samples shows quantized steps (see Fig. A.1). The fundamental unit of resistance is h/e^2 (about 25.7 kΩ in SI units), and the steps occur at values of this constant divided by an integer i. Since 1990, the SI standard of ohm is defined using the $i = 4$ step in the Hall resistance of a semiconductor sample. Given the robust nature of the steps, it is easy to reproduce the resistance and its value is determined only by physical constants. The standard volt and ohm in the SI system have thus been successfully tied to fundamental constants of nature. The definition of current (ampere) follows from Ohm's law $I = V/R$.

5. Summary

We have thus seen a trend where fundamental constants play an increasingly important role in eliminating artifact standards and in deriving some units from others. Our faith in physical laws makes us believe that these constants are truly constant, and do not vary with time. Ultimately, we would like to find enough fundamental constants that we are really left with only one defined standard and all others are derived from it, just like we have been able to do for length and time by specifying the value of c, so that there are only two standards for mass and time. All indications are that when we understand the force of gravity from a quantum mechanical perspective, mass (or inertia) and spacetime will be linked in a fundamental way. When this happens, the unit of mass will most likely be fundamentally related to the unit of time. We will be left with just one arbitrary definition for "second" which sets the scale for expressing all the laws of physics. Someday, if we were to meet an intelligent alien civilization and would like to compare their scientific knowledge with ours, we would only need to translate their time standard to ours. Will this ever happen? "Time" alone will tell!

6. Additional items

(i) A brief history of time-keeping

As mentioned in the main text, the oldest clocks in the world are sundials. Many designs for sundials exist. They have evolved from simple designs of flat horizontal or vertical plates to more elaborate forms that compensate for the motion of the sun in the sky during the course of the year. For example, one design has a bowl-shaped depression cut into a block of stone, with a central vertical gnomon (pointer) and scribed with sets of hour lines for different seasons.

The oldest clocks that did not rely on observation of celestial bodies were water clocks. They were designed to either drip water from a small hole or fill up at a steady rate. Elaborate mechanical accessories were added to regulate the rate of flow of water and display the time. But the inherent difficulty in controlling the flow of water led to other approaches for time keeping. The first mechanical clocks appeared in 14th century Italy, but they were not significantly more accurate. Accuracy improved only when the Dutch scientist Huygens made the first pendulum clock in 1656. Around 1675, Huygens also developed the balance wheel and spring assembly, which is still found in mechanical wrist watches today. In the early 18th century, temperature compensation in pendulum clocks made them accurate to better than 1 second per day.

John Harrison, a carpenter and self-taught clockmaker, refined temperature compensation techniques and added new methods of reducing friction. He constructed many "marine chronometers"— highly accurate clocks that were used on ships to tell the time from the start of the voyage. A comparison of *local noon* (that is, the time at which the sun was at its highest point) with the time on the clock (which would give the time of the noon at the starting point) could be used to determine with precision the longitude of the ship's current position. The British government had instituted the Longitude Prize so that ships could navigate on transatlantic voyages without getting lost; specifically, the prize was for a clock capable of determining longitude to within half a degree at the end of a voyage from England to Jamaica. Harrison's prize-winning design (in 1761) kept time on board a rolling ship nearly as well as a pendulum clock on land, and was only 5.1 seconds slow after 81 days of rough sailing, about 10 times better than required.

The next improvement was the development of the nearly free pendulum at the end of the 19th century with an accuracy of one hundredth of a second per day. A very accurate free-pendulum clock called the Shortt clock was demonstrated in 1921. It consisted of two pendulums, one a slave and the other a master. The slave pendulum gave the master pendulum gentle pushes needed to maintain its motion, and also drove the hands of the clock. This allowed the master pendulum to remain free from mechanical tasks that would disturb its regularity.

Time keeping was revolutionized by the development of quartz crystal clocks in the 1920s and 30s. Quartz is a piezoelectric material, meaning that it generates an electric field when mechanical stress is applied, and changes shape when an electric field is applied. By cutting the crystal suitably and applying an electric field, the crystal can be made to vibrate like a tuning fork at a constant frequency. The vibration produces a periodic electrical signal that can be used to operate an electronic display. Quartz crystal clocks are far superior than mechanical clocks because they have no moving parts to disturb their regular frequency. They dominate the commercial market due to their phenomenal accuracy and low cost. For less than a dollar, you can get a watch that is accurate to about 1 second per year! And you do not have to worry about winding the clock every day or replacing the battery more than once a year or so.

Despite this success, quartz crystal clocks ultimately rely on a mechanical vibration whose frequency depends critically on the crystal's size and shape. Thus, no two crystals can be precisely alike and have exactly the same frequency. That is why they cannot be used as primary time standards. And atomic clocks, which *are* used as primary standards, are exactly reproducible (when operated under the prescribed conditions). In Fig. A.2, we see that the accuracy of clocks in the last millennium has increased exponentially *on a logarithmic scale.* Extrapolating into the near future, we expect *optical* atomic clocks to be 1000 times more accurate than the current radio-frequency standard.

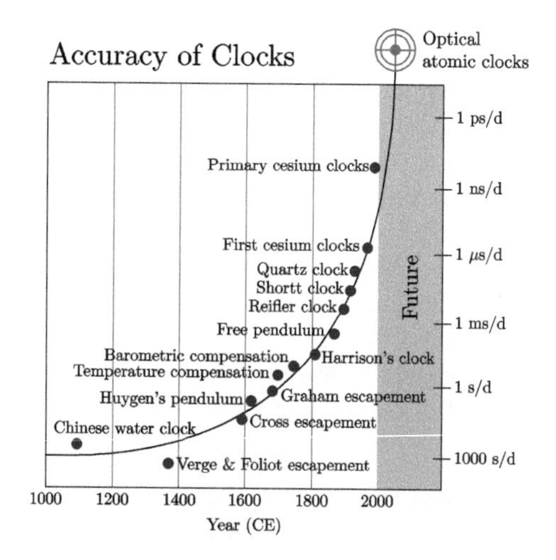

Figure A.2: Exponential improvement in clock accuracy over the last millennium. Note the logarithmic scale of the y axis.

(ii) Frequency measurements

Measuring the frequency in an atomic clock is similar to pushing a child on a swing. The child has a natural oscillation frequency on the swing, and can be thought of as the atom. The person pushing the child periodically can be thought of as the laboratory oscillator. If the two frequencies are identical, the person will give a push to the child each time the child reaches her and will keep the child swinging. On the other hand, if the two frequencies are not identical, the pushes will come at random times, sometimes slowing the child and sometimes speeding it up. The average effect will be lower and will tend to zero as the mismatch between the two frequencies increases.

In the separated-oscillatory-fields (SOF) technique, invented by Norman Ramsey in 1949, there is a period when the system evolves in the "dark," unperturbed by the driving field. In the example above, it would be like giving a push to the child, and then letting the child swing unperturbed for some time before giving another push. During the time between the two pushes, the frequency difference between the two oscillators builds up as a phase difference, so that after sufficient dark time the person pushing is exactly out of phase with the child and the second push brings the child to a complete halt. Even a small frequency difference can be built up to a large phase difference by increasing the dark period. Thus the frequency (or more precisely, the frequency mismatch) can be measured more and more precisely by waiting for longer and longer times between the two pushes. This is the advantage in the latest cesium clocks—by using laser cooled atoms in a fountain arrangement the dark period is about 1 second, whereas in older atomic clocks with thermal beams the dark period was a few milliseconds.

(iii) Fun with dimensions

It is somewhat instructive to play with dimensions of fundamental constants to see how they result in phenomena that can be used for defining standards. For electrical phenomena at the quantum level, there are two constants—the charge on an electron e and Planck's constant h. In the SI system, e has units of coulombs (C), and h has units of joules/hertz (J/Hz). It can be verified that h/e has the dimensions of volts/hertz (V/Hz), which is the quantum mechanical voltage-to-frequency conversion factor. The actual factor in the Josephson effect is $h/2e$, but the factor of 2 can be understood when we remember that the basic charge carriers in a superconductor are paired electrons, which thus carry charge of $2e$. Similarly h/e^2 has dimensions of ohms (Ω), and is the basic unit of quantum resistance in a semiconductor sample.

Continuing with this theme, we expect a deeper understanding of inertia when we develop a satisfactory theory of relativistic quantum gravity. The fundamental constants that are expected to be important in such a theory are c (for relativistic), h (for quantum), and Newton's constant G (for gravity). These constants have values 3×10^8 m/s, 6.6×10^{-34} J/Hz, and 6.7×10^{-11} Nm^2/kg^2, respectively. The three constants can be combined in various ways to get other constants that have dimensions of mass, length, and time. Together the new constants set the scale at which we expect quantum gravity effects to become significant. This is called the Planck scale. Thus the Planck length $\sqrt{hG/c^3}$ is 1.6×10^{-35} m, the Planck time $\sqrt{hG/c^5}$ is 5.4×10^{-44} s, and the Planck mass $\sqrt{hc/G}$ is 2.2×10^{-8} kg. It looks as though only the Planck mass scale is currently accessible to human technology, and perhaps we will not understand quantum gravity until we can access the Planck length and time scales.

Appendix B

What Is a Photon?

All the fifty years of conscious brooding have brought me no closer to answer the question, "What are light quanta?" Of course today every rascal thinks he knows the answer, but he is deluding himself.

— Albert Einstein

OUR current understanding of the photon is based on quantum electrodynamics (QED), which is arguably the pinnacle of any quantum theory to date. Feynman, who along with Wheeler is the coauthor of the seminal paper on the absorber theory of radiation, was instrumental in developing QED. He later shared the Nobel prize for this work, and called QED "the strange theory of light and matter." Strange indeed, but also immensely successful. In fact, it can be called our most successful theory since its prediction of the anomalous magnetic moment, $(g-2)$, of the electron has been verified to an unprecedented accuracy of 12 digits! Its success means that it has managed to capture some inherent description of the workings of nature, so that any future theory has to at least reproduce its quantitative results. Indeed, it is our best example of a quantum field theory, with its naturally occurring creation and annihilation operators. It is used as a model for formulating field theories for other interactions—a *canonical* field theory if ever there was one.

The idea of this essay, which represents a personal journey into the nature of the photon, is to show that there are alternate, *more understandable* ways of looking at light. Feynman gave up on the absorber theory because he could not quantize it in a satisfactory manner so as to get the experimental results of the $(g-2)$ of the electron and the Lamb shift of the hydrogen atom. But, this failure may reflect our current way of doing physics, rather than any fundamental problem. In the immortal words of Wheeler:

> *Behind it all is surely an idea so simple, so beautiful, that when we grasp it—in a decade, a century, or a millennium—we will all say to each other, how could it be have been otherwise? How could we have been so stupid?*

I too think that we will find such a beautiful idea to explain the puzzles of light, simple enough that it can be explained to high-school students.

Light is a propagating disturbance of the electromagnetic field. It appears as the solution of a wave equation resulting from the four Maxwell's equations in source-free region. Not surprisingly, it was treated as a classical wave, and seemed to have all the properties that one associates with a wave—interference, diffraction, reflection and refraction, coherence, etc. Then came the mystery of blackbody radiation spectrum, which was inexplicable from this classical wave picture. In a stroke of genius, Max Planck (in 1900) made the *ad hoc* proposal that the energy of the emitted radiaton is *quantized* in units of the frequency ($E = h\nu$), and with this assumption, he could explain all the features of the spectrum. This ushered in the "quantum era," and caused, in the terminology of Thomas Kuhn, a paradigm shift in our understanding of nature. But the quantization of light was only an implicit idea in Planck's theory. The explicit nature of the light quantum, or photon as it is called now, came with its use by the *young* Einstein (in 1905) in explaining the *photoelectric effect*. He went on to win the Nobel Prize for this work, because this explanation firmed up the photon concept in the thinking of scientists, and the (additional) particle nature of light came to be accepted. Things came a full circle when de Broglie (in 1924) introduced the idea of wave nature for particles of matter, showing that *wave-particle duality* is a fundamental property of everything in nature, matter and its interactions.

A century later, most of us know how to work with photons. The advent of *lasers*, a coherent source of photons, has put an indispensable tool in the hands of scientists and engineers. Lasers are used everywhere today—in your computer hard drive, in bar-code scanners in shops, in laser pointers, in DVD players, in all kinds of surgery including delicate surgery of the eye, in metal cutting, in the modern research laboratory, to name a few. We, in our atomic physics laboratory, also use lasers all the time. We use them for laser cooling, to cool atoms down to a temperature of a millionth of a degree above absolute zero. We use lasers as optical tweezers, to trap micron-sized beads and cells. We use lasers in high-resolution spectroscopy, to understand the structure of atoms and validate fundamental theories.

In short, we know how to use photons, and how to use them well. But do we understand them? Perhaps not. We certainly have a useful mental picture of a straight-line beam of particles traveling at the speed of light c. In fact, we believe that we can actually see a laser beam—think of the familiar red line coming out of a laser pointer. But a moment's introspection will make us realize that what we are "seeing" is actually those photons that scatter into our eye from the ever-present dust particles in the room. Indeed, it is quite illuminating (pun intended) to see a light beam entering a vacuum chamber through a window—it seems to disappear after the window because there are no particles inside the vacuum chamber to scatter the light. A simpler experiment can be done if you have access to a plane polarized beam. If the polarization axis is oriented in the vertical direction, then you will not see the beam if you view it from the top. This is because the scattering probability is exactly *zero* along the polarization axis. Therefore, when we say we "see" something, what we are talking about is that some photons have reached our retina.

Consider the phenomenon of spontaneous emission. One learns that an atom in an excited state "wants" to go to the lowest-energy ground state. What do you mean "wants"? Atoms do not have feelings. The excited state is as good a solution of the Hamiltonian of the atom as the ground state. Every state is a stable *stationary* solution (called an eigenstate), just that they all have different energies. So why is a lower energy better, and the lowest energy the best? What, in fact, *causes* spontaneous emission, i.e. *induces* the atom to go from a higher to a lower state? Our latest and most-successful theory to date—quantum electrodynamics (QED)—says that spontaneous emission is actually stimulated emission, but one where the stimulation is from the vacuum modes. This may be a clever way of doing calculations, but it is unsatisfactory because the total energy in the vacuum modes (called the zero-point energy) is *infinity.* This is one of several infinities that plague QED; we know how to work around these infinities, but it still leaves a bad taste in the mouth.

Equally puzzling is the phenomenon of *photon recoil,* also known as *radiation reaction.* This is the momentum kick that an atom* receives when it emits a photon, similar to the recoil that you feel when you fire a gun. The bullet is a real particle that carries momentum, and the recoil kick is just a consequence of momentum conservation. But the photon recoil is due to the momentum transferred by a massless particle of *interaction.* This recoil effect is real, in fact Einstein used it in his 1917 paper† to show that maintaining the Maxwell–Boltzmann velocity distribution in a gas in thermal equilibrium is crucially dependent on this recoil. And the same momentum transfer is used for the well-known phenomenon of laser cooling. But, unless the photon is given *independent* reality, the mechanism by which the momentum of the atom changes cannot be understood.

Which brings us to the question—is the photon independently real? Let us not forget that light is an interaction between electrical charges. The big-bang model of cosmology says that there was a time in the early universe when only photons were present. It seems illogical to say that the early universe was full of interactions, but had no matter between which the interactions could occur. It is like saying there is a room full of conversations, but no people to converse between. Conversation is an interaction between people. No people, no conversation.

---

*I use the word *atom* to mean any piece of matter.

†A. Einstein, "Quantentheorie der Strahlung (On the Quantum Theory of Radiation)," *Phys. Z.* **18**, 121–128 (1917). English translation of the paper available in *Sources of Quantum Mechanics*, B. L. Van Der Waerden, ed. (Dover, New York, 1968).

Enter Wheeler and Feynman, and their paper titled "Interaction with the absorber as the mechanism of radiation."* They show that the photon is *not* independently real, and give a satisfactory answer to all of the above puzzles. In fact, the puzzles—especially that of radiation reaction—were known for a long time, and many scientists (like Fokker and Schwarzchild) had proposed solutions. The idea that Wheeler and Feynman developed was based on an earlier proposal by Tetrode,† a fact that was pointed out to them by Einstein. As they write in a footnote in the paper:

When we gave a preliminary account of the considerations which appear in this paper, we had not seen Tetrode's paper. We are indebted to Professor Einstein for bringing to our attention the ideas of Tetrode and also of Ritz, who is cited in this article. An idea similar to that of Tetrode was subsequently proposed by G. N. Lewis:‡ "I am going to make the ... assumption that an atom never emits light except to another atom, and to claim that it is as absurd to think of light emitted by one atom regardless of the existence of a receiving atom as it would be to think of an atom absorbing light without the existence of light to be absorbed.§ I propose to eliminate the idea of mere emission of light and substitute the idea of *transmission*, or a process of exchange of energy between two definite atoms or molecules." Lewis went nearly as far as it is possible to go without explicitly recognizing the importance of other absorbing matter in the system, a point touched upon by Tetrode, and shown below to be essential for the existence of the normal radiative mechanism.

The idea of Tetrode also is to abandon the concept of electromagnetic radiation as an elementary process and to interpret it as a consequence of an *interaction* between a source and an absorber. His exact words are worth repeating:

The sun would not radiate if it were alone in space and no other bodies could absorb its radiation .... If for example I observed through my telescope yesterday evening that star which let us say is 100 light years away, then not only did I know that the light which it allowed to reach my eye was emitted 100 years ago, but also the star or individual atoms of it knew already 100 years ago that I, who then did not even exist, would view it yesterday evening at such and such a time. One might accordingly adopt the opinion that the amount of material in the universe determines the rate of emission. Still this is not necessarily so, for two competing absorption centers will not collaborate but will presumably interfere with each other. If only the amount of matter is great enough and is distributed to some extent in all directions, further additions to it may well be without influence.

Radiation reaction was well known from the fact that a charged particle on being accelerated loses energy by emitting radiation. This loss can be interpreted as being caused by a force acting on the particle given in

*J. A. Wheeler and R. P. Feynman, "Interaction with the absorber as the mechanism of radiation," *Rev. Mod. Phys.* **17**, 157–181 (1945).

†H. Tetrode, "Über den Wirkungszusammenhang der Welt. Eine Erweiterung der klassischen Dynamik (About the causal relationship in the world. An extension of classical dynamics)," *Z. Phys.* **10**, 317–328 (1922).

‡G. N. Lewis, "The nature of light," *P. Natl. Acad. Sci. USA* **12**, 22–29 (1926).

§I would add that it is equally "absurd" to think of a universe with only light and no atoms to emit or absorb it, *apropos* my previous comment.

magnitude and direction by the expression

$$\frac{2(\text{charge})^2(\text{time rate of change of acceleration})}{3(\text{velocity of light})^3}$$

when the particle is moving slowly. Wheeler and Feynman take up the proposal of Tetrode to get two results: the above expression for radiation reaction, and that the fields we are familiar with from experience are all time retarded. For this, they give his idea the following definite formulation:

1. An accelerated point charge in otherwise charge-free space does not radiate electromagnetic energy.
2. The fields which act on a given particle arise only from other particles.
3. These fields are represented by one-half the retarded plus one-half the advanced Lienard-Wiechert solutions of Maxwell's equations. This law of force is symmetric with respect to past and future.* In connection with this assumption we may recall an inconclusive but illuminating discussion carried on by Ritz and Einstein in 1909, in which Ritz treats the limitation to retarded potentials as one of the foundations of the second law of thermodynamics, while Einstein believes that the irreversibility of radiation depends exclusively on considerations of probability. Tetrode, himself, like Ritz, was willing to assume elementary interactions which were not symmetric in time. However, complete reversibility is assumed here because it is an essential element in a unified theory of action at a distance. In proceeding on the basis of this symmetrical law of interaction, we shall be testing not only Tetrode's idea of absorber reaction, but also Einstein's view that the one-sidedness of the force of radiative reaction is a purely statistical phenomenon. This point leads to our final assumption:
4. Sufficiently many particles are present to absorb completely the radiation given off by the source.

As mentioned in point 3, this is a theory of action at a distance, but not the kind of *instantaneous* action at a distance envisaged by Newton for his theory of gravitation. It is action propagated at a finite velocity, in this case the velocity of light.

In this picture, the absorber is the *cause* of radiation. When the absorber receives the photon, it *moves*, or more correctly accelerates. Therefore,

*We now have some evidence (based on the decay of the neutral kaon particle) that the fundamental laws of physics violate such time-reversal symmetry. One consequence of this would be the existence of a permanent electric dipole moment (EDM) in an atom or molecule, though none has been found so far. Therefore, EDM searches are among the most important experiments in physics today.

processes such as spontaneous emission and radiation reaction are caused by the advanced field of this movement appearing at the source. The half-advanced field is essential so that this cause appears at the exact instant of radiation—the recoil felt by the atom is simultaneous with the emission of the photon. If the retarded and advanced fields due to acceleration of the source are F_{ret} and F_{adv} respectively, then the total field emanating from the source is

$$\frac{F_{\text{ret}}}{2} + \frac{F_{\text{adv}}}{2}$$

Wheeler and Feynman show that the total field near the source due to all the absorbers is

$$\frac{F_{\text{ret}}}{2} - \frac{F_{\text{adv}}}{2}$$

This field was called the "radiation field" by Dirac, and its form was *assumed* by him in order to get the correct expression for the radiation reaction. Now, we have an explanation for its origin. Moreover, the complete field diverging from the source that would be felt by a test particle (which is just the sum of the above two terms), is *the full retarded field*, as required by experience.

We see that the above picture gives a self-consistent explanation of radiation. To quote from the paper:

> Our picture of the mechanism of radiation is seen to be self-consistent. Any particle on being accelerated generates a field which is half-advanced and half-retarded. From the source a disturbance travels outward into the surrounding absorbing medium and sets into motion all the constituent particles. They generate a field which is equal to half the retarded minus half the advanced field of the source. In this field we have the explanation of the radiation field assumed by Dirac. The radiation field combines with the field of the source itself to produce the usual retarded effects which we expect from observation, and such retarded effects only. The radiation field also acts on the source itself to produce the force of radiative reaction. What we have said of one particle holds for every particle in a completely absorbing medium. All advanced fields are concealed by interference. Their effects show up directly only in the force of radiative reaction. Otherwise we appear to have a system of particles acting on each other via purely retarded forces.

Wheeler and Feynman next show that the irreversibility of radiation is not due to electrodynamics itself but due to the statistical nature of absorption, *à la* Einstein. To understand this, it is enlightening to compare radiation with heat conduction. Both processes convert ordered into disordered motion although every elementary interaction involved is microscopically reversible. In heat conduction, an initially hot body cools off with time because the *probability* for cooling is overwhelmingly greater than the chance for it to grow hotter. Similarly, if we start with a charged particle whose energy is large in comparison to the surrounding absorber particles, then there is an overwhelming probability that the particle will lose energy to the absorber (at a rate in close accord with the law of radiative damping).

Take the classic example of the irreversible breaking of an egg. If we could choose the initial conditions so that the millions of particles involved had exactly the reverse of the motion acquired during breakage, we would see an egg forming from its constituent pieces. It is just that the probability of this happening is negligible.*

The expression for the force of radiation reaction shows that it is proportional to the first derivative of acceleration, or the third derivative of position. This means that a charge starts to move before the arrival of the disturbance; and e^2/mc^3 seconds ahead of the time when it attains a velocity comparable with its final speed. This has been termed *pre-acceleration.* Since the disturbance in this case is the advanced field of the absorber, we have to give up the notion that the movement of a particle at a given instant is completely determined by the motions of all other particles at earlier moments. Pre-acceleration can be hence viewed as an influence of the future on the past, i.e. the distinction between past and future is blurred on time scales of the order of e^2/mc^3. In other words, *those phenomena which take place in times shorter than this figure require us to recognize the complete interdependence of past and future in nature, an interdependence due to an elementary law of interaction between particles which is perfectly symmetrical between advanced and retarded fields.*

The absorber picture of radiation seems "repugnant to our notions of causality,"† in the sense that we can (at least in principle) change the process of emission by intervening suitably—by blocking the path from emitter to absorber for example. Without bringing notions of human free will and philosophical complications involving life, let us imagine a simple intervention scheme where a shutter is designed to (automatically) block the path of the photon halfway between the source and absorber. Does the photon go back to the source and re-excite it because the path to the absorber is now blocked? No. The correct solution, which comes out of the absorber theory, is that the *advanced* field of the shutter tells the atom not to radiate in the first place. That is why the advanced field is so important to this theory, it gives a consistent solution irrespective of the distance between the absorber and emitter—your eye and the light from a distant star millions of light years away, for example.

The theory also gives a satisfactory explanation for the well-known phenomenon of *photonic bandgap.* This is a system where a periodic array of dielectric materials is used to create a bandgap for light—a situation where the system does not allow the propagation of light waves with certain energies or wavelengths. This is akin to the bandgap for electrons in a crystal, where the periodic array of nuclei creates a (Bragg-scattering) condition so

*The phenomenon of "spin echo," discussed in Chapter 8, "Coherence," is an example of such re-formation, which is induced by a pi-pulse that acts to reverse the randomization of a collection of spins.

†G. N. Lewis, "The nature of light," *P. Natl. Acad. Sci. USA* **12**, 22–29 (1926).

that certain electron waves cannot propagate. One can therefore suppress spontaneous emission from an atom in the excited state by placing it within a photonic-bandgap material, with a band gap in the correct range. This is easily understood in the absorber picture as arising due to the fact that the field of the absorber is not allowed to reach the atom. In the conventional picture, this is explained by saying the bandgap material creates a "better vacuum," one where the vacuum modes are suppressed.

The above analysis shows that the process of emission is *nonlocal*—the states of the emitter and absorber are coupled no matter how far apart they are. Nonlocality is an inherent part of quantum mechanics, and John Bell showed that it can be experimentally tested using what are now called Bell's inequalities. Most of these tests are done with entangled photons. And experimenters try to enforce locality by changing the (polarization) state of the detector while the photons are in flight. But the absorber picture tells us that the emitting atom "knows" the final state of the detector in advance. So there is no possibility of a "delayed choice"—a phrase coined by Wheeler to indicate changes made after a particle has chosen one of two paths in an interference experiment. The correct way to test quantum nonlocality is to use entangled pieces of matter (two atoms dissociating from a paired singlet state, for example), and not photons.

The one puzzle that remains in the above picture is the phenomenon of *pair creation*, a process where a photon of suitable energy gets converted into matter consisting of a particle and antiparticle pair. Here, suitable energy means that it is at least equal to $2mc^2$, the total rest mass energy of the pair. The theory of relativity tells us that matter and energy are equivalent: the famous relation $E = mc^2$, which gives the above requirement for the minimum photon energy for pair production. But to say that a photon (remember that it is a particle of interaction) can be converted into matter seems absurd. Or to say that matter can be created out of pure energy. A better solution is that the particle-antiparticle pair was always there, but in a bound state that did not interact with other matter (or was "invisible"). This state then absorbed a photon and disassociated into its constituent pair. No matter is ever created or destroyed, just that a photon takes the matter from being invisible to visible. Of course, this picture is valid only if such a non-interacting state is shown to exist. Note that such a state is different from the ground state of the well-known *positronium* atom formed using an electron and a positron (anti-electron), which is like a hydrogen atom but with the proton replaced by the positron. The positronium atom can absorb photons of much smaller energy because it has excited states that are analogous to the excited states of hydrogen.

I met Wheeler at a conference in his honor at the University of Maryland in 1994. I asked him why the seemingly good theory of absorber interaction was not widely accepted. He said that they (he and Feynman) felt the theory was incorrect because they had looked empirically for complete absorption in all directions in the sky, and not found it. In their paper, Wheeler and Feynman do discuss the consequences of incomplete absorption. The problems depend to a large extent on the model of the universe, and the description of electromagnetism in curved spacetime. Anyway, I think (in agreement with Lewis) that emission without absorption is not possible, so there is no question of partial absorption. Certainly, the present model of the photon and radiation has many puzzling features that make it unsatisfactory. To paraphrase Einstein, perhaps *we are deluding ourselves into thinking that we know the photon.*

Appendix C

Einstein as Armchair Detective: The Case of Stimulated Radiation

EINSTEIN was in many ways like a detective on a mystery trail, though in his case he was on the trail of nature's mysteries and not some murder mystery! And like all good detectives he had a style. It consisted of taking facts that he knew were correct, and forcing nature into a situation that would contradict this established truth. In this process she would be forced to reveal some new truths. Einstein's 1917 paper on the quantum theory of radiation is a classic example of this style, and enabled him to predict the existence of stimulated radiation starting from an analysis of thermodynamic equilibrium between matter and radiation.

Einstein is rightly regarded as one of the greatest scientific geniuses of all time. Perhaps the most amazing and awe-inspiring feature of his work was that he was an "armchair" scientist, not a scientist who spent long hours in a darkened laboratory conducting delicate experiments, but one who performed *gedanken* (thought) experiments while sitting in his favorite chair that nevertheless advanced our understanding of nature by leaps and bounds. Two of his greatest contributions are the special theory of relativity and the general theory of relativity, both abstract creations of his remarkable intellect. They stand out as scientific revolutions that completely changed our perceptions of nature—of space and time in the case of the special theory, and of gravity in the case of the general theory. It might be argued that the special theory of relativity was necessitated by experimental facts such as the constancy of the speed of light, but the general theory was almost completely a product of Einstein's imagination. For a person to have achieved one revolution in his lifetime is great enough, but two revolutions seems quite supernatural.

But is it really so magical? While it is certain that Einstein was a one-of-a-kind genius, is it at least possible to understand the way in which his mind tackled these problems? I think the answer is yes, because deep inside Einstein was like a detective hot on a mystery trail, of course not one solving murder mysteries but one trying to unravel the mysteries of nature. Any keen follower of murder mysteries knows that there are two types of detectives: those who get down on their hands and knees looking for some microscopic piece of clinching evidence at the scene of the crime, and the second type of "armchair detectives" who seem to arrive at the solution just by thinking logically about the possibilities. Einstein was most certainly of the second kind, and true to this breed, he had his own *modus operandi.* In simple terms, his technique was to imagine nature in a situation where she contradicted established truths, and revealed new truths in the process. As a case in point, we will look at Einstein's 1917 paper* titled "On the Quantum Theory of Radiation" where he predicted the existence of stimulated emission. While Einstein will always be remembered for his revolutionary relativity theories, his contributions to the early quantum theory are certainly of the highest caliber, and the 1917 paper is a classic.

It is useful to first set the paper in its historical perspective. By the time Einstein wrote this paper, he had already finished most of his work on the relativity theories. He had earlier done his doctoral thesis on Brownian motion and was a pioneer of what is now called statistical mechanics. He was thus a master at using thermodynamic arguments. He was one of the earliest scientists to accept Planck's radiation law and its quantum hypothesis. He had already used it in 1905 for his explanation of the photoelectric effect. He was also aware of Bohr's theory of atomic spectra and Bohr's

*English translation of the paper available in *Sources of Quantum Mechanics*, B. L. Van Der Waerden, ed. (Dover, New York, 1968).

model of the atom, which gave some explanation for why atoms emitted radiation in discrete quanta. What he did **not know** in 1917 was any of the formalism of quantum mechanics, no Schrödinger equation and not the de Broglie hypothesis for wave nature of particles that we learn in high school these days. Despite this, Einstein was successful in predicting many new things in this paper.

Let us now see what Einstein's strategy in this paper is. He is attempting to understand the interaction between atoms and radiation from a quantum mechanical perspective. For this, he imagines a situation where a gas of atoms is in thermal equilibrium with radiation at a temperature T. The temperature T determines both the Maxwell-Boltzmann velocity distribution of atoms and the radiation density ρ at different frequencies through Planck's law. He assumes that there are two quantum states of the atom Z_n and Z_m, whose energies are ε_n and ε_m respectively, and which satisfy the inequality $\varepsilon_m > \varepsilon_n$. The relative occupancy W of these states at a temperature T depends on the Boltzmann factor as follows

$$\begin{aligned} W_n &= p_n \exp(-\varepsilon_n/k_B T) \\ W_m &= p_m \exp(-\varepsilon_m/k_B T) \end{aligned} \tag{C.1}$$

where p_n is a number, independent of T and characteristic of the atom and its nth quantum state, called the degeneracy or "weight" of the particular state. Similarly, p_m is the weight of the mth state.

Einstein then makes the following basic hypotheses about the laws governing the absorption and emission of radiation.

1. Atoms in the upper state m make a transition to the lower state n by spontaneous emission. The probability dW that such a transition occurs in the time dt is given by

   $$dW = A_m^n dt$$

   A_n^m in modern terminology is called the Einstein A coefficient. Since this process is intrinsic to the system and is not driven by the radiation field, it has no dependence on the radiation density.

2. Atoms in the lower state make a transition to the upper state by absorbing radiation. The probability that such a transition occurs in the time dt is given by

   $$dW = B_n^m \rho dt \tag{C.2}$$

   B_m^n is now called the Einstein B coefficient. The absorption process is driven by the radiation field, therefore the probability is directly proportional to the radiation density ρ at frequency ν.

3. The two postulates above seem quite reasonable. Now comes his new postulate, that there is a third process of radiative transition from the upper state to the lower state, namely stimulated emission, *driven by the radiation field.* By analogy with the probability for absorption, the probability for stimulated emission is

$$dW = B_m^n \rho dt \tag{C.3}$$

Einstein calls the processes in both 2 and 3 as "changes of state due to irradiation." We will see below how he is forced to include postulate 3 in order to maintain thermodynamic equilibrium.

The main requirement of thermodynamic equilibrium is that the occupancy of atomic levels given by Eq. (C.1) should not be disturbed by the absorption and emission processes postulated above. Therefore the number of absorption processes (type 2) per unit time from state n into state m should equal the number of emission processes (type 1 and 3 combined) out of state m into state n. This is called detailed balance. Since the number of processes from a given state occurring in a time dt is given by the occupancy of that state times the probability of a transition, the detailed balance condition is written as

$$p_n \exp(-\varepsilon_n/k_B T) B_n^m \rho = p_m \exp(-\varepsilon_m/k_B T)(B_m^n \rho + A_m^n) \tag{C.4}$$

Notice the importance of the third hypothesis about stimulated emission to make the equation consistent. If one does not put that in, the equation becomes

$$p_n \exp(-\varepsilon_n/k_B T) B_n^m \rho = p_m \exp(-\varepsilon_m/k_B T) A_m^n$$

which clearly will not work. At high temperatures, when the Boltzmann factor makes the occupancy of the two levels almost equal, the rate of absorption on the LHS increases with temperature as the radiation density increases. But the rate of emission on the RHS does not increase because spontaneous emission is independent of the radiation density. Thermodynamic equilibrium will therefore not be maintained. This is vintage Einstein: he imagines a situation that forces a contradiction with what he "knows," namely thermal equilibrium, and uses it to obtain a new result, namely stimulated emission during radiative transfer.

With the grace and confidence of an Olympic hurdler, Einstein now moves on to make quantitative predictions based on the bold new hypothesis. First, he uses the high temperature limit to derive a relation between the coefficients for absorption and stimulated emission. Under the reasonable assumption that $\rho \to \infty$ as $T \to \infty$, the spontaneous emission term on the RHS of Eq. (C.4) can be neglected at high temperatures. From this, it follows that

$$p_n B_n^m = p_m B_m^n$$

By substituting this result in Eq. (C.4), Einstein obtains a new, simple derivation of Planck's law

$$\rho = \frac{A^n_m / B^n_m}{\exp[(\varepsilon_m - \varepsilon_n)/k_B T] - 1}$$

Notice that he will not get the correct form of this law if he did not have the stimulated emission term in Eq. (C.4). Another reason for him to be confident that his three hypotheses about absorption and emission are correct. He then compares the above expression for ρ with Wien's displacement law

$$\rho = \nu^3 f(\nu/T)$$

to obtain

$$\frac{A^n_m}{B^n_m} = \alpha\nu^3 \qquad \text{and} \qquad \varepsilon_m - \varepsilon_n = h\nu$$

with constants α and h. The second result is well known from the Bohr theory of atomic spectra. Einstein is now completely sure that his three hypotheses about radiation transfer are correct since he has been able to derive both Planck's law and Bohr's principle based on these hypotheses.

Einstein does not stop here. He now considers how interaction with radiation affects the atomic motion in order to see if he can predict new features of the momentum transferred by radiation. Earlier he had argued that thermal equilibrium demands that the occupancy of the states remain undisturbed by interaction with radiation. Now he argues that the Maxwell-Boltzmann velocity distribution of the atoms should not be disturbed by the interaction. In other words, the momentum transfer during absorption and emission should result in the same statistical distribution of velocities as obtained from collisions. From kinetic theory, we know that the Maxwell velocity distribution results in an average kinetic energy along each direction given by

$$\frac{1}{2} M \left\langle v^2 \right\rangle = \frac{1}{2} k_B T \tag{C.5}$$

This result should remain unchanged by the interaction with radiation.

To calculate the momentum change during radiative transfer, Einstein brings into play his tremendous insight into Brownian motion. As is now well known from the Langevin equation, he argues that the momentum of the atom undergoes two types of changes during a short time interval τ. One is a frictional or damping force arising from the radiation pressure that systematically opposes the motion. The second is a fluctuating term arising from the random nature of the absorption-emission process. It is well known from Brownian motion theory that the atoms would come to rest from the damping force if the fluctuating term were not present. Thus, if the initial momentum of the atom is Mv, then after a time τ, the momentum will have the value

$$Mv - Rv\tau + \Delta$$

where the second term is the damping term and the last term is the fluctuating term. If the velocity distribution of the atoms at temperature T is to remain unchanged by this momentum transfer process, the average of the above quantity must be equal to Mv, and the mean values of the squares of these quantities must also be equal

$$\left\langle (Mv - Rv\tau + \Delta)^2 \right\rangle = \left\langle (Mv)^2 \right\rangle$$

Since we are only interested in the systematic effect of v on the momentum change due to interaction with radiation, v and Δ can be regarded as independent statistical processes and the average of the cross term $v\Delta$ can be neglected. This yields

$$\left\langle \Delta^2 \right\rangle = 2RM \left\langle v^2 \right\rangle \tau$$

To maintain consistency with kinetic theory, the value of $\left\langle v^2 \right\rangle$ in the above equation must be the same as the one obtained from Eq. (C.5). Thus

$$\frac{\left\langle \Delta^2 \right\rangle}{\tau} = 2Rk_BT \tag{C.6}$$

This is the equation that will tell Einstein if his hypotheses about momentum transfer are correct. In other words, he assumes that the radiation density is given by Planck's law, and calculates R and $\left\langle \Delta^2 \right\rangle$ based on some hypotheses about momentum transfer during radiative processes. If the hypotheses are valid, the above equation should be satisfied identically in order not to contradict thermal equilibrium.

His main hypothesis about momentum transfer is that, if the photon behaves like a localized packet of energy E, it must also carry directional momentum of E/c. Without going into the details, I just outline the approach he uses for calculating R and $\left\langle \Delta^2 \right\rangle$. For calculating R, he uses the following argument. In the laboratory frame in which the atom has a velocity v, the radiation is isotropic. But in the rest frame of the atom, the radiation is anisotropic because of the Doppler shift. This gives rise to a velocity-dependent radiation density and a velocity-dependent probability of absorption and stimulated emission [from Eqs. (C.2) and (C.3)]. The average momentum transferred to the atom is calculated from the modified rates of absorption stimulated emission, thus yielding R. R does not depend on the rate of spontaneous emission because spontaneous emission occurs independently of the radiation field and is therefore isotropic in the rest frame of the atom. Calculating $\left\langle \Delta^2 \right\rangle$ is relatively simpler. If each absorption or emission process gives a momentum kick of E/c in a random direction, the mean square momentum after ℓ kicks is simply $\ell \times (E/c)^2$. ℓ is equal to twice the number of absorption processes taking place in the time τ since each absorption process is followed by an emission process. Using this approach, Einstein calculates R and $\langle \Delta^2 \rangle$. He shows that Eq. (C.6) is satisfied identically when these values are substituted, which implies that the velocity distribution from kinetic theory is not disturbed if and only if momentum exchange with radiation occurs in units of E/c in a definite direction.

He thus concludes the paper with the following observations. There must be three processes for radiative transfer, namely absorption, spontaneous emission, and **stimulated emission**. Each of these interactions is quantized and takes place by interaction with a single radiation bundle. The radiation bundle (which we today call a photon) carries not only energy of $h\nu$ but also momentum of $h\nu/c$ in a well defined direction. The momentum transferred to the atom is in the direction of propagation for absorption and in the opposite direction for emission. And finally, ever loyal to his dislike for randomness in physical laws ("God does not play dice ...!"), he concludes that one weakness of the theory is that it leaves the duration and direction of the spontaneous emission process to "chance." However, he is quick to point out that the results obtained are still reliable and the randomness is only a defect of the "present state of the theory."

What far reaching conclusions starting from an analysis of simple thermodynamic equilibrium. This is a truly great paper in which we see two totally new predictions. First, he predicts the existence of stimulated emission. And to top that, for the first time, he shows that each light quantum carries discrete momentum, in addition to discrete energy. He shows that the directional momentum is present even in the case of spontaneous emission. Thus an atom cannot decay by emitting "outgoing radiation in the form of spherical waves" with no momentum recoil.

Today his conclusions about momentum transfer during absorption and emission of radiation have been abundantly verified. Equally well verified is his prediction of stimulated emission of radiation. Stimulated emission is the mechanism responsible for operation of the laser, which is used in everything from home computers and CD players to long-distance communication systems. Stimulated emission, or more correctly stimulated scattering, underlies our understanding of the phenomenon of Bose-Einstein condensation. It plays an important role in the explanation of superconductivity and superfluidity. The two predictions, momentum transfer from photons and stimulated emission, are particularly close to my heart because they play a fundamental role in one of my areas of research, namely laser cooling of atoms. In laser cooling, momentum transfer from laser photons is used to cool atoms to very low temperatures of a few millionths of a degree above absolute zero. Perhaps fittingly, it is the randomness or "chance" associated with the spontaneous emission process (which he disliked so much) that is responsible for the entropy loss associated with cooling. In other words, as the randomness from the atomic motion gets reduced by cooling, it gets added to the randomness in the radiation field through the spontaneous emission process, thus maintaining consistency with the second law of thermodynamics.

In conclusion, we have seen how Einstein was able to use the principle of thermodynamic equilibrium to imagine a situation where radiation and matter were in dynamical equilibrium and from that predict new features of the radiative transfer process. As mentioned before, this was a recurring theme in his work, a kind of *modus operandi* for the great "detective." In his later writings, he said that he always sought one fundamental governing principle from which he could derive results through these kind of arguments. He found such a principle for thermodynamics, namely the second law of thermodynamics, which states that it is impossible to build a perpetual motion machine. He showed that the second law was a necessary and sufficient condition for deriving all the results of thermodynamics. His quest in the last four decades of his life was to geometrize all forces of nature. In this quest, he felt that he had indeed found the one principle that would allow him to do this uniquely, and this was the *principle of relativity*

> **the laws of physics must look the same to all observers no matter what their state of motion.**

He had already used this principle to geometrize gravity in the general theory of relativity. His attempts at geometrizing electromagnetic forces remained an unfulfilled dream.

Examples of *gedanken* experiments

We present two examples of *gedanken* experiments that illustrate the Einstein technique for arriving at new results. Both of these experiments yield results associated with the general theory of relativity, but are so simple and elegant that they can be understood without any knowledge of the complex mathematical apparatus needed for the general theory. The first experiment is due to Einstein himself, while the second is due to Hermann Bondi.

(i) Need for curved spacetime for gravity

This is a thought experiment devised by Einstein to arrive at the conclusion that the general theory of relativity is an extension of the special theory which requires curved spacetime, or spacetime in which the rules of plane (Euclidean) geometry do not apply. The "known" facts are the results of special theory of relativity applicable to inertial systems, and the equivalence principle which states that inertial mass is exactly equal to gravitational mass. Einstein's argument proceeds as follows.

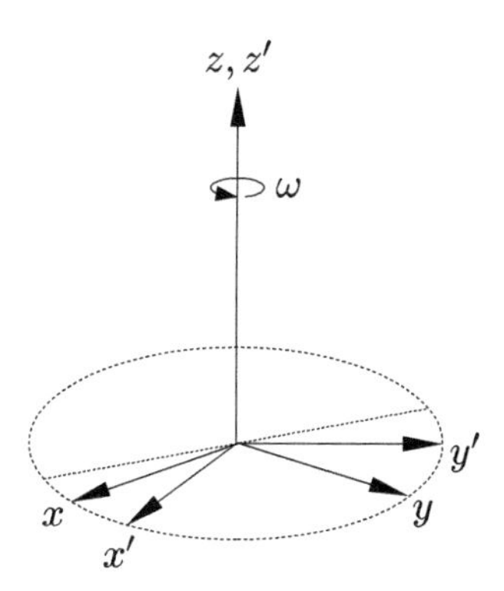

Figure C.1: Coordinate systems with relative rotation between them.

Imagine two observers or coordinate systems O and O$'$. Let the z' axis of O$'$ coincide with the z axis of O, and let the system O$'$ rotate about the z axis of O with a constant angular velocity, as shown in Fig. C.1. Thus O is an inertial system where the laws of special relativity apply, while O$'$ is a non-inertial system. Imagine a circle drawn about the origin in the x'–y' plane of O$'$ with some given diameter. Imagine, further, that we have a large number of rigid rods, all identical to each other. We lay these rods in series along the circumference and the diameter of the circle, at rest with respect to O$'$. If the number of rods along the circumference is U and the number of rods along the diameter is D, then, if O$'$ does not rotate with respect to O, we have (from plane geometry)

$$\frac{U}{D} = \pi$$

However, if O′ rotates, we get a different result. We know from special relativity that, relative to O, the rods on the circumference undergo Lorentz contraction while the rods along the diameter do not undergo this contraction (the relative motion is perpendicular to the diameter). Therefore, we are led to the unavoidable conclusion that

$$\frac{U}{D} > \pi$$

i.e. the laws of configuration of rigid bodies with respect to O′ are not in accordance with plane geometry. If, further, we place two identical clocks, at rest with respect to O′, one at the periphery and one at the center of the circle, then with respect to O the clock at the periphery will go slower than the clock at the center from special relativity. A similar conclusion will be reached by O′, i.e. the two clocks go at different rates.

We thus see that space and time cannot be defined with respect to O′ as they were defined in special theory of relativity for inertial systems. But, according to the equivalence principle, O′ can also be considered a system at rest with respect to which there is a gravitational field (corresponding to the centrifugal force field and the Coriolis force field). We therefore arrive at the following remarkable result: the gravitational field influences and even determines the geometry of the space-time continuum, and this geometry is not Euclidean. From this conclusion, Einstein goes on to develop a curved spacetime theory of gravitation.

(ii) Gravitational redshift

This example illustrates the use of a thought experiment to calculate the difference in rates between two clocks placed at different gravitational potentials, called the gravitational redshift. We have already seen in the first example how the rate of the clock at the periphery differs from the rate of the clock at the center. Here, we derive a quantitative value for this difference using an Einstein-like *gedanken* experiment, first conceived by Bondi. The "known" things are the second law of thermodynamics and the special relativistic energy-mass relationship, $E = mc^2$. The argument proceeds as follows.

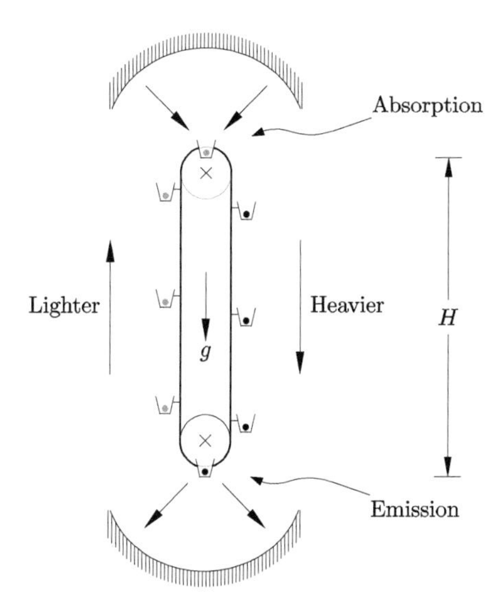

Figure C.2: Bondi's perpetual motion machine. The buckets on the right side contain atoms that have higher energy and are thus heavier than the atoms on the left side. When a bucket reaches the bottom, the atom inside emits a photon which is absorbed by the corresponding atom in the top bucket. The heavier buckets on the right keep falling down in the gravitational field and their gravitational energy can be converted to useful work. The resolution to the paradox is that the photon absorbed at the top has a lower frequency than the photon emitted at the bottom.

Imagine a series of buckets on a frictionless pulley system, as shown in Fig. C.2. Each bucket contains an atom capable of absorbing or emitting a photon of energy $h\nu$. The system is in a uniform gravitational field with acceleration g. If the photon frequency were unaffected by the gravitational field, we can operate the system as a perpetual motion machine in the following way. Imagine that the pulleys rotate clockwise and that all the atoms on the left are in the ground state and the atoms on the right are in the excited state. The lifetime in the excited state is such that, on average, every time a bucket reaches the bottom the atom inside decays to the ground state and emits a photon. Suitable reflectors direct this photon to the corresponding bucket at the top so that the atom inside absorbs the photon and goes into the excited state. All the excited state atoms on the right have more energy and, from the relation $E = mc^2$, are therefore heavier by an amount $\Delta m = h\nu/c^2$. The heavier masses are accelerated down by the gravitational field and the system remains in perpetual motion. The excess gravitational potential energy can be converted to unlimited useful work, in violation of the second law of thermodynamics.

The solution to the paradox lies in postulating that the frequency of the photon emitted by the atom at the bottom is not the same as the frequency of the photon when it reaches the top. Let the two frequencies be ν and ν' respectively. Then the additional mass for the atom at the top by absorbing a photon of frequency ν' is $h\nu'/c^2$, and the potential energy of this excess mass at a height H between the two buckets is $h\nu'/c^2 \times gH$. To maintain consistency with the second law of thermodynamics, this excess energy should exactly compensate for the loss in energy of the photon as its frequency changes from ν to ν'

$$\frac{h\nu'}{c^2} gH = h(\nu - \nu')$$

which yields

$$\frac{\nu' - \nu}{\nu'} = -\frac{gH}{c^2}$$

i.e. the relative frequency shift is given by gH/c^2 and is negative (redshift) at the location where the gravitational potential is higher. The shift can be understood from the fact that the photon is also affected by the gravitational field and therefore loses energy as it climbs up the potential. Since the photon always travels at the speed c, it loses energy by changing its frequency. This result explains why, in the first example, the clock at the center goes slower than the clock in the periphery according to O′. With respect to O′, there is a gravitational field (corresponding to the centrifugal force) pointing away from the center. The clock at the center is at a higher gravitational potential and hence goes slower.

The gravitational redshift on the surface of the earth is very tiny at any reasonable height, but it was experimentally verified in a remarkable experiment by Pound and Rebka in 1959. They measured the frequency shift of a recoilless Mössbauer transition between the top and bottom of a building at Harvard University, a height difference of about 25 m. The relative frequency shift measured was a tiny 3 parts in 10^{14}, consistent with the above calculation!

Appendix D

Frequency Comb

IN recent times, one atomic transition that has inspired many advances in high resolution spectroscopy and optical frequency measurements is the 1S $\rightarrow$ 2S resonance in hydrogen, with a natural linewidth of only 1 Hz. Measurement of the frequency of this transition is important as a test of QED and for the measurement of fundamental constants. However, the wavelength of this transition is 121 nm, corresponding to a frequency of 2.5×10^{15} Hz. Since the SI unit of time is defined in terms of the cesium radio frequency transition at 9.2×10^{9} Hz, measuring the optical frequency with reference to the atomic clock requires spanning six orders of magnitude! You can think of this as having two shafts whose rotation speeds differ by a factor of one million, and you need to measure the ratio of their speeds accurately. If we use a belt arrangement to couple the two shafts, then there is a possibility of errors in the ratio measurement due to phase slip. Instead, one would like to couple them through a gearbox mechanism with the correct teeth ratio so that there is no possibility of slip. The idea is shown schematically in Fig. D.1, and is precisely what is achieved by the frequency comb. This measurement technique was pioneered by John Hall of the National Institute of Standards and Technology (NIST) in Boulder, Colorado, USA, and Theodor Hänsch of the Max-Planck Institute for Quantum Optics in Garching, Germany. They shared the 2005 Nobel Prize in physics for this work.

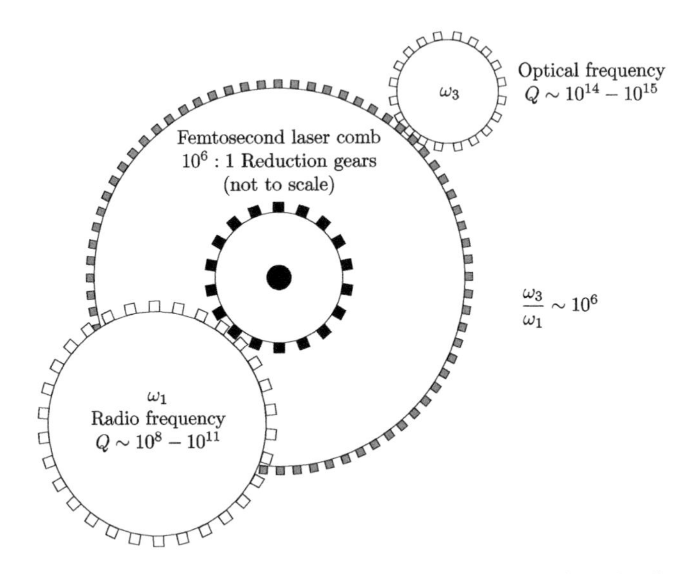

Figure D.1: Schematic of a gearbox mechanism to couple radio frequencies to optical frequencies. The span of 6 orders of magnitude is what is achieved by the frequency comb.

The basic idea of the comb technique, shown in Fig. D.2, is that periodicity in time implies periodicity in frequency through the Fourier transform. Thus, if you take a pulsed laser that produces a series of pulses that are uniformly spaced in time (corresponding to a fixed repetition rate), then the frequency spectrum of the laser will consist of a set of uniformly spaced peaks on either side of a central peak. The central peak is at the optical frequency of each laser pulse (**carrier frequency**), and the peaks on either side are spaced by the inverse of the repetition period (**sidebands**), i.e. if the repetition period is τ_r then the sideband spacing is $\Delta = 1/\tau_r$.

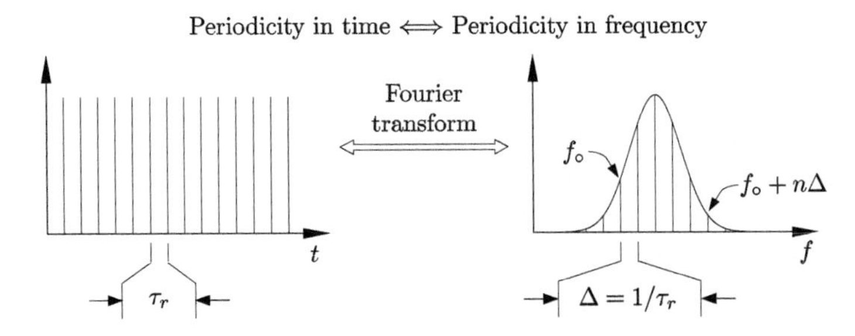

Figure D.2: Periodicity in time implies periodicity in frequency through the Fourier transform. The spacing Δ in frequency domain is the inverse of the repetition period τ_r in time domain.

You can produce such a spectrum by putting the laser through a nonlinear medium such as a nonlinear fiber. The larger the nonlinearity, the more the number of sidebands. Around the year 1999, there was a major development in making nonlinear fibers; fibers with honeycomb microstructure were developed which had such extreme nonlinearity that the sidebands spanned almost an octave. If you sent a pulsed laser (operating near 800 nm) through such a fiber, you would get a near continuum of sidebands spanning the entire visible spectrum. The series of uniformly spaced peaks stretching out over a large frequency range looks like the teeth of a comb, hence the name optical frequency comb. The beautiful part of the technique is that the comb spacing is determined solely by the repetition rate, thus by referencing the repetition rate to a cesium atomic clock, the comb spacing can be determined as precisely as possible. In 1999, Hänsch and coworkers showed that the comb spacing was uniform to 3 parts in 10^{17}, even far out into the wings.

Thus the procedure to produce a frequency comb is now quite straightforward. One starts with a mode-locked pulsed Ti-sapphire laser and sends its output through 20–30 cm of nonlinear fiber. The pulse repetition rate is referenced to an atomic clock, and determines the comb spacing. The carrier frequency is controlled independently, and determines the comb position. But how does one measure an optical frequency using this comb? This can be done in two ways. One way is to use a reference transition whose frequency $f_\circ$ is previously known. We now adjust the comb spacing Δ so that the reference frequency $f_\circ$ lies on one peak, and the unknown frequency f lies on another peak that is n comb lines away, i.e. $f = f_\circ + n\Delta$.* Thus by measuring n, the number of comb lines in between, and using our knowledge of $f_\circ$ and Δ, we can determine f. This was the method used by Hänsch in 1999 to determine the frequency of the D_1 line in cesium (at 895 nm). The measurement of this frequency can be related to the fine structure constant α, which is one of the most important constants in physics because it sets the scale for electromagnetic interactions and is a fundamental parameter in QED calculations.

However, the above method requires that we already know some optical frequency $f_\circ$. If we want to determine the absolute value of f solely in terms of the atomic clock, the scheme is slightly more complicated. In effect, we take two multiples (or harmonics) of the laser frequency, and use the uniform comb lines as a precise ruler to span this frequency difference. Let us say we align one peak to $3.5f$, and another peak that is n comb lines away to $4f$, then we have determined

$$4f - 3.5f = n\Delta \qquad \Longrightarrow \qquad f = 2n\Delta$$

*It is not necessary that the comb peak aligns perfectly with the laser frequency. A small difference between the two can be measured easily since the beat signal will be at a sufficiently low frequency.

so that we have f in terms of the comb spacing. In 2000, Hänsch and coworkers used this method to determine the frequency of the 1S $\rightarrow$ 2S resonance in hydrogen with an unprecedented accuracy of 13 digits. This was the first time that a frequency comb was used to link a radio frequency to an optical frequency.

Currently, one of the most important questions in physics is whether fundamental constants of nature are really constant, or are changing with time. For example, is the fine structure constant α constant throughout the life of the universe, or is it different in different epochs? Now, if you want to measure a very small rate of change $\dot{\alpha}$ ($= d\alpha/dt$), then you can do it in two ways. You can take a large dt so that the integrated change in α is very large. This is what is done in astronomy, where looking at the light from a distant star is like looking back millions of years in time. You can then compare atomic spectra from distant stars to spectra taken in the laboratory today. Alternately, if you want to do a laboratory experiment to determine $\dot{\alpha}$, then you have no choice but to use a small dt. Therefore, you have to improve the accuracy of measuring α so that even small changes become measurable. This is what has been done by Hänsch and his group. By measuring the 1S $\rightarrow$ 2S resonance in hydrogen over a few years, they have been able to put a limit on the variation of α. Similar limits have been put by other groups using frequency comb measurements of other optical transitions. The current limit (as of 2014) on $\dot{\alpha}/\alpha$ from both astronomy and atomic physics measurements is about 10^{-15} per year.

In the last few years, several optical transitions have been measured using frequency combs. The primary motivation is to find a suitable candidate for an optical clock to replace the microwave transition used in the current definition. An optical clock will "tick" a million times faster, and will be inherently more accurate. However, since the cesium atomic clock has an accuracy of 10^{-15}, one has to measure the candidate optical transition to this accuracy to make sure it is consistent with the current definition. The race is on to find the best candidate among several alternatives such as laser-cooled single ions in a trap, ultracold neutral atoms in an optical lattice, or molecules. The applications for more precise clocks of the future range from telecommunications and satellite navigation to fundamental physics issues such as measurement of pulsar periods, tests of general relativity, and variation of physical constants.

Let me conclude this piece with a personal anecdote. I attended a small reception in honor of John Hall after he won the Nobel Prize. In his speech, he mentioned that the thing he enjoyed most about being at NIST was that the management allowed him complete freedom to play with the latest "toys and gadgets," pleasures that he has carried from his childhood. I remember that, as a child, I too was fascinated by mechanical and electrical gadgets, and the precision with which they were engineered. I think many of us take to experimental research precisely for this reason, that it allows us to take our childhood pleasures of playing with toys into adulthood, and even make a living out of this enjoyment!

Index